Electrical, Electronic and Control Engineering

Electrical, Electronic and Control Engineering

Edited by **Dewayne Hopper**

WILLFORD PRESS

New York

Published by Willford Press,
118-35 Queens Blvd., Suite 400,
Forest Hills, NY 11375, USA
www.willfordpress.com

Electrical, Electronic and Control Engineering
Edited by Dewayne Hopper

International Standard Book Number: 978-1-68285-051-0 (Hardback)

Printed in the United States of America.

Contents

Preface

The innovations and advancements in the field of science and technology have brought a tremendous shift in the field of engineering. Electrical, electronic and control engineering are some of the major engineering disciplines that have developed at a rapid pace. This book explores all the important aspects of these disciplines in the present day scenario with respect to topics such as power and electrical engineering, information and communication technologies, mechatronics, control engineering, design of new devices, etc. Those in search of information to further their knowledge will be greatly assisted by this comprehensive book.

All of the data presented henceforth, was collaborated in the wake of recent advancements in the field. The aim of this book is to present the diversified developments from across the globe in a comprehensible manner. The opinions expressed in each chapter belong solely to the contributing authors. Their interpretations of the topics are the integral part of this book, which I have carefully compiled for a better understanding of the readers.

At the end, I would like to thank all those who dedicated their time and efforts for the successful completion of this book. I also wish to convey my gratitude towards my friends and family who supported me at every step.

Editor

Impact of the Patient Torso Model on the Solution of the Inverse Problem of Electrocardiography

Milan TYSLER, Jana LENKOVA, Jana SVEHLIKOVA

Institute of Measurement Science, Slovak Academy of Sciences, Dubravska cesta 9, 841 04 Bratislava, Slovak Republic

tysler@savba.sk, umermacu@savba.sk, umersveh@savba.sk

Abstract. *Cardiac diagnostics based on a solution of the inverse problem of electrocardiography offers new tools for visual assessment of cardiac ischemia. The accuracy of the inverse solution is influenced by fidelity of the patient torso model. As optimum, an individual torso model with real heart shape and position obtained from CT or MRI is desirable. However, imaging is not always available in clinical practice, hence we investigated, if a generic torso shape individually adjusted according to patient's chest dimensions, with a simplified heart model placed to a vertical position obtained from inverse localization of the early ventricular activation can result in an inverse solution close to the result obtained with an accurate torso model. Simulated inverse localization of 18 ischemic lesions for 9 subjects showed that the use of individually adjusted generic torso instead of real torso shape led to an acceptable increase of the lesion localization error from 0.7±0.7 cm to 1.1±0.7 cm when accurate heart model was used. However, if simplified heart model was used and placed in a vertical position according to the V2 lead level, the lesion localization error increased to 3.5±0.9 cm. Moving the simplified heart model to a position estimated by the inverse solution decreased the vertical heart positioning error from 1.6±2.3 cm to 0.2±1.2 cm but without adjusting the heart shape and rotation the lesion localization error did not improve and reached 3.7±1.0 cm.*

Keywords

Individual torso shape model, inverse problem of electrocardiography, inversely estimated heart position.

1. Introduction

Solution of the inverse problem of electrocardiography and topographical visualization of an cardiac electrical generator is promising tool for assessment of various cardiac disorders including local ischemic lesions or arrhythmogenic substrates. For an accurate inverse solution it is necessary to have an individual torso model with internal structures representing at least the main electrical inhomogeneities, such as lungs and ventricular cavities filled with blood [1], [2]. Another important issue discussed in the literature is the large variability of the heart position that can vary by several centimeters, namely in the vertical direction [3], [4]. Missing information on the exact heart position can strongly influence the result of the inverse solution [5]. As optimum, the real heart position should be used in the torso model rather than usually assumed position relatively to anatomical landmarks, such as the fourth intercostal space.

To obtain a faithful model of the patient torso, the use of computed tomography (CT) or magnetic resonance imaging (MRI) technique is preferable. However, in clinical practice these techniques are not always available for cardiac patients. Hence it is desirable to search other methods how to create enough accurate patient specific torso model without the need of imaging techniques.

In this simulation study an approach based on the use of a generic model of the human torso containing simplified model of the ventricular myocardium was attempted. Using several anthropometric measures the torso shape was adjusted to match with the torso of an individual subject. To estimate the vertical heart position, measured ECG data were used for solving a simplified inverse problem and finding the location of early ventricular activation that was supposed in the upper part of the septum. The aim of the study was to verify whether the use of the individually adjusted generic torso model and a simplified heart model placed in the estimated vertical position allow inverse solution with sufficient accuracy.

2. Methods and Material

2.1. Simulation of Body Surface Potentials

A simplified model of ventricular myocardium was used to simulate normal ventricular activation and activation in ventricles containing single ischemic region with changed repolarization [8], [9]. The geometry of the model was defined using several ellipsoids and its volume consisted of $1 \times 1 \times 1$ mm cubic elements. Each model element was assigned realistically shaped action potential (AP) and the ventricular activation process was simulated by a cellular automaton. In each time step of the activation, elementary dipole moments were computed from the differences between APs of adjacent model elements, thus the equivalent cardiac electrical generator was represented by a multiple-dipole model. Using the boundary element method, body surface potentials (BSPs) $\mathbf{p}(t)$ were computed in points representing electrode positions on the surface of an inhomogeneous torso model:

$$\mathbf{p}(t) = \mathbf{A}\,\mathbf{s}(t), \qquad (1)$$

where $s(t)$ is a multiple dipole source in the ventricular myocardium model and matrix $\mathbf{A}$ represents the influence of the torso as an inhomogeneous volume conductor.

From the simulated BSP maps the QRST integral map (IM) $\mathbf{i}$ was computed using the formula

$$\mathbf{i} = \int\limits_{QRST} \mathbf{p}(t)\mathrm{d}t = \int\limits_{QRST} \mathbf{A}\,\mathbf{s}(t)\mathrm{d}t = \mathbf{A} \int\limits_{QRST} \mathbf{s}(t)\mathrm{d}t = \mathbf{A}\,\mathbf{s}, \quad (2)$$

where $\mathbf{i}$ is the vector of integrals of BSPs and s is an integral of multiple dipole source of the cardiac electrical field.

To mimic the local repolarization changes in the ischemic lesions, 18 small areas were modeled in the ventricular myocardium, one at a time. They were formed as spherical caps with varying diameter and height, and placed in 3 typical regions supplied by the main coronary arteries: anterior - in the region supplied by the left descending artery, posterior – in the region supplied by the left circumflex artery, and inferior - in the region supplied by the right coronary artery. In each region, 3 endocardial and 3 epicardial lesions of different sizes were modeled. In the model elements within the ischemic lesions, the AP was shortened by 20 % to simulate the changed repolarization.

For each ischemic lesion the difference QRST integral map (DIM) $\mathbf{\Delta i}$ was calculated by subtracting the IM computed for the normal activation from the IM computed in the presence of the particular lesion as

$$\mathbf{\Delta i} = \mathbf{i_i} - \mathbf{i_n} = \mathbf{As_i} - \mathbf{As_n} = \mathbf{A}(\mathbf{s_i} - \mathbf{s_n}) = \mathbf{A\Delta s}, \quad (3)$$

where $\mathbf{i_i}$ and $\mathbf{i_n}$ represent the vectors of QRST integrals of BSPs in case of ischemia and during normal activation, $\mathbf{\Delta s}$ represents the difference between the integral multiple dipole source under normal conditions and during ischemia. The DIM thus represents the topographical changes in the surface cardiac electrical field due to the local ischemia.

2.2. Inverse Localization of an Ischemic Lesion

To identify the ischemic lesion by an inverse solution, equivalent integral generator representing the original multiple dipole generator $\mathbf{\Delta s}$ should be determined. Because this inverse problem is generally ill-posed, additional constraints are needed for its unique solution. The constraint used in this study was the assumption that the equivalent integral generator representing the small ischemic area can be represented by a single dipole. The magnitude, orientation and position of the dipole can be searched as parameters of a "moving dipole", what yields a nonlinear problem. In this study another approach was used: only dipole magnitude and orientation were determined for dipoles in predefined possible positions. In this way the problem was converted to a linear one, however, the parameters of an equivalent integral dipole (EID) had to be computed for many positions within the ventricular myocardium and then the proper position had to be selected. To achieve sufficient resolution for the dipole localization, the mean distance between the neighboring possible dipole positions less than 1 cm was selected. For every predefined position j, the dipole moment $\mathbf{d}_j$ was computed as

$$\mathbf{d}_j = \mathbf{A}_j^{+}\mathbf{\Delta i}, \qquad (4)$$

where $\mathbf{A}_j^{+}$ is the pseudo-inverse of a submatrix $\mathbf{A}_j$ of the matrix $\mathbf{A}$ that represents the relation between the EID placed in the position j and the DIM. The equation (4) is overdetermined and its unique solution exists for each position of the EID. To compute the pseudo-inverse, singular value decomposition was applied to the submatrix $\mathbf{A}_j$.

To find the best representative generator of the lesion, for each position j the surface map $\mathbf{q}_i$ was computed using corresponding EID as the generator. This map was compared with the input DIM using the relative root mean square difference $RMSDIF_j$:

$$RMSDIF_j = \sqrt{\sum_k (\mathbf{q}_{j,k} - \mathbf{\Delta i}_k)^2} \Big/ \sqrt{\sum_k (\mathbf{\Delta i}_k)^2}, \quad (5)$$

where k is the number of electrodes on the torso surface. The EID in a position that produced the map with smallest $RMSDIF_j$ was selected as the best representative of the lesion.

The distance between the selected EID position and the gravity center of the simulated lesion was defined as the lesion localization error (LE) and was used to evaluate the accuracy of the inverse lesion localization.

2.3. Vertical Heart Position Estimation

From the observed high variability of the vertical heart position relatively to the anatomically fixed electrode positions (Fig. 1) it is apparent that adjustment of the vertical heart position is highly desirable.

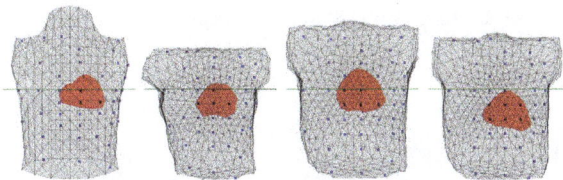

Fig. 1: Generic torso model (left) and 3 examples of real chest models of subjects used in the study. Dots indicate electrode positions, vertical position of ECG lead V2 is marked by a horizontal line.

In Fig. 1 the inter-individual variability of the vertical distance between the heart position and the level of ECG lead V2 defined in the 4th intercostal space is demonstrated. If the same generic torso and heart model (Fig. 1 left) is used for all subjects, this vertical distance is assumed to be zero what apparently may not be correct.

The possibility to estimate the individual vertical position of the heart by inverse localization of the early ventricular activation was studied using real ECG signals measured in 9 subjects (7 men, 2 women) published in [3]. The ECG signals in each subject were recorded by 62 leads of the Amsterdam lead system. Realistic torso models, as well as the electrode positions for these subjects, were obtained from MRI scans. For each subject ECG signals were recorded for 10 seconds with a sampling rate of 1000 Hz. Low-pass filter with 50 Hz stop-band was applied and the signals were time averaged to create representative signal for one heart cycle in each lead [11]. Finally, the baseline of averaged signals was adjusted by setting the mean potential of the PQ interval to zero. The time instant of the QRS onset was set manually from rms signal computed from all measured leads.

The integral map (IM) for the first 20 ms of the ventricular depolarization (from the QRS onset) was computed for each subject and used as the representative of the cardiac electrical generator during the early ventricular activation that normally occurs in the upper part of the left endocardial septum [11].

Site of the initial ventricular depolarization was estimated from the IM using the inverse solution in homogeneous torso model. Similar approach as described in section 2.2. was applied. The region activated during the early depolarization was assumed to be small enough to be represented by single EID that was searched in the whole modeled ventricular myocardium volume in predefined positions placed in regular 3 mm grid. For each subject the position j of the early activated area was determined as the site in which the RMSDIFj between the IM and the map generated by the EID was minimal. The heart model was then vertically shifted so that the site of the early ventricular activation vertically coincided with the anatomically determined area in the upper part of the left endocardial septum. The vertical errors between the real heart position and the inversely estimated heart position, as well as the standard heart position (representing the situation with no individual information about the heart position), were then evaluated.

The described method for inverse estimation of the vertical heart position was used in this study to create one type of the individual torso models for each subject.

2.4. Torso Models Used in the Study

Torso models of 9 healthy subjects introduced in section 2.3. obtained from MRI scans and described by triangulated surfaces of torso, lungs and ventricular myocardium were used in the study. The positions of 62 ECG electrodes were also included in the torso models.

Modified Dalhousie torso [6] containing the simplified ventricular myocardium model [8] described in section 2.1. and placed in anatomically defined standard position was used as the generic model of a human torso.

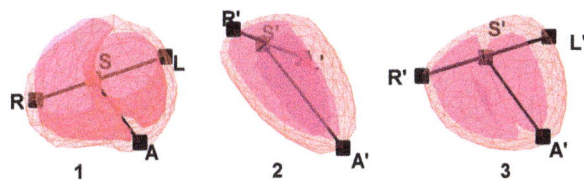

Fig. 2: Individual adjustment of the simplified heart model geometry to the heart of a particular subject (frontal view): 1 – subject's heart geometry, 2 – simplified heart model in standard position, 3 – simplified heart model rotated along the long (A-S) and short (L-R) axis and scaled along the long axis for best correspondence to the subject's heart geometry.

To have comparable heart anatomy for forward simulations in all subjects, the simplified ventricular my-

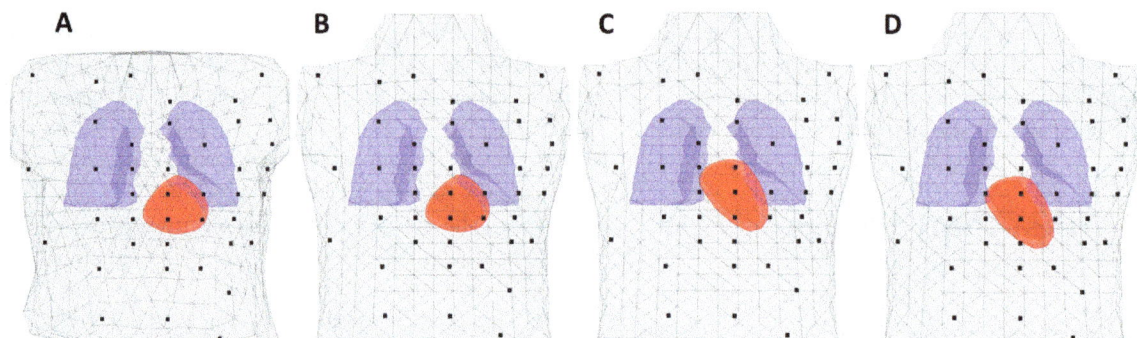

Fig. 3: Four types of torso models used for each subject in the study: A - original torso model obtained from MRI with a simplified model of ventricles placed, rotated and scaled to correspond to the subject's real heart. B - generic torso model adjusted for best match with subject's torso shape with the ventricles as in case A. C – adjusted generic torso shape as in case B but with a simplified model of ventricles in standard position, D – adjusted generic torso shape as in cases B and C but with a simplified model of ventricles vertically shifted to a vertical position based on the inversely estimated site of early ventricular activation.

ocardium model described in section 2.1. but adjusted to the heart geometry of each subject was used (Fig. 2). For each subject the long heart axis (from apex A to point S in the septum) and short heart axis (from point L in the left ventricular free wall to point R in the right ventricular free wall) were defined. Then the simplified ventricular model was positioned so that its axes coincided with the axes in the heart of the real subject and was properly scaled along its long axis.

For all 9 subjects body surface potential maps (BSPMs) corresponding to normal activation as well as to activation in case of 18 modeled ischemic lesions were simulated and single DIM was computed for each case. Realistic inhomogeneous torso models based on subjects' MRI scans, containing lungs and heart cavities filled with blood were used in the simulations. Individually adjusted simplified geometrical models of the ventricles described above were inserted into each torso. Respective electrical conductivities assigned to the lungs and heart cavities were 4 times lower and 3 times higher than the average conductivity of the rest of the torso. The DIMs were computed from 62 simulated leads placed on the torso surface according the Amsterdam lead system and used as input for the inverse solutions.

To study the impact of the torso model shape and heart position on the noninvasive inverse localization of ischemic lesions, several types of torso models were used in the inverse computations (Fig. 3).

Model A – the same torso model as used in the forward simulations. It consists of realistic outer torso shape and electrode positions based on MRI scan, lungs, and individually adjusted simplified heart model that was placed, oriented and scaled for best correspondence with the subject's heart model obtained from MRI.

Model B – torso shape created from the generic torso model by adjusting its shape according to 10 anthropometric measures of the subject (see Fig. 4) as proposed in [7]. The same individually adjusted simplified heart model as in torso model A was used.

Model C – the same adjusted generic torso shape with electrodes as in model B but with a generic heart model placed and oriented in a standard way – as if no knowledge about the heart position, orientation and size was available. The vertical position of the heart model is in the level of the standard ECG lead V2.

Model D – the same adjusted generic torso shape with electrodes as in models B and C but with a generic heart model vertically shifted according to the result of the inverse estimation of the vertical heart position.

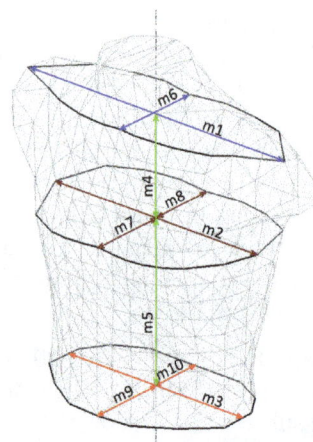

Fig. 4: Selected 10 anthropometric measures for subject-specific adjustment of the generic torso shape.

The errors of the inverse localization of all 18 modeled lesions were evaluated for each of the 9 subjects and each type of the torso model. The results for different torso model types were compared.

3. Results

3.1. Vertical Heart Position Estimation

The inversely estimated sites of early ventricular depolarization were found in the upper septal area for all 9 investigated subjects. Their positions (transformed to a single standard simplified ventricular model) are depicted in Fig. 5 together with their mean position (larger marker) computed as the gravity center of the results for individual subjects. The average spatial distance between the individual positions of the early depolarization sites from their mean position was 1.6±0.6 cm and the standard deviation of the vertical position of results for individual subjects was ±1.3 cm. These numbers indicate the possible error range when assuming that the early activation site should serve as a reference point for adjustment of the vertical heart position.

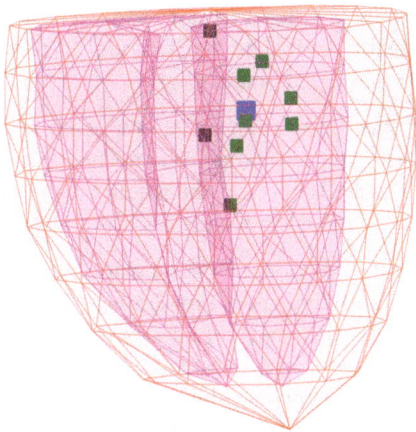

Fig. 5: The estimated sites of early ventricular depolarization for 9 studied subjects (small markers) and the gravity center of the positions (large marker) depicted in standard simplified ventricular model.

While the average vertical error (for all 9 subjects) between the real position of the heart ventricles and position of the standard ventricular model (marked as "stand. pos.") was 1.6±2.3 cm, the average vertical error between the real position of the ventricles and the position of ventricles estimated from the site of early ventricular activation (marked as "early dep.") dropped to 0.2±1.2 cm. The results for all 9 subjects are shown in Fig. 6. These results indicate that despite

the vague definition of the site of the early ventricular activation as a reference point, its inverse estimation can improve the vertical positioning of the heart model if no other information on the heart position in the torso is available.

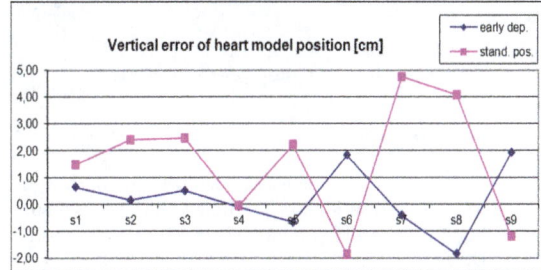

Fig. 6: Errors of vertical position of the heart models for all studied subjects: Diamonds – errors between real positions of ventricles and positions estimated from the early ventricular activation. Squares - errors between real position of ventricles and the position of standard simplified ventricular model.

3.2. Impact of Approximate Torso Shape on the Inverse Solution

To study the impact of the use of an approximate torso shape created by adjusting a generic torso shape according 10 anthropometric parameters of the subject, results of the inverse solutions with torso models A and B were compared. As it can be seen in Fig. 7, the LE values obtained with approximate torso model B (from 0.6±0.4 cm to 1.6±1.1 cm) with the mean LE for all subjects of 1.1±0.7 cm were only slightly worse than the LE values obtained with the torso model A created from MRI scans (from 0.5±0.3 cm to 0.8±0.9 cm) with mean LE for all subjects of 0.7±0.7 cm. In all subjects, the mean LE was slightly worse when torso model B was used, with the exception of subject s7, where the values were equal.

This result suggests that the use of individually adjusted generic torso shape in the inverse solution can be acceptable if no imaging data are available.

3.3. Impact of the Vertical Heart Position Estimation on the Inverse Solution

To study the impact of the vertical positioning of the heart model, results of the inverse lesion localization with torso models C and D were evaluated and compared also with results with torso model B. In all these models individually adjusted generic torso shape was used. When the torso model C with the heart model

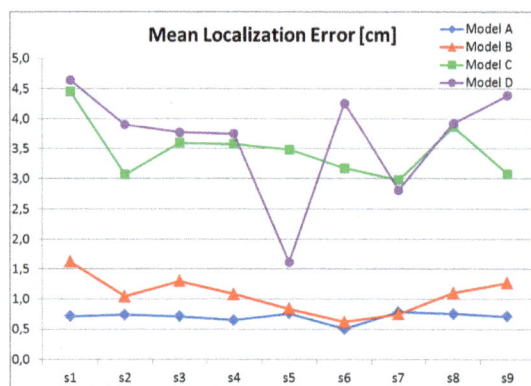

Fig. 7: Mean errors of the inverse lesion localization computed for all 18 modeled lesions in 9 studied subjects (s1–s9) and using 4 torso and heart model configurations (models A, B, C, D).

located in standard vertical position given by the level of ECG lead V2 was used in the inverse solution, the mean LE values varied from 3.0 ± 0.5 cm to 4.5 ± 0.7 cm, with the mean LE for all subjects of 3.5 ± 0.9 cm. For torso model D, where the heart model was vertically moved to the inversely estimated position, the LE values ranged from 2.8 ± 1.2 cm to 4.6 ± 0.5 cm, with a noticeably lower value of 1.6 ± 1.4 cm for the subject s5. The LE averaged for all subjects was 3.7 ± 1.0 cm.

These results show, that despite the improved vertical positioning of the heart model in torso model D in comparison with torso model C, in all but two subjects (s5 and s7) the results with model D were even worse than those with torso model C. Comparison with much better results obtained with torso model B indicates that merely positioning of the heart model without its proper rotation and scaling does not yield acceptable errors of the inverse lesion localization.

4. Discussion

The experimental inverse localization of the early ventricular activation in 9 subjects indicated that the site of the initial activation can be estimated within about 1.6 ± 0.6 cm. In all studied subjects the found positions were in agreement with Durrer's findings [11] that the ventricular activation in healthy subjects starts in endocardial areas of the left ventricular cavity near septum. Although the achieved average error of the estimated vertical position was only 0.2 cm, its standard deviation of ±1.2 cm is quite large and errors of almost 2 cm were found in subjects s6, s8 and s9 (Fig. 6). In spite of this, the method generally improved the vertical positioning of the heart model in comparison with the error of 1.6 ± 2.3 cm in a situation when no information on the heart position was used and the ventricular model was positioned with the use of anatomical

landmarks. The reason for the remaining inaccuracy of the estimated vertical heart position could be the individual variability of the normal ventricular activation sequence as well as neglect of torso inhomogeneities in the inverse computations. However, serious limitation of this method is the imperative of normal initial ventricular depolarization.

From the results with the torso model B in the second part of the study it implies that adjustment of a generic torso shape according to individual anthropometric measures of the subject and maintaining real electrode positions is a promising way how to obtain subject-specific torso geometry accurate enough for the inverse solution. However, from the results obtained with torso models C and D the great impact of the used heart model on precision of the inverse solution is also apparent.

In the third part of the study the method for individual assessment of the vertical heart position using the inverse localization of early ventricular activation was used in 9 subjects to create their individual torso models (model D). From the graph for model D in Fig. 7 it is apparent that the improvement of vertical position of the heart, without its additional adjustment by proper rotation and scaling did not decrease the lesion localization error in the inverse solution. Individual positioning of the heart model in 3D space along all three coordinates based on the estimated site of the early ventricular activation was not used because it was not always possible to fit the heart model in the torso without additional heart scaling.

The importance of information about heart size and rotation suggests the necessity of some heart imaging (e.g. USG, CT or MRI) even if the whole torso imaging is not available. This issue should be studied further.

The limitation of the forward simulations used in the study was the simplified model of the heart activation and cardiac electrical generator. However, it was sufficient to demonstrate the importance of individually adjusted torso and heart models used in the inverse solution for each examined subject.

The principal limitation of the presented inverse method is the need of BSPMs measured during the ischemia (with changed repolarization phase of the myocytes AP) and also in a situation without the ischemia manifestation. Both measurements in the same subject are necessary for computation of the DIM that is used as the input for the inverse solution. To have such data for a patient admitted with acute myocardial infarction would be extremely difficult. However, such data can be obtained when ischemia is evoked in controlled conditions, e.g. before and after the exercise stress test or by repeated examinations. Another possible application of the inverse method could be in revealing of regions responsible for transient beat-to-beat changes in ECG, e.g. those expressed as changes of the non-dipolarity index in integral BSPMs reported in [12].

5. Conclusion

From the results obtained in this study it is apparent that the use of a generic torso model with patient-specifically adjusted torso shape and with electrode positions defined in accordance with their real placement allows acceptable inverse localization of pathological cardiac events based on a dipole model of the cardiac electric generator. Accuracy of the solution is only slightly worse than that obtained with individual torso model created from MRI scans. However, the use of reasonably accurate heart model is still necessary.

The use of information from measured ECG signals can improve the individual positioning of the heart model in the torso in comparison to the standard heart position based on the ECG lead V2 level. However, in spite of this result, such information without proper rotation and scaling of the heart model does not lead to improved accuracy of the inverse solution. Hence some heart imaging allowing the creation of a patient/specific heart model seems unavoidable even if the whole torso imaging is not available.

Acknowledgment

The authors thank to Dr. Hoekema and prof. van Oosterom for providing the measured ECG data and MRI based real torso models used in this study.

The present study was supported by the research grant 2/0131/13 from the VEGA Grant Agency and by the grant APVV-0513-10 from the Slovak Research and Development Agency.

References

[1] HUISKAMP, G. and A. VAN OOSTEROM. Tailored versus realistic geometry in the inverse problem of electrocardiography. *IEEE Transactions on Biomedical Engineering*. 1989, vol. 36, iss. 8, pp. 827–35. ISSN 0018-9294. DOI: 10.1109/10.30808.

[2] BRUDER, H., B. SCHOLZ and K. ABRAHAM-FUCHS. The influence of inhomogeneous volume conductor models on the ECG and the MCG. *Physics in Medicine and Biology*. 1994, vol. 39, iss. 11, pp. 1949–1968. ISSN 0031-9155. DOI: 10.1088/0031-9155/39/11/010.

[3] HOEKEMA, R., G. J. UIJEN, L. VAN ERNING and A. VAN OOSTEROM. Interindividual variability of multilead electrocardiographic recordings: influence of heart position. *Journal of Electrocardiology*. 1999, vol. 32, iss. 2, pp. 137–148. ISSN 0022-0736. DOI: 10.1016/S1053-0770(99)90050-2.

[4] HOEKEMA, R., G. J. UIJEN and A. VAN OOSTEROM. Geometrical aspects of the interindividual variability of multilead ECG recordings. *IEEE Transactions on Biomedical Engineering*. 2001, vol. 48, iss. 5, pp. 551–559. ISSN 0018-9294. DOI: 10.1109/10.918594.

[5] CHENG, L. K, J. M. BODLEY and A. J. PULLAN. Effects of experimental and modeling errors on electrocardiographic inverse formulations. *IEEE Transactions on Biomedical Engineering*. 2003, vol. 50, iss. 1, pp. 23–32. ISSN 0018-9294. DOI: 10.1109/TBME.2002.807325.

[6] HORACEK, B. M. Numerical model of an inhomogeneous human torso. *Advances in Cardiology*. 1974, iss. 10, pp. 51–57. ISSN 0065-2326.

[7] LENKOVA, J., J. SVEHLIKOVA and M. TYSLER. Individualized model of torso surface for the inverse problem of electrocardiology. *Journal of Electrocardiology*. 2012, vol. 45, iss. 3, pp. 231–236. ISSN 0022-0736. DOI: 10.1016/j.jelectrocard.2012.01.006.

[8] SZATHMARY, V. and R. OSVALD. An interactive computer model of propagated activation with analytically defined geometry of ventricles. *Computers and Biomedical Research*. 1994, vol. 27, iss. 1, pp. 27–38. ISSN 0010-4809. DOI: 10.1006/cbmr.1994.

[9] TYSLER, M., M. TURZOVA and J. SVEHLIKOVA. Modeling of Heart Repolarization Using Realistic Action Potentials. *Measurement Science Review*. 2003, vol. 3, pp. 37–40, ISSN 1335-8871.

[10] OOSTENDORP, T. F. and A. VAN OOSTEROM. Source parameter estimation in inhomogeneous volume conductors of arbitrary shape. *IEEE Transactions on Biomedical Engineering*. 1989, vol. 36, iss. 3, pp. 382–391. ISSN 0018-9294. DOI: 10.1109/10.19859.

[11] ONDRACEK, O., J. PUCIK and E. COCHEROVA. Filters for ECG Digital Signal Processing. In: *Trends in Biomedical Engineering. Proceedings of International Conference*. Zilina: University of Zilina, 2005. pp. 91–96. ISBN 80-8070-443-0.

[12] DURRER, D., R. T. VAN DAM, G. E. FREUD, M. J. JANSE, F. L. MEIJLER and R. C. ARZBAECHER. Total excitation of the isolated human heart. *Circulation*. 1970, vol. 41, iss. 6, pp. 899–912. ISSN 1941-3149. DOI: 10.1161/01.CIR.41.6.899.

[13] METTINGVANRIJN, A. C., A. P. KUIPER, A. C. LINNENBANK and C. A. GRIMBERGEN. Patient isolation in multichannel bioelectric recordings by digital transmission through a single optical fiber. *IEEE Transactions on Biomedical Engineering*. 1993, vol. 40, iss 3, pp. 302–308. ISSN 0018-9294. DOI: 10.1109/10.216416.

[14] KOZMANN, G., K. HARASZTI and I. PREDA. Beat-to-beat interplay of heart rate, ventricular depolarization, and repolarization. *Journal of Electrocardiology*. 2010, vol. 43, iss. 1, pp. 15–24. ISSN 0022-0736. DOI: 10.1016/j.jelectrocard.2009.08.003.

About Authors

Milan TYSLER was born in Prague, Czech Republic. He received his M.Sc. in Computer Science from Faculty of Electrical Engineering, Slovak Technical University in Bratislava in 1974, Ph.D. degree from the Institute of Measurement Theory, Slovak Academy of Sciences in 1982 and became associate professor of Technical University in Kosice in 2006. His research interests include biosignal processing, modeling of biological processes oriented to the human cardiovascular system and development of intelligent biomedical instrumentation.

Jana LENKOVA was born in Presov, Slovakia. She received her M.Sc. in Biomedical Engineering from Faculty of Electrical Engineering, University of Zilina in 2009. Currently she finished her Ph.D. study in the Institute of Measurement Science, Slovak Academy of Sciences. Her research interests include cardiac electrical field modeling and research of the role of individual torso geometry in the forward and inverse problem of electrocardiography.

Jana SVEHLIKOVA was born in Bratislava, Slovakia. She received her M.Sc. in Biocybernetics from Faculty of Electrical Engineering, Slovak Technical University in Bratislava in 1986 and her Ph.D. degree from the Institute of Measurement Science, Slovak Academy of Sciences in 2011. Her research interests include modeling of the heart electrical activity, forward and inverse problem of electrocardiography and real-time biosignal measurement.

Development of the Real Time Situation Identification Model for Adaptive Service Support in Vehicular Communication Networks Domain

Mindaugas KURMIS[1,2], *Arunas ANDZIULIS*[2], *Dale DZEMYDIENE*[3],
Sergej JAKOVLEV[2], *Miroslav VOZNAK*[4], *Darius DRUNGILAS*[2]

[1]Institute of Mathematics and Informatics, Vilnius University, Akademijos St. 4, LT-08663 Vilnius, Lithuania
[2]Department of Informatics Engineering, Faculty of Marine Engineering Klaipeda University,
Bijunu St. 17, LT-91225 Klaipeda, Lithuania
[3]Institute of Communication and Informatics, Faculty of Social Policy, Mykolas Romeris University,
Ateities St. 20, LT-0830 Vilnius, Lithuania
[4]Department of Telecommunications, Faculty of Electrical Engineering and Computer Science,
VSB–Technical University of Ostrava, 17. listopadu 15, 708 33 Ostrava, Czech Republic

mindaugask01@gmail.com, arunas.iik.ku@gmail.com, daledz@mruni.eu, s.jakovlev.86@gmail.com,
voznak@ieee.org, dorition@gmail.com

Abstract. *The article discusses analyses and assesses the key proposals how to deal with the situation identification for the heterogeneous service support in vehicular cooperation environment. This is one of the most important topics of the pervasive computing. Without the solution it is impossible to adequately respond to the user's needs and to provide needed services in the right place at the right moment and in the right way. In this work we present our developed real time situation identification model for adaptive service support in vehicular communication networks domain. Our solution is different from the others as it uses additional virtual context information source - information from other vehicles which for our knowledge is not addressed in the past. The simulation results show the promising context exchange rate between vehicles. The other vehicles provided additional context source in our developed model helps to increase situations identification level.*

Keywords

Adaptive service support, context aware, situation identification, vehicular communication networks.

1. Introduction

Today, the vehicle is a very important component of human life, so its combination with the intelligence based software and hardware equipment can improve the level of travel safety and comfort. At the moment one of the most interesting and developed mobile technologies is the vehicular cooperation networks. In these networks vehicles communicating with each other and it open new opportunities for the vehicle industry and mobile service providers (Fig. 1). Differently from other pervasive computing devices vehicles have specific requirements and does not have strict energy constraints so it can be equipped with powerful computational resources, wireless transmitters and various sensors [1]. The vehicle must not distract drivers attention during driving, it must provide user with services autonomous and without user intervention. To provide the necessary services at the right time in the right place and in the right way it is necessary to adapt services and their support to user needs.

To solve these problems the vehicle must understand its environment and to identify current, past and possible future situations. For the situation awareness it can be used data from various sensors but this data is complex (different modality, huge amount with complex dependencies between sources), dynamic (real-time update, critical ageing) and different confidence [2]. Situation identification system must be able to recognize many different situations, to understand their relationships and context, and to control these situations. The system must be aware of simultaneous different situations and of that it cannot occur at the same moment. Considering the complex environment of system operation, high level of dynamics, imprecise data of sensors and other circumstances it is very difficult challenge to achieve a high level of situation identification.

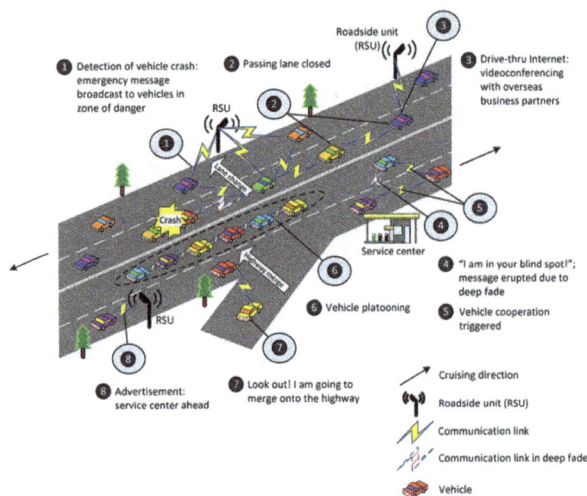

Fig. 1: Illustration of vehicular communication networks [1].

The aim of this work is to develop the real time situation identification model for adaptive service support in vehicular communication networks domain.

2.   Related Works

Despite the fact that it is a huge amount of researches in situation identification in pervasive computing domain but there is very small number of researches where it is addressed the specifics of vehicular communication and cooperation.

One of the first examples of the situation reasoning in the vehicular communication networks domain was presented in [3]. This paper proposes an approach for context and situation reasoning in V2V environment. The context and situations modelling are based on Context Spaces and it is integrated with the Dempster-Shafer rule of combination for situation reasoning. This approach was applied to a context middleware framework that aims to facilitate context and situation reasoning to provide cooperative applications in V2V environment. This work does not assess the real-time requirements and virtual context sources in the proposed framework.

Other example analyzing the context and intra-vehicular context can be found in [4]. In this work authors presents an on-board system which is able to perceive certain characteristics of the intra-vehicular context of its EgoV. It was defined a formal representation of the intra-vehicular context. The proposed system fusions the data from different vehicle sensors by means of a CEP approach to perceive two characteristics of the vehicular context, the occupancy and the places or landmarks of the itineraries usually covered by the EgoV. The real-time constrains also was not addressed in this work.

The survey of context modelling and reasoning techniques can be found at [5]. Authors described the state-of-the-art in context modelling and reasoning that supports gathering, evaluation and dissemination of context information in pervasive computing. It was showed that the existing approaches to context information modelling differ in the expressive power of the context information models, in the support they can provide for reasoning about context information, and in the computational performance of reasoning. Unfortunately authors did not take into account the specifics of vehicular communication networks domain. Most of the analyzed methods are applicable to healthcare and other pervasive systems.

3.   Specifics of Vehicular Communication Networks

Vehicular communication networks have special characteristics and features that distinguish it from the other kinds of mobile communication networks. The main unique characteristics:

- High energy reserve.

- Huge mass and size of the vehicle.

- Moving by the templates.

The vehicles have much bigger reserve of the energy comparing to ordinary mobile device. The energy can be obtained from batteries and it can be recharged by the gasoline, diesel or alternative fuel engine. The vehicles are many times larger and heavier compared to ordinary mobile networks clients and it can support much larger and powerful computational and sensor components. The computers can be provided by powerful processors, huge amount of memory and fast wireless connections (3G, LTE, WiMAX, 802.11p, etc.). Vehicles can move at high speed (160 km·h^{-1}) or even more so it is difficult to maintain constant V2V or V2I connection and to provide necessary services. However existing statistical data about traffic such as moving together by some templates or moving in the rush hours can be used to identify some types of situations and sequences of situation occur. The situation identification is also influenced by the scenario of vehicles movement. In the rural areas there are fewer obstacles and interferences but the driving speed is higher and the number of information sources is lower. In the city there is a high level of interference and obstacles however the driving speed is lower and number of information sources is higher (see Tab. 1).

Tab. 1: Influence of different vehicles movement scenarios.

Parameter Scenario	Rural	Town	City	Highway
Average movement speed	Average	Low	Very low	Very high
Node density	Low	Average	Very high	Average/ low
Interference	Low	Average	Very high	Low
Number of radio obstacles	Low	Average	Very high	Low

3.1. Sensors for the Situation Identification

To identify the situation in the vehicular environment it can be used various sensors and other sources of information. The raw data can be acquired from physical

Tab. 2: Proposed sensors for the situation recognition in the vehicular communication networks environment.

Sensor	Update rate	Information source	Data exchange
Physical			
GPS	High	Vehicle	inV
Speed	High	Vehicle	inV
Accelerometer	High	Vehicle	inV
Temperature	Low	Vehicle	inV
Fuel quantity	Low	Vehicle	inV
No of passengers	Low	Vehicle	inV
Vision	High	Vehicle	inV
Voice commands	Average	Vehicle	inV
Radar (Milimetre wave radar system)	High	Vehicle	inV
WSN	Average	Environment	V2I
Wireless interface info	Low	Wireless equipment	inV
Virtual			
Calls	Low	Smartphone	V2M
Calendar	Low	Smartphone	V2M
Reminders	Low	Smartphone	V2M
User preferences	Low	Smartphone	V2M
Road information	High	Other vehicles government environment	V2I V2V V2M
Warnings	High	Other vehicles government environment	V2I V2V V2M
Interaction with other vehicles	Average	Environment	V2I V2V V2M

sensors deployed in the vehicle: video cameras, GPS, microphones, movement dynamics, vehicle parameters, etc. and from virtual sensors: user preferences, data from Smartphone/tablet (calendar, reminders, social networks) and from other vehicles data (warnings, road information, etc.). This collected data makes vehicle user context. The context of an entity is a collection of measured and inferred knowledge that describe the state and environment in which an entity exists

or has existed [6]. This definition includes two types of knowledge: facts that can be measured by sensors (physical or virtual) and inferred data using machine learning, reasoning or applying other methods of artificial intelligence to the current of past context. Due to discussed specifics of vehicular communication networks sensors used in the vehicles covers much broader spectrum than used in the traditional ubiquitous environment. Table 2 shows the proposed sensors for the situation recognition in the vehicular communication networks environment.

Due to not strict requirements of energy consumption it can be used more different sensors (physical and virtual) and it can be acquired and analyzed more data. In this way using the methods of artificial intelligence the situations can be recognized more accurate and faster. In the Tab. 2 can be seen the update rate of the data, information source and data exchange ways: inV (in vehicle), V2I (vehicle to infrastructure), V2M (vehicle to mobile device), V2V (vehicle to vehicle). Different sensors provide different data types: binary, numerical and features so the software and hardware must be able to deal with all types of data.

4. Proposed Model

Our proposed system model (Fig. 2) for situation identification in the vehicle communication networks domain infers situations and associates it with other situations in the system. The system acquires the data from different sources (physical and virtual) of sensors, then it is performed the information pre-processing and de-noising procedures. The processed data is transferred to the reasoning engine which employing different methods of artificial intelligence (logic rules, expert system, ontologies) associates the context with the data from different sensors. In this way the reasoning engine infers the current vehicle situation. By using the knowledge of the current, past and possible future situation the system selects best services, adopts it to the user and provides it.

5. Experimental Results

It was created a program to test the developed system (see Fig. 3) in LabVIEW graphical programming environment. It acquires data from various sensors with the cRIO Real-Time controller. It was used 4 AI modules to acquire 16 channels of context data at 40 kS·s^{-1} per channel. A DMA FIFO was used to pass the data to the real-time controller, which then, via an RT FIFO, passes the data to a TCP/IP consumer loop and streams the data over the network to a host PC.

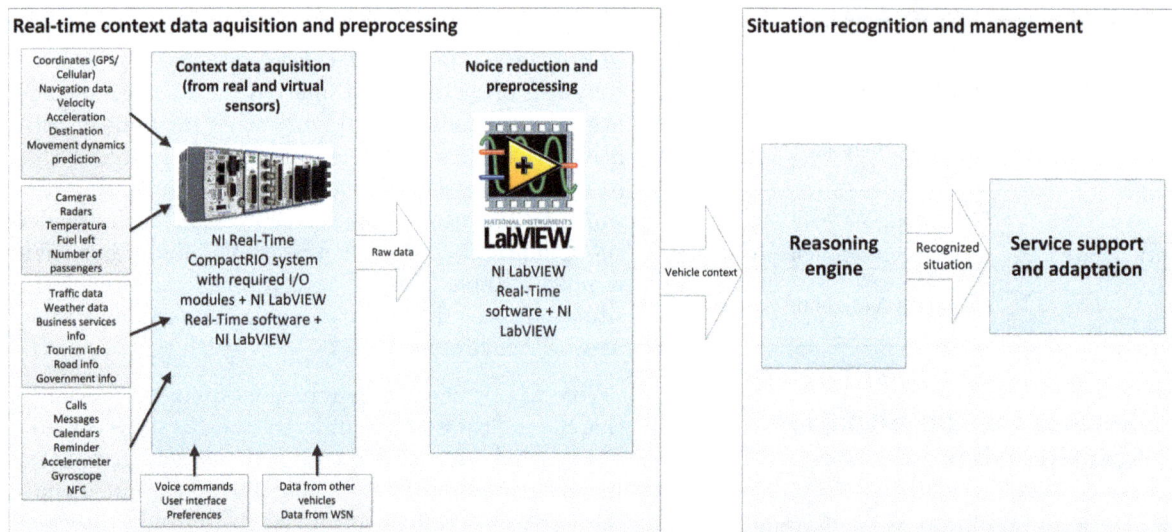

Fig. 2: Architecture of the proposed real-time situation identification model.

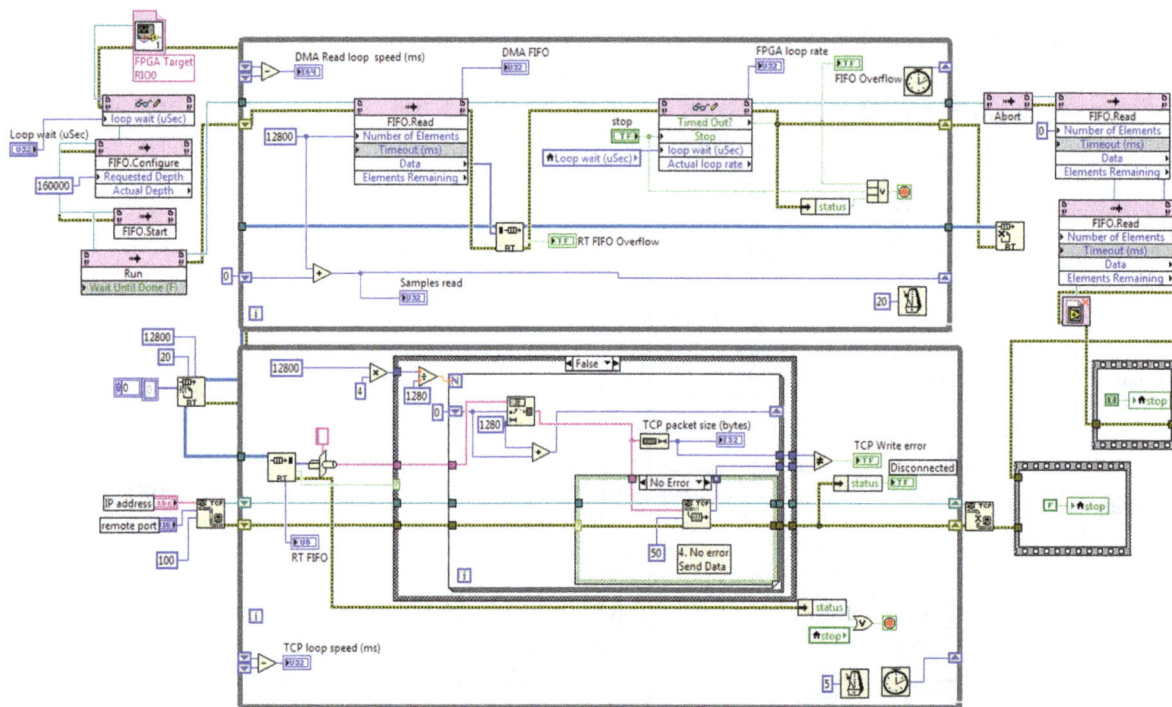

Fig. 3: LabVIEW environment program for the signal acquisition from sensors.

The PC Host program was used to get the context data to the host PC over the network (see Fig. 4). The data was pre-processed and transferred to the reasoning engine. Then the context data was transferred to the emulated mobile nodes (vehicles) in the ESTINET simulation environment. It was investigated the context data transfer capabilities in the mobile network.

The emulation were carried out in the simulation environment ESTINET 8.0 [7]. The environment was chosen as it uses the existent Linux TCP/UDP/IP protocols stack, it provides high-accuracy results; it can be used with any actual Unix application on a sim-

ulated node without additional modifications; it supports 802.11a/b/p communication networks and vehicle mobility modelling, user-friendly user interface, and it is capable of repeating the simulation results. In the experimental scenario the context data was sent from one vehicle to the other. Communication is provided via 801.11b standard interface and is used multi-hop data transmission method.

The experiment was carried out when the number of nodes in the network is from 10 to 100 - simulating different traffic congestion to determine the impact of the vehicle's number for the data-transfer ef-

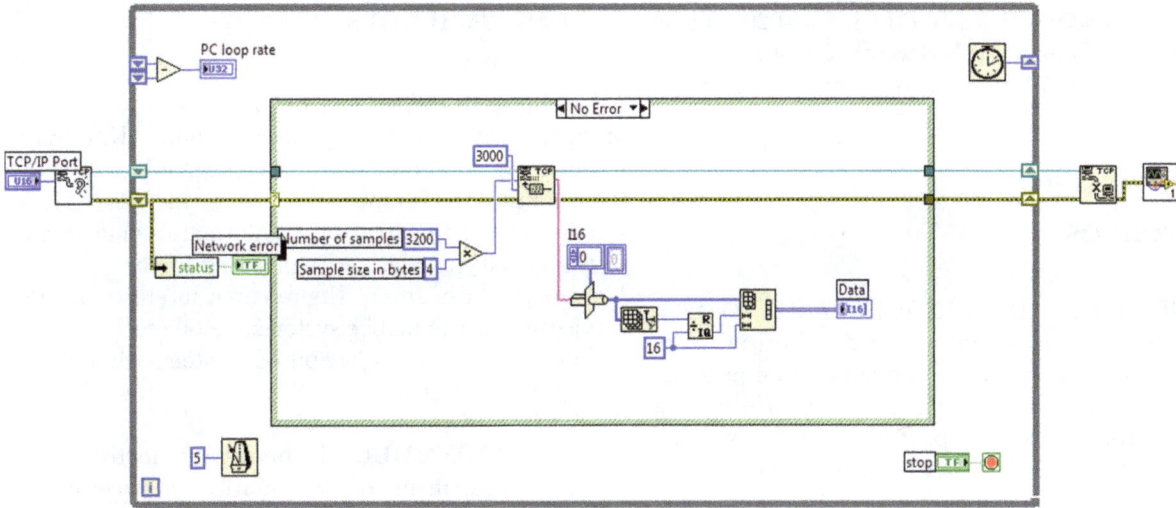

Fig. 4: PC Host program to get the context data to the host PC over the network.

ficiency. Senders and receiver's nodes are moving at high speed (130 km·h^{-1}) in the opposite directions. The remaining vehicles are moving at different speeds from 90 km·h^{-1} to 150 km·h^{-1}, and their speed and directions of movement are spread evenly. These parameters are chosen to simulate the realistic movement of cars on highway conditions.

During the experiments the average data uplink and downlink throughput was measured (Fig. 5). In this case, the highest mean transfer rate achieved by the network operating 20 vehicles, while the meanest - 30. The maximum average data rate of downlink - 100 vehicles, while the meanest - 50. The data rate is sufficient for the real implementation of the solution.

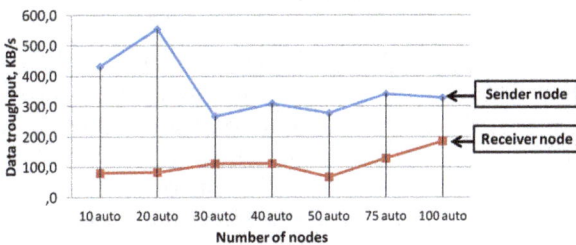

Fig. 5: The average context data downlink and uplink throughput with a different number of vehicles on the network.

Also it was found investigated collision's dependence on sender and receiver nodes with a different number of vehicles (Fig. 6). Collision rate is directly proportional to the number of vehicles. Up to 40 vehicles, collisions rate at the receiver and sender nodes is similar, but from 50 vehicles, collision is greater in sender node because of unsuitable channel access mechanisms.

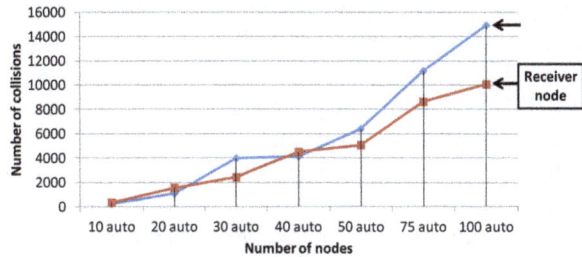

Fig. 6: Collisions rate dependence on receiver and sender nodes with a different number of vehicles on the network.

6.   Conclusion

In this work we present our developed real time situation identification model for adaptive service support in vehicular communication networks domain. We have tested our solution during various experiments. The results showed that the solution is able to work in the real-time vehicular communication networks environment. There are still many problems in the field of routing protocols with a huge number of nodes. Future plans are to extend the study and to test the system in real life vehicular environment.

Acknowledgment

Authors would like to thank the Project LLIV-215 "JRTC Extension in Area of Development of Distributed Real-Time Signal Processing and Control Systems" for the opportunity to complete a scientific research. Also this work has been supported by the European Social Fund within the project "Development and application of innovative research methods and solutions for traffic structures, vehicles and their flows", code VP1-3.1-SMM-08-K-01-020 and by the European Community's Seventh Framework Programme (FP7/2007-2013) under grant agreement no. 218086.

References

[1] CHENG, H. T., H. SHAN and W. ZHUANG. Infotainment and road safety service support in vehicular networking: From a communication perspective. *Mechanical Systems and Signal Processing.* 2011, vol. 25, iss. 6, pp. 2020–2038. ISSN 0888-3270. DOI: 10.1016/j.ymssp.2010.11.009.

[2] YE, J., S. DOBSON and S. MCKEEVER. Situation identification techniques in pervasive computing: A review. *Pervasive and Mobile Computing.* 2012, vol. 8, iss. 1, pp. 36–66. ISSN 1574-1192. DOI: 10.1016/j.pmcj.2011.01.004.

[3] WIBISONO, W., A. ZASLAVSKY and S. LING. Improving situation awareness for intelligent on-board vehicle management system using context middleware. In: *IEEE Intelligent Vehicles Symposium, 2009.* Xian: IEEE, 2009, pp. 1109–1114. ISBN 978-1-4244-3503-6. DOI: 10.1109/IVS.2009.5164437.

[4] TERROSO-SAENZ, F., M. VALDES-VELA, F. CAMPUZANO, J. A. BOTIA and A. F. SKARMETA-GOMEZ. A Complex Event Processing Approach to Perceive the Vehicular Context. *Information Fusion.* 2012, vol. 13, iss. 2, pp. 1–23. ISSN 1566-2535. DOI: 10.1016/j.inffus.2012.08.008.

[5] BETTINI, C., O. BRDICZKA, K. HENRICKSEN, J. INDULSKA, D. NICKLAS, A. RANGANATHAN and D. RIBONI. A survey of context modelling and reasoning techniques. *Pervasive and Mobile Computing.* 2010, vol. 6, iss. 2, pp. 161–180. ISSN 1574-1192. DOI: 10.1016/j.pmcj.2009.06.002.

[6] LATRE, S., J. FAMAEY, J. STRASSNER and F. DE TURCK. Automated context dissemination for autonomic collaborative networks through semantic subscription filter generation. *Journal of Network and Computer Applications.* 2013, vol. 36, iss. 1, pp. 1–13. ISSN 1084-8045. DOI: 10.1016/j.jnca.2013.01.011.

[7] WANG, S.-Y. and Y.-M. HUANG. NCTUns distributed network emulator. *Internet Journal.* 2012, vol. 4, no. 2, pp. 61–94. ISSN 1937-3805.

About Authors

Mindaugas KURMIS is Ph.D. student at Institute of Mathematics and Informatics, Vilnius University (Lithuania). He graduated M.Sc. degree in Informatics Engineering, Klaipeda University in 2011. He is lecturer at the Department of Informatics engineering of the Faculty of Marine Engineering of the Klaipeda University (Lithuania). His research interests include service support in mobile systems, mobile technologies, computer networks, distributed systems, intelligent transport systems.

Arunas ANDZIULIS is professor, doctor, head of the Department of Informatics engineering of the Faculty of Marine Engineering of the Klaipeda University (Lithuania). He holds a diploma with honour of physics in 1968 of Vilnius University, Ph.D. in electronic engineering in 1968 of Kaunas Institute of Technology, habilitation doctor procedure in the field of transport engineering sciences in 2007, and long time works as a head, designer and project manager. His research interests include operation research, nanotechnology and modelling, optimization of technical systems, intelligent transportation/logistic systems.

Dale DZEMYDIENE is doctor, professor of the Institute of Communication and informatics, Mykolas Romeris University (Lithuania). She holds a diploma of Applied mathematics in 1980 of Kaunas University of Technologies, Ph.D. in mathematics - informatics in 1995 of Vilnius University, habilitation doctor procedure in the field of management and administration in 2004. Her research interests include information systems, intelligent transportation systems, service support in mobile networks.

Sergej JAKOVLEV is Ph.D. student at Department of Informatics Engineering, Klaipeda University (Lithuania). He graduated M.Sc. degree in informatics engineering, Klaipeda University in 2011. His research interests include intelligent transportation systems, mobile technologies, wireless sensor networks.

Darius DRUNGILAS is Ph.D. student at Institute of Mathematics and Informatics, Vilnius University (Lithuania). He graduated M.Sc. degree in computer science, Klaipeda University in 2009. He is lecturer at the Department of Informatics engineering of the Faculty of Marine Engineering of the Klaipeda University (Lithuania). His research interests include methods of artificial intelligence, agent-based modelling, digital adaptive control, machine learning and data mining.

Miroslav VOZNAK holds position as an associate professor with Department of Telecommunications, Technical University of Ostrava, since 2013 he is a department chair. He received his M.Sc. and Ph.D. degrees in telecommunications, dissertation thesis "Voice traffic optimization with regard to speech quality in network with VoIP technology" from the Technical University of Ostrava, in 1995 and 2002, respectively. The topics of his research interests are Next Generation Networks, IP telephony, speech quality and network security.

3

Analysis of Fixing Nodes Used in Generalized Inverse Computation

Pavla HRUSKOVA

Department of Applied Mathematics, Faculty of Electrical Engineering and Computer Science,
VSB–Technical University Ostrava, 17. listopadu 15, 708 33 Ostrava, Czech Republic

pavla.hruskova@vsb.cz

Abstract. *In various fields of numerical mathematics, there arises the need to compute a generalized inverse of a symmetric positive semidefinite matrix, for example in the solution of contact problems. Systems with semidefinite matrices can be solved by standard direct methods for the solution of systems with positive definite matrices adapted to the solution of systems with only positive semidefinite matrix. One of the possibilities is a modification of Cholesky decomposition using so called fixing nodes, which is presented in this paper with particular emphasise on proper definition of fixing nodes. The generalised inverse algorithm consisting in Cholesky decomposition with usage of fixing nodes is adopted from paper [1]. In [1], authors choose the fixing nodes using Perron vector of an adjacency matrix of the graph which is only a sub-optimal choice. Their choice is discussed in this paper together with other possible candidates on fixing node. Several numerical experiments including all candidates have been done. Based on these results, it turns out that using eigenvectors of Laplacian matrix provides better choice of fixing node than using Perron vector.*

Keywords

Cholesky decomposition, fixing nodes, generalised inverse, Laplacian matrix, spectral graph theory, stiffness matrix.

1. Introduction

The motivation of this paper is a computation of a generalized inverse of a symmetric positive semidefinite (SPS) matrix which arises for example in solution of contact problems. One of the possibilities is a modification of Cholesky decomposition.

For the purposes of this paper, we will consider the meshes arising during discretization of numerical problems from graph theory point of view and we will analyse these meshes as graphs.

The particular purpose of this analysis is to find certain nodes (further defined as "fixing nodes") in meshes to reduce numerical instability in computation of a generalized inverse of semidefinite stiffness matrix.

Algorithm presented in this paper is taken from [1]. In [1], authors use the Perron vector of an adjacency matrix for the computation of fixing nodes, which is only a sub-optimal choice.

In this paper, the proper definition of fixing nodes is presented. Several candidates to fixing nodes are provided based on (spectral) graph techniques together with experimental results. Based on these results, it turns out that using eigenvectors of Laplacian matrix provides better choice of fixing node than using Perron vector (eigenvector of the adjacency matrix). The best choice of fixing node is analysed at the end of this paper.

2. Generalized Inverse Algorithm

Let us consider a system of consistent linear equations:

$$\mathbf{A}x = b, \tag{1}$$

with an SPS matrix $\mathbf{A}$. In case of FETI method, matrix $\mathbf{A}$ is called the stiffness matrix.

If $\mathbf{A} \in \mathbb{R}^{n \times n}$ and $b \in \operatorname{Im} \mathbf{A}$, where $\operatorname{Im} \mathbf{A}$ denotes the range of $\mathbf{A}$, then a solution $x = \mathbf{A}^{+}b + \mathbf{R}c$ of the system of linear equations Eq. (1) can be expressed by means of a (left) generalized inverse matrix, $\mathbf{A}^{+} \in \mathbb{R}^{n \times n}$, $\mathbf{R}$ is the matrix of rigid body modes, and $\vec{c}$ is a vector of coefficients.

There are several known algorithms how to compute generalized inverse matrix. There can be used either standard direct methods or some of iterative methods such as Cholesky decomposition, singular value decomposition, and their combinations.

In paper *Cholesky decomposition with fixing nodes to stable computation of a generalized inverse of the stiffness matrix of a floating structure* [1], authors Dostal, Kozubek, Kovar, Markopoulos, and Brzobohaty present their algorithm of computation of generalized inverse matrix based on Cholesky decomposition, adapted to the solution of systems with SPS matrix.

Generalized inverse algorithm [1] consists in a decomposition of the SPS matrix $\mathbf{A} \in \mathbb{R}^{n \times n}$:

$$\mathbf{PAP}^T = \begin{bmatrix} \widetilde{A}_{JJ} & \widetilde{A}_{JI} \\ \widetilde{A}_{IJ} & \widetilde{A}_{II} \end{bmatrix} =$$
$$= \begin{bmatrix} L_{JJ} & O \\ L_{IJ} & I \end{bmatrix} \begin{bmatrix} L_{JJ}^T & L_{IJ}^T \\ O & S \end{bmatrix}, \quad (2)$$

where $\widetilde{A}_{JJ} \in \mathbb{R}^{r \times r}$ is well-conditioned regular part of $\mathbf{A} \in \mathbb{R}^{n \times n}$, $L_{JJ} \in \mathbb{R}^{r \times r}$ is a lower factor of the Cholesky factorization of $\widetilde{A}_{JJ}$, $L_{IJ} \in \mathbb{R}^{s \times r}$, $L_{IJ} = \widetilde{A}_{IJ} L_{JJ}^{-T}$, $\mathbf{S} \in \mathbb{R}^{s \times s}$ is a singular matrix, $s = n - r$ is the number of displacements corresponding to the fixing nodes, and $\mathbf{P}$ is a permutation matrix.

Then the generalised inverse $\mathbf{A}^+$ is computed as:

$$\mathbf{A}^+ = \mathbf{P}^T \begin{bmatrix} L_{JJ}^{-T} & -L_{JJ}^{-T} L_{IJ}^T S^\dagger \\ O & S^\dagger \end{bmatrix} \cdot$$
$$\cdot \begin{bmatrix} L_{JJ}^{-1} & O \\ -L_{IJ} L_{JJ}^{-1} & I \end{bmatrix} \mathbf{P}, \quad (3)$$

where $S^\dagger$ denotes the Moore–Penrose generalized inverse. Permutation matrix $\mathbf{P}$, $\mathbf{P} = \mathbf{P}_2 \mathbf{P}_1$, is computed in two steps. First, matrix $\mathbf{P}_1$ is found in the form:

$$\mathbf{P}_1 \mathbf{A} \mathbf{P}_1^T = \begin{bmatrix} \widetilde{A}_{JJ} & \widetilde{A}_{JI} \\ \widetilde{A}_{IJ} & \widetilde{A}_{II} \end{bmatrix}, \quad (4)$$

where $\widetilde{A}_{JJ}$ is nonsingular and $\widetilde{A}_{II}$ corresponds to the degrees of freedom of the M fixing nodes.

Second, reordering algorithm on $\mathbf{P}_1 \mathbf{A} \mathbf{P}_1^T$ is applied to get a permutation matrix $\mathbf{P}_2$ which leaves the part $\widetilde{A}_{II}$ without changes and enables the sparse Cholesky factorization of $\widetilde{A}_{JJ}$.

3. Fixing Nodes and Center-like Points

3.1. Fixing Nodes

Finding fixing nodes effectively and accurately is the main ingredient of the generalized inverse algorithm.

The active choice is made in sense of permutation matrix $\mathbf{P}_1$ in Eq. (4) such that the rows/columns corresponding to those vertices marked as fixing nodes

are permuted at the end of the original stiffness matrix (bottom-right block). Also, the rigid body motions corresponding to degrees of freedom of selected fixing nodes are removed. Let us rewrite the definition of fixing node (first written in [2]).

Definition 1. *Fixing node*: Let $\mathbf{A}x = b$ be a system of linear equations arising from a finite element or finite difference discretization of the problem, such that $\mathbf{A}$ has one-dimensional kernel (i.e. the singular part $\widetilde{A}_{II}$ in Eq. (4) is formed by one zero element).

The one–fixing node is the node that makes the regular part $\widetilde{A}_{JJ}$ of the stiffness matrix $\mathbf{A}$ produced by the permutation $\mathbf{P}$ nonsingular and well conditioned, i.e. permutation of this node to the last row/column of the matrix $\mathbf{A}$ makes the condition number of the regular part $\widetilde{A}_{JJ}$ finite and sufficiently small.

The best choice of one–fixing node is the node k for which the regular part $\widetilde{A}_{JJ}$ of the stiffness matrix $\mathbf{A}$ produced by the permutation $\mathbf{P}$ has the minimal condition number over all $\widetilde{A}_{\widehat{kk}}$ (Symbol $\widetilde{A}_{\widehat{kk}}$ denotes the principal submatrix arising by removing k-th row and column from original matrix $\mathbf{A}$):

$$\text{cond}(\widetilde{A}_{JJ}) = \min_{k=\{1,\dots,n\}} \text{cond}(\widetilde{A}_{\widehat{kk}}). \quad (5)$$

Condition number $\text{cond}(\widetilde{A}_{\widehat{kk}})$ is computed as:

$$\text{cond}(\widetilde{A}_{\widehat{kk}}) = \frac{\lambda_{max}(\widetilde{A}_{\widehat{kk}})}{\lambda_{min}(\widetilde{A}_{\widehat{kk}})}, \quad (6)$$

where $\lambda_{max}(\widetilde{A}_{\widehat{kk}})$ and $\lambda_{min}(\widetilde{A}_{\widehat{kk}})$ denote the largest and the smallest eigenvalue of the non-singular matrix $\widetilde{A}_{\widehat{kk}}$, respectively. Definition of more fixing nodes is composed accordingly.

Definition 2. *i-fixing nodes*: The i-fixing nodes (if the number i of i-fixing nodes is not important, we use only the term "fixing nodes" instead of the term "i-fixing nodes") are the set of i nodes that make the regular part $\widetilde{A}_{JJ}$ of the stiffness matrix A produced by the permutation $\mathbf{P}$ nonsingular and well conditioned, i.e., permutation of these nodes to the bottommost rows, rightmost columns respectively, of the matrix $\mathbf{A}$ makes the condition number of regular part $\widetilde{A}_{JJ}$ finite and sufficiently small (see Eq. (4)).

The best choice of i-fixing nodes is the set of i-fixing nodes for which the regular part $\widetilde{A}_{JJ}$ of the stiffness matrix $\mathbf{A}$ produced by the permutation $\mathbf{P}$ has the minimal condition number over all such permutations.

Further, we show techniques how to find one fixing node. One fixing node can stabilize well the solution of problems with one-dimensional kernel. Problem of finding more fixing nodes can be reduced to the problem of finding one fixing node by decomposition of the mesh corresponding to given problem by a suitable software into several parts. The fixing nodes are then found one in each part.

In this case, it is guaranteed that we can obtain the best choice of one–fixing node in each part in sense of Definition 1 but the resulting set of i-fixing nodes of the whole problem do not need to be the best choice of i-fixing nodes in sense of Definition 2.

Definition 1 operates on condition numbers of residual matrices, thus it is not suitable to evaluate the fixing node in a reasonable time. Therefore, a fast heuristic has been looked for to provide a good approximation of the fixing node.

As the fixing nodes should lie near the "center" of the mesh [3], there are several possibilities how to define center-like points of mesh, or corresponding graph respectively.

3.2.   Center-like Points

One of the approaches to find an approximation of the best choice of one–fixing node is based on the mechanical interpretation of the problem. It follows the idea of choice of the fixing node near the "center" of mesh. This translates directly into choosing one of the vertices in the center of the graph. Finding graph centers is not suitable for the numerical solution of large problems but it provides a good referential point to the other methods. This approach has been published in [3].

Another idea how to identify the fixing node is based on spectral approach, namely on eigenvector of the adjacency matrix of the graph.

Definition 3. *Capital vertex*: Let the capital vertex of a graph G be a vertex v_i that corresponds to the index i of the highest value in the Perron vector of the corresponding adjacency matrix of graph G.

The capital vertex does not identify the center of the graph, rather a vertex from which most walks of a given length can be realized. By choosing the capital vertex to be the fixing node we expect to achieve a stable numerical solution. Moreover, identifying the capital vertex is fast.

Authors Dostal, Kozubek, Kovar, Markopoulos, and Brzobohaty use exactly the Perron vector for fixing nodes computation in their generalized inverse algorithm [1].

In Fig. 1 overtaken from [1], we can see that the position of the fixing node in the rightmost side subdomain differs from the position where we naturally expect it should be (i.e. closer to the "center" of given subdomain). One can see that the highest value of the Perron

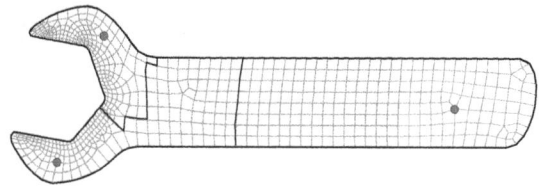

Fig. 1: Mesh with wrong position of fixing nodes.

vector arises at the vertex with the higher degree rather than in one with the mean value of the degree.

The last presented approach how to obtain a good approximation of the best choice of one–fixing node mentioned in this text is based on the eigenvectors of the Laplacian matrix.

The Laplacian matrix can be set up in dependence on the boundary conditions. As standard Laplacian matrix corresponds to problem with the Neumann boundary, it is more suitable to represent the problem that we are interesting in.

According to M. Fiedler theory [4], [5], the eigenvector corresponding to the second smallest eigenvalue of the Laplacian matrix is used to graph partitioning [6], [7]. It is known, that the so-called Fiedler cut defines the cut in certain coordinate direction. The interesting thing is that in two-dimensional case, the eigenvector corresponding to the third smallest eigenvalue defines the cut in the second coordinate direction. Because the Fiedler cut is known to be somehow "optimal", it is reasonable to expect a good approximation of one–fixing node near the crossing of both cuts.

For purposes of definition of a generalized eigenvector cut, the notation based on "Fiedler's tree theorem" is used, especially notation of the characteristic vertex and the characteristic edge.

Using the other Fiedler's results, the Fiedler cut is formed by the characteristic set (i.e. set of characteristic vertices or characteristic edges). In general, characteristic set can be defined for arbitrary eigenvector of the Laplacian matrix and for arbitrary type of graph.

Definition 4. *k-level cut*: Let $G = (V, E)$ be a graph on n vertices, labelled $1, 2, \ldots, n$, with Laplacian matrix $\mathbf{L}(G)$. Let v_k be an eigenvector of $\mathbf{L}(G)$ associated with the k-th eigenvalue ($v_k[i]$ denotes i-th element of v_k).

The k-level cut is defined as a characteristic set, i.e. either set of characteristic vertices or set of characteristic edges, where:

- characteristic vertices are the vertices i that satisfy $v_k[i] = 0$,

- characteristic edges are the edges ij such that i and j are adjacent in G, $v_k[i] > 0$ and $v_k[j] < 0$ (or $v_k[i] < 0$, $v_k[j] > 0$).

Following this notation, the 2-level cut is exactly the Fiedler cut. As the spectrum of graph (or meshes respectively), respects the physical properties of given objects, i.e. the space-dimension, also the cross-eigenvector center has to be defined with respect to this characteristic (dimension of the mesh).

Definition 5. *Cross-eigenvector center*: Let $G = (V, E)$ be a graph on n vertices:

- For one-dimensional mesh, the cross-eigenvector center is the vertex, edge respectively, that lies on the 2-level cut.

- For two-dimensional mesh, the cross-eigenvector center is the vertex, edge or 2D element respectively, that lies on the crossing of the 2-level cut and the 3-level cut.

- For three-dimensional mesh, the cross-eigenvector center is the vertex, edge, face or 3D element respectively, that lies on the crossing of the 2-level cut, 3-level cut and 4-level cut.

In computational arithmetic, i.e. in presence of rounding errors, there is usually problem with zero identification. The cross-eigenvector center is always assigned to the vertex with the smallest value (in absolute value). The exact position of the cross-eigenvector center can differ no more than half edge, half element respectively.

In Section 5. there is presented the sketch of the proof that the cross-eigenvector center delivers the best choice of one–fixing node (instead of the other center-like points), which is its biggest advantage. In opposite to the capital vertex, the eigenvectors corresponding to the second, third, and fourth smallest eigenvalue of the Laplacian matrix are harder to compute than the eigenvector corresponding to the largest eigenvalue of the adjacency matrix, which is a disadvantage.

4. Experiments

The results of positioning of fixing nodes based on listed definitions are presented in this section. The criterion to measure quality of the approximation of one–fixing node is the condition number of the regular part $\tilde{A}_{JJ}$ of the matrix $\mathbf{A}$ of given problem Eq. (4) after permutation of the row and column corresponding to the fixing node at the end of the matrix.

Given examples are quite small (less than 121 vertices and 100 edges) because the finding of the best choice of one–fixing node according to Definition 1 yields to removing all vertices, computing the condition number of the regular part of the matrix (i.e. the largest and the smallest eigenvalue) for each removal,

and choosing the minimal one which is a huge time-consuming process.

In figures the following symbols appear:

- The best choice of one–fixing node satisfying Definition 1 is drawn as a circle ○.

- The graph centers are drawn as a square ■.

- The capital vertex satisfying Definition 3 is drawn as a triangle △.

- The cross-eigenvector center satisfying Definition 5 is drawn as a star ✳.

Let us have a look in the rightmost subdomain in Fig. 2. As the cross-eigenvector center (color cyan) almost the same vertex as the best choice of one–fixing node (color green) has been detected, meanwhile the capital vertex (color red) has been assigned to the vertex with higher degree which is far from the best choice of one–fixing node.

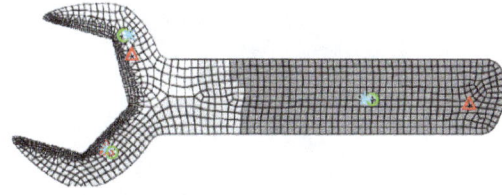

Fig. 2: Comparison of two approaches.

Let us briefly present the result of this phenomenon.

At first, let us have a look on the behaviour of center-like points of the meshes with regular elements, see Fig. 3(a). In two-dimensional space, the element corresponds to the graph face (region bounded by edges). Our examples consist of quadrilateral type elements only. As we can see from Fig. 3(a) and from the Tab. 1, almost all approaches deliver a good approximation of one–fixing node.

One can suppose that the behaviour of all center-like points is similar. But this holds only for meshes with uniform elements, or similarly, for meshes with inner vertices of the same degree. Let us have a look on the mesh with mixed elements (quadrilateral and triangle type) for further testing. This means that the degree of inner vertices differs from four to six.

For well understanding of the problem, several eigenvector of testing meshes are plotted (compare Fig. 3, Fig. 4). In Fig. 3(b), Fig. 4(b), the eigenvectors of the adjacency matrix corresponding to the highest eigenvalue are plotted. In Fig. 3(c), Fig. 4(c), the eigenvectors of the Laplacian matrix corresponding to the second smallest eigenvalue are plotted, and in Fig. 3(d), Fig. 4(d), the eigenvectors of the Laplacian matrix corresponding to the third smallest eigenvalue are plotted.

(a) Example 1

(b) A(G), v_1

(c) L(G), u_2

(d) L(G), u_3

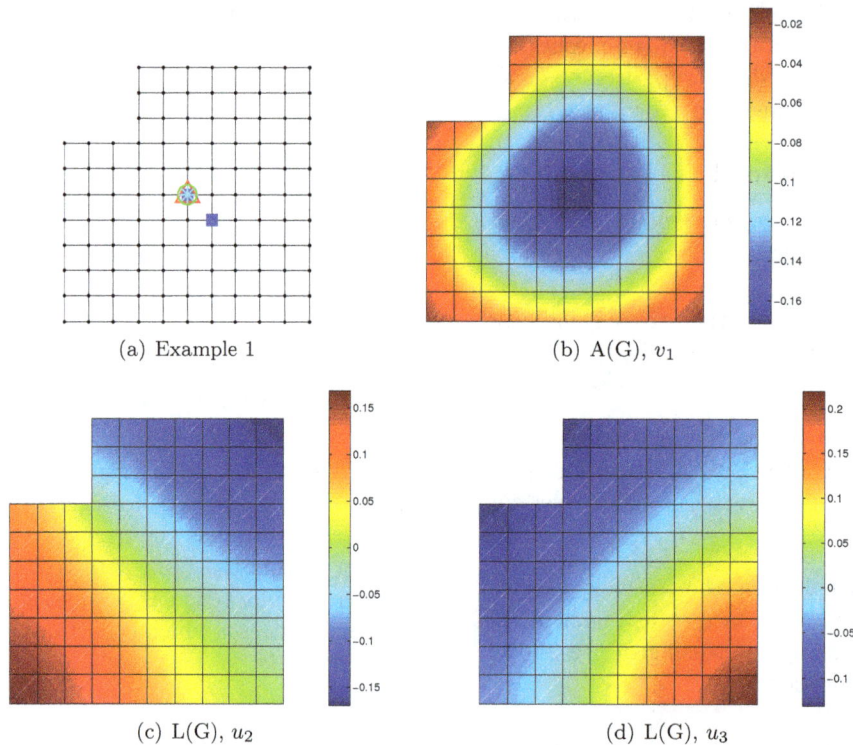

Fig. 3: Example 1 with plotted eigenvectors.

The testing mesh (Example 1) is similar to the Cartesian product of two paths $P_{11} \times P_{11}$ and also, the spectrum of this mesh is similar to the spectrum of Cartesian product $P_{11} \times P_{11}$. The mesh Example 1 can be understood as an approximate Cartesian product with deleted both edges and vertices. The missing left upper corner has no essential influence on the behaviour of the eigenvectors.

Another situation arises if the edges are added instead of deleted (see Fig. 4). The behaviour of the eigenvector of the adjacency matrix corresponding to the highest eigenvalue does not correspond to the classical behaviour of the spectrum of Cartesian products. Therefore, the capital vertex (based on the adjacency matrix) do not approximate the best choice of one–fixing node well.

As we can see in Fig. 4(c), Fig. 4(d), the eigenvectors of the Laplacian matrix are still indifferent to change of the structure of the mesh. The cross-eigenvector center still approximates the best choice of one–fixing node well.

For better comprehension, Tab. 1 is presented. The first column represents the name of the test example. The regular condition number of the matrix $\mathbf{A}$ in the second column is computed by:

$$\overline{\mathrm{cond}}(\mathbf{A}) = \frac{\lambda_{max}(\mathbf{A})}{\overline{\lambda}_{min}(\mathbf{A})}, \tag{7}$$

where $\lambda_{max}(\mathbf{A})$ and $\overline{\lambda}_{min}(\mathbf{A})$ correspond to the largest and to the smallest non-zero eigenvalue of $\mathbf{A}$. The regular condition number is used as a reference value, because it represents the smallest boundary to the condition number of the generalized inverse (the condition number of the generalized inverse can be never smaller than the condition number of the original matrix). The third column represents the condition number of the matrix $\widetilde{A}_{JJ}$ considered in Definition 1, i.e. the minimal possible condition number of the regular part $\widetilde{A}_{JJ}$ after removing the best choice of one–fixing node. The condition number of the regular part $\widetilde{A}_{JJ}$ after removing the graph center (without definition) is written in the fourth column.

If more vertices satisfy the definition, the vertex that causes the minimal condition number has been chosen. The fifth column represents the condition number of the matrix $\widetilde{A}_{JJ}$ when the capital vertex satisfying Definition 3 is removed. The last column represents the condition number of the matrix $\widetilde{A}_{JJ}$ when the cross-eigenvector center satisfying Definition 5 is removed.

We can see that choosing the capital vertex as the approximation of the best choice of one–fixing node instead the cross–eigenvector center debases the size of the condition number from 426.5 (best choice of one–fixing node, cross-eigenvector center) to 537.1, i.e. by 26 %.

Tab. 1: Approximation of the best choice of one–fixing node.

No.	$\overline{\text{cond}}(A)$ (regular)	$\kappa(\widetilde{A}_{JJ})$ (1-fixing node)	$\text{cond}(\widetilde{A}_{JJ})$ (graph center)	$\text{cond}(\widetilde{A}_{JJ})$ (Adjacency m.)	$\text{cond}(\widetilde{A}_{JJ})$ (Laplace m.)
1	107.7	379.8	379.8	379.8	379.8
2	99.2	426.5	426.5	**537.1**	426.5

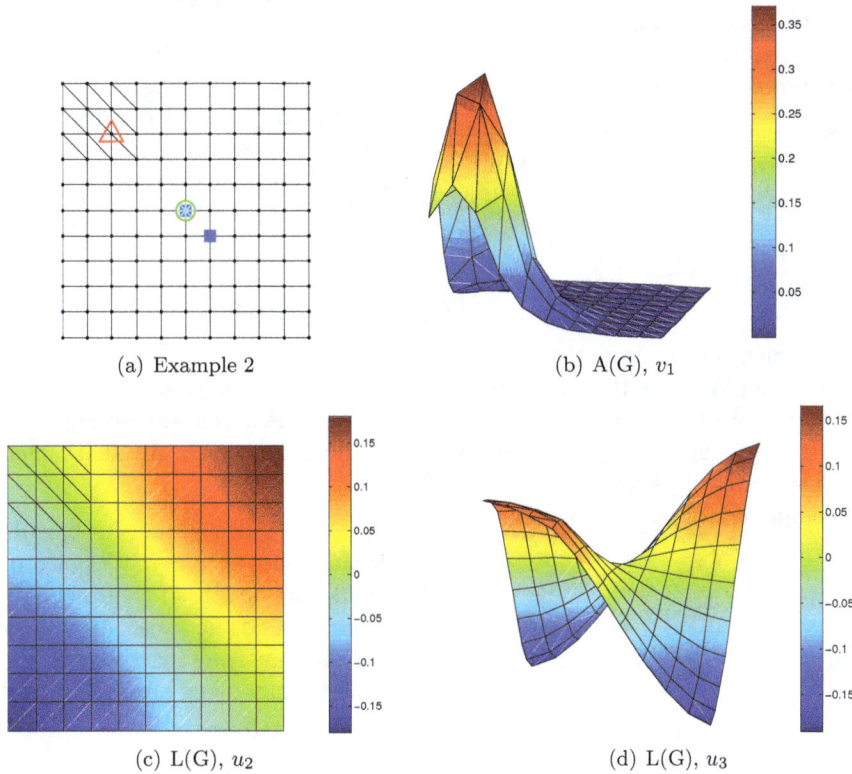

(a) Example 2

(b) A(G), v_1

(c) L(G), u_2

(d) L(G), u_3

Fig. 4: Example 2 with plotted eigenvectors.

5. The Best Choice of One–Fixing Node

Based on the experiments, it is seemed that the best approximation of the best choice of one–fixing node is the cross-eigenvector center. In this section, the particular theorem is presented.

Theorem 6. *The best choice of one–fixing node: The cross-eigenvector center (according to Definition 5) is the best choice of one–fixing node (according to Definition 1).*

I.e. if we remove the row and column corresponding to the cross-eigenvector center from the original matrix **A**, the remaining principal submatrix has the best condition number over all principal submatrices.

Remark that we restrict on the matrices with one-dimensional kernel, i.e. the resulting submatrix has deleted one row and one corresponding column. In

above definitions, we have worked with some general stiffness matrix **A**. As we consider the Laplacian matrix of graph, we will use the symbol **L** hereinafter. Symbols with $\sim$ as $\widetilde{\mathbf{L}}$, $\widetilde{\lambda}$, etc. will further denote variables of corresponding reduced problem. If we could emphasize that the i-th row/column is removed, we assign the corresponding variables as $\widetilde{\mathbf{L}}^i$, $\widetilde{\lambda}^i$, etc.

The condition number is considered in the form:

$$\text{cond}(\widetilde{\mathbf{L}}) = \frac{\widetilde{\lambda}_{max}(\widetilde{\mathbf{L}})}{\widetilde{\lambda}_{min}(\widetilde{\mathbf{L}})}. \tag{8}$$

From one of the variant of Cauchy-like interlacing theorems (see i.e. [8]), removing the fixing node does not change the $\widetilde{\lambda}_{max}$ so much as the $\widetilde{\lambda}_{min}$. Therefore, the minimization of the condition number corresponds to the maximization of the $\widetilde{\lambda}_{min}$. Theorem 6 can be paraphrased into following form:

Theorem 7. *Maximization of $\widetilde{\lambda}_{min}$: Removing the vertex k corresponding to the cross-eigenvector center* in Laplacian matrix $\mathbf{L}$, the maximal value of $\widetilde{\lambda}_{min}^{\star}$ of $\widetilde{\mathbf{L}}$ is obtained over all $\widetilde{\lambda}_{min}^{i}$ of all principal submatrices $\widetilde{\mathbf{L}}^{i}$, i.e., the condition number $\widetilde{\mathbf{L}}^{i}$ is minimized over all i.

Finding the maximal value of $\widetilde{\lambda}_{min}^{\star}$ can be written:

$$\widetilde{\lambda}_{min}^{\star} = \max_{i=\{1,2,\ldots,n\}} \widetilde{\lambda}_{min}^{i}. \tag{9}$$

As the whole proof of this theorem consists in comprehensive analysis of all cases, we restrict on the main ideas only. The complete proof can be found in [2].

Proof. In our considerations, we come out from the Rayleigh's principle [9]. The smallest eigenvalue of the reduced matrix $\widetilde{\mathbf{L}}$ can be computed as:

$$0 \neq \widetilde{\lambda}_{min} = \min \frac{x^T \widetilde{\mathbf{L}} x}{x^T x} = \min_{\|x\|=1} x^T \widetilde{\mathbf{L}} x. \tag{10}$$

As we would like to find which vertex i from the original matrix $\mathbf{L}$ should be removed to obtain the maximal value of $\widetilde{\lambda}_{min}$, we have to rewrite the above equation using original matrix $\mathbf{L}$. Remark that the removing of the vertex i in original matrix $\mathbf{L}$ corresponds to setting the corresponding component of vector x to zero, which can be written as:

$$\widetilde{\lambda}_{min}^{i} = \min_{\substack{\|x\|=1 \\ x_i=0}} x^T \mathbf{L} x =$$
$$= \min_{\substack{\|x\|=1 \\ x_i=0 \sim x^T e_i=0 \sim e_i^T x=0}} x^T \mathbf{Q} \Lambda \mathbf{Q}^T x. \tag{11}$$

In the last equation, we have substitute the Laplacian matrix by its spectral decomposition, where Λ is a diagonal matrix of eigenvalues of $\mathbf{L}$ ordered in ascending order and $\mathbf{Q}$ is the matrix, whose columns are eigenvectors of $\mathbf{L}$ ordered accordingly.

After some manipulations (and substituting $y = \mathbf{Q}^T x$, $y^T = x^T \mathbf{Q}$ respectively) we get:

$$\widetilde{\lambda}_{min}^{i} = \min_{\substack{\|y\|=1 \\ r_i(Q)y=0}} y^T \Lambda y =$$
$$= \min_{\substack{\|y\|=1 \\ r_i(Q)y=0}} \sum_{i=1}^{n} \lambda_i, \tag{12}$$

where $r_i(\mathbf{Q})$ denotes the i-th row of matrix $\mathbf{Q}$ (and it is not the eigenvector).

For given n we solve following system of equations:

$$\widetilde{\lambda}_{min}^{i} = \min \left(\lambda_1 y_1^2 + \lambda_2 y_2^2 + \cdots + \lambda_n y_n^2 \right), \tag{13}$$

subject to:

$$y_1^2 + y_2^2 + \cdots + y_n^2 = 1, \tag{14}$$
$$q_{i1} y_1 + q_{i2} y_2 + \cdots + q_{in} y_n = 0. \tag{15}$$

Here, q_{ij} denotes the ij-th entry of matrix $\mathbf{Q}$, i.e. $q_{i\star}$ denotes again the i-th row of matrix $\mathbf{Q}$.

Thanks to Cauchy Interlacing Theorem [8]:

$$0 = \lambda_{1=min} \leq \widetilde{\lambda}_{1=min} \leq \lambda_2 \leq \ldots \leq \widetilde{\lambda}_{n-1} \leq \lambda_n, \tag{16}$$

we get the upper bound for $\widetilde{\lambda}_{min}^{i} \leq \lambda_2$.

Considering the particular example of regular rectangle mesh with known eigenvalues and eigenvectors, it can be shown, that we obtain exactly the upper bound for i such that $q_{i2} = 0$, [2]:

$$\widetilde{\lambda}_{min}^{i} = \min \left(\lambda_1 y_1^2 + \lambda_2 y_2^2 + \cdots + \lambda_n y_n^2 \right) = \lambda_2. \tag{17}$$

Notice that in this notation, the q_{i2} corresponds to the i-th entry of the second smallest eigenvector of the Laplacian matrix which corresponds exactly to 2-level cut according to Definition 4.

6.   Conclusion

In this paper, we have presented a modification in stabilization of generalised inverse algorithm used in [1]. Our modification consists in another choice of fixing node.

Several numerical experiments have be done to observe the influence of different choice of fixing nodes on the condition number of resulting matrix. Based on these experiments, it has been shown that using eigenvectors of the Laplacian matrix provides better choice of fixing node than using eigenvector of the adjacency matrix (used in [1]).

We have presented the main ideas of the analytical proof that the capital vertex (defined using eigenvectors of the Laplacian matrix) provides the best choice of one-fixing node. The whole proof is based on knowledge of eigenvalues and eigenvectors of the particular example of regular rectangle mesh and it can be found in [2].

References

[1] BRZOBOHATY, T., Z. DOSTAL, T. KOZUBEK, P. KOVAC and A. MARKOPOULOS. Cholesky decomposition with fixing nodes to stable computation of a generalized inverse of the stiffness matrix of a floating structure. *International Journal for Numerical Methods in Engineering*. 2011, vol. 88, iss. 5, pp. 493–509. ISSN 0029-5981. DOI: 10.1002/nme.3187.

[2] KABELIKOVA, P. *Implementation of Non-Overlapping Domain Decomposition Techniques for FETI Methods*. Ostrava, 2012. Dissertation thesis. VSB–Technical University of Ostrava. Supervisor Vit Vondrak.

[3] KABELIKOVA, P. Graph centers used for stabilization of matrix factorizations. *Discussiones Mathematicae Graph Theory Journal.* 2010, vol. 30, iss. 2, pp. 249–259. ISSN 1234-3099.

[4] FIEDLER, M. Algebraic connectivity of graphs. *Czechoslovak Mathematical Journal.* 1973, vol. 23, iss. 98, pp. 298–305. ISSN 0011-4642.

[5] FIEDLER, M. A property of eigenvectors of nonnegative symmetric matrices and its application to graph theory. *Czechoslovak Mathematical Journal.* 1975, vol. 25, iss. 100, pp. 619–633. ISSN 0011-4642.

[6] BARNARD, S. T. and H. D. SIMON. A Fast Multilevel Implementation of Recursive Spectral Bisection for Partitioning Unstructured Problems. *Concurrency: Practice and Experience.* 1992, vol. 6, iss. 2, pp. 101–117. ISSN 1532-0634.

[7] POTHEN, A., H. D. SIMON and K. LIOU. Partitioning Sparse Matrices with Eigenvectors of graphs. *SIAM Journal on Matrix Analysis and Applications.* 1990, vol. 11, iss. 3, pp. 430–452. ISSN 1095-7162. DOI: 10.1137/0611030.

[8] HORN, R. A. and Ch. R. Johnson. *Matrix Analysis.* Cambridge: Cambridge University Press, 1985. ISBN 978-0-521-38632-6.

[9] QUENELL, G. T. Eigenvalue comparisons in graph theory. *Pacific Journal of Mathematics.* 1996, vol. 176, iss. 2, pp. 619–633. ISSN 0030-8730.

About Authors

Pavla HRUSKOVA was born in Cesky Tesin as Pavla Kabelikova. She received her M.Sc. from Computer Science and Technology in 2006 and her Ph.D. from Computational and Applied mathematics in 2012, both at VSB–Technical University of Ostrava. Her research interests include graph theory with particular emphasis on the spectral graph theory, numerical mathematics and numerical modelling, and high performance computing. She has married in 2013 and she adopted surname Hruskova. At present, she works at the Department of Applied Mathematics of Faculty of Electrical Engineering and Computer Science of VSB–Technical University of Ostrava.

4

ADMISSION CONTROL IN IMS NETWORKS

Erik CHROMY, Marcel JADRON, Matej KAVACKY, Stanislav KLUCIK

Institute of Telecomunications, Faculty of Electrical Engineering and Information Technology, Slovak University of Technology Bratislava, Ilkovicova 3, Bratislava 812 19, Slovakia

chromy@ut.fei.stuba.sk, marcel.jadron@gmail.com, kavacky@ut.fei.stuba.sk, klucik@ut.fei.stuba.sk

Abstract. *In our paper there is an emphasis on simulations of admission control methods in MATLAB environment. The main task of admission control method is to make a decision if the connection requiring network access should be accepted to the network or the access should be rejected. If the connection is accepted to the network, the admission control has to ensure that Quality of Service of this connection will be satisfied, as well as Quality of Service of all other existing connections. We have observed several Measurement based admission control algorithms and the result is the identification of the suitable algorithm which can estimate the required bandwidth.*

Keywords

Admission control, IMS, MBAC algorithms, Quality of Service.

1. Introduction

Trends in telecommunication networks and services tend to IP Multimedia Subsystem networks (IMS). IMS network guarantees Quality of Service (QoS) and Admission control methods are one of applied QoS mechanisms [1].

Within the scope of the project "Support of Center of Excellence for SMART Technologies, Systems and Services II" funded by structural funds of European union we have built the most modern IMS lab at the Institute of Telecommunications. In this lab we can also conduct research aimed to admission control methods, which are necessary for Quality of Service providing for real-time services. The IMS architecture also contains the Resource Admission Control Subsystem (RACS) in which the admission control methods can be applied.

2. IP Multimedia Subsystem

IP Multimedia Subsystem architecture was developed by 3GPP group. This architecture allows to providers to offer multimedia services such as IPTV, VoIP and many others. IMS is not dedicated only to new services, but it must also support legacy services and should be ready for development of new services. Telecommunication providers can deliver their services to customers irrespective of their location, access to technology or terminals. IMS defines architecture which allows convergence of voice, video, data through IP based infrastructure [2], [3], [4], [5], [6], [7], [8].

RACS is one of the important IMS components for interaction between control layer and transmission functions for control of resources including resource reservations, admission control and QoS support. Therefore RACS component ensures QoS in IMS networks. Admission control block receives requests for QoS resources via reference point, e.g. bandwidth requirement. AC uses information from QoS for admission control, i.e. AC checks if the required QoS resources are available and sends decision if the request is fulfilled or not via reference point [9], [10].

3. Admission Control Methods

QoS in the network must be guaranteed in order to support real-time requests and real-time applications. Three QoS classes are defined for Integrated services. The first is Best-effort class. In the network with this class all connections are permitted. Network sends data of these connections with its maximal transmission rate. Each connection needs some network resources therefore there is not QoS assurance. Due to this fact admission control for this class of service is needed. The second class is Guaranteed services. This class ensures that packet in the network will be not lost and guarantees bounded end-to-end delay. This guaranty needs particular bandwidth reservation. The last class is represented by services with controlled traffic. Guaranteed services with controlled traffic require some grade of QoS, therefore they need admission control for estimation of a number of connection for which they can ensure QoS. Admission control makes a decision if the incoming connection will be accepted or rejected. The measure is provision of QoS of incoming connection and preservation of QoS of existing connections.

The field of admission control is divided into Parameter based admission control (PBAC) and Measurement based admission control (MBAC). PBAC methods regard traffic characteristics of all connections, such as peak transmission rate. This method determines required network resources for all connections based on such parameter. MBAC methods are aimed to measurement of actual traffic in the network. This method accepts incoming connections upon realized measurements.

Admission control is necessary for admission of new connection. It is possible to design the model of admission control which ensures QoS by use of admission control methods. Created models can be used separately or in combination for achievement of better QoS. The main task of admission control methods is to estimate required bandwidth for incoming data flow and to decide if this bandwidth can be allocated. Admission control methods are used mainly for services sensitive for the delay and jitter or for real-time services. There are various admission control methods and they differ mainly in different traffic types and method of realization. Some of the AC methods are based on mathematical calculations and statistical markers while others are based on traffic measurements [1].

3.1. Conditions for AC Methods

For QoS provision admission control methods must fullfill following conditions:

- provide QoS for incoming connection while existing connections are not affected,

- fast decision (in order to prevent delays),

- efficient capacity utilization and effective bandwidth allocation for particular flows,

- simple applicability into the system,

- adaptation for new service [1].

Admission control is important mainly in access network whereas nodes in backbone networks have high transmission rates and information about bandwidth calculations they send to edge nodes. The network is utilized mostly at the edge what is also a reason for centralization of admission control to edge nodes. RACS is one of the IMS components that ensures required QoS [9].

3.2. PBAC Methods

PBAC can be preferred due to their simple implementation. They work with parameters such as peak or effective bandwidth of incoming flow instead of values measured in the network. Through PBAC methods we can limit constraints caused by measurements and network monitoring [11].

3.3. MBAC Methods

MBAC methods use measurement of actual traffic in the network for decision about admission of new data flow. MBAC methods make the decision process based on measurements and QoS parameters. In the case of measurement and network monitoring the more efficient network resources utilization for aggregate data flows or for lower transmission rate than the peak rate is possible compared to PBAC methods. In this scenario remaining bandwidth can be used for other data flows [12].

3.4. MBAC Algorithms

Various measurement based algorithms are known. In the paper we deal with following algorithms:

- Simple Sum [13],

- Measured Sum [14],

- Predicted Sum [15],

- Hoeffding Bound [16] and

- Acceptance Region [17].

1) Simple Sum Algorithm

The algorithm through the sum of existing flows simple ensures itself against not exceeding of available bandwidth. It accepts new data flow only if the following condition is fulfilled:

$$v + r_\alpha < C, \tag{1}$$

where v is a sum of reserved transmission rates [kbit·s^{-1}], C is link capacity [kbit·s^{-1}], α is index for incoming flow and r_α is transmission rate of new data flow [kbit·s^{-1}].

Measured sum algorithm is the simplest admission control algorithm. Therefore it is the most implemented algorithm in routers and switches. It is often used in combination with WFQ method [18] in order to ensure low enqueue delay. WFQ allocates individual service queue for each data flow, so data bursts are separated from each other.

2) Measured Sum Algorithm

This algorithm ensures that sum of peak transmission rates of new flows and actual traffic is lower than targeted link capacity utilization. It is expressed through the condition:

$$v + p \leq u \cdot C, \qquad (2)$$

where v is measured actual traffic of existing connections [kbit·s^{-1}], p is peak rate of incoming data flow [kbit·s^{-1}], C is link capacity [kbit·s^{-1}] and u is link utilization parameter.

Actual measured traffic will not be usable for future connection admissions, because new connection will occur in the network and traffic vary. Link utilization parameter is set to a value lower than 1 in order to guarantee the QoS for all connections. Measuring mechanism makes measurements required for measured sum and it is based on window size of this algorithm.

This mechanism sets the fixed time interval for sample period and a longer time interval for window period which is multiple of sample period. Link utilization is measured at the end of each sample period and the highest value of traffic in one window period is defined as the end of the window period. End of the previous window period is regarded as measured traffic. At the end of each window the window closing period is reset to initial value.

3) Predicted Sum

Proposed algorithm samples data during defined intervals. Such acquired samples represent input data for prediction. After each sampling the prediction of the next data traffic is done. Such prediction will be then used for future admission control for admission or rejection of connection. Algorithm of predicted sum makes a decision based on the condition:

$$\hat{x}(n+1) + p_\alpha \leq uC, \qquad (3)$$

where α is index for incoming data flow, $x(n+1)$ is predicted aggregated traffic used in next sampling $n+1$ [kbit·s^{-1}], p_α is peak transmission rate of new flow [kbit·s^{-1}], C is link capacity [kbit·s^{-1}] and u is source utilization.

4) Hoeffding Bound

This method uses Hoeffding bound for estimation of link traffic. Hoeffding bound sets the higher bound of traffic for connections in the networks according to equation:

$$C_H(v, \{p_i\}1 \leq i \leq n, \varepsilon) = avg + \sqrt{\frac{\ln(\frac{1}{\varepsilon}) \sum_{i=1}^{n} p_i^2}{2}}, \qquad (4)$$

where avg is total traffic of all connections [kbit·s^{-1}], p_i is peak rate of i-th connection [kbit·s^{-1}] and ε is prediction that traffic will exceed link capacity (probability of packet losses).

Hoeffding bound algorithm makes admission decision based on the equation:

$$C_H + p \leq C, \qquad (5)$$

where p is peak transmission rate of new connection [kbit·s^{-1}] and C is link capacity [kbit·s^{-1}].

If the sum of Hoeffding bound of all existing connections and peak rate of new connection is lower than available link capacity the admission control accepts new connection into network. On the contrary, if this sum is higher than available link capacity, connection will be rejected. Compared to Measured sum algorithm, the Hoeffding bound algorithm will not reserve the above capacity for a short-term raised traffic, because Hoeffding bound is adapted for this case. Mechanism of measurement used in this algorithm uses exponential averaging. Firstly, average rate is measured, then the exponential average is calculated and finally Hoeffding bound C_H is estimated.

5) Acceptance Region

This algorithm estimates region in which the link utilization is maximized at the expense of packet losses. Acceptance region can be estimated on the basis of following parameters: given bandwidth, memory space of switches, parameters of buffer stack filter, data bursts of data flow and probability that actual traffic will exceed the acceptance bound. We suppose the Poisson distribution of incoming independent requests in acceptance region calculation. In the case of parameter values variation the algorithm behavior is not precise. Version of the measurement process of this algorithm ensures that sum of measured traffic and transmission rate of new data flow will not exceed the acceptance region.

$$C(s) = \frac{1}{s}\log\left[1 + \frac{v}{p}(e^{sp} - 1)\right], \qquad (6)$$

where $C(s)$ is estimated bandwidth for aggregated traffic [kbit·s^{-1}], v is average transmission rate of traffic [kbit·s^{-1}], p is peak transmission rate [kbit·s^{-1}] and s

is space parameter which value is from interval from 0 to 1.

4. Simulations

This chapter deals with simulations of AC methods. Network topology used in simulations is shown in the Fig. 1.

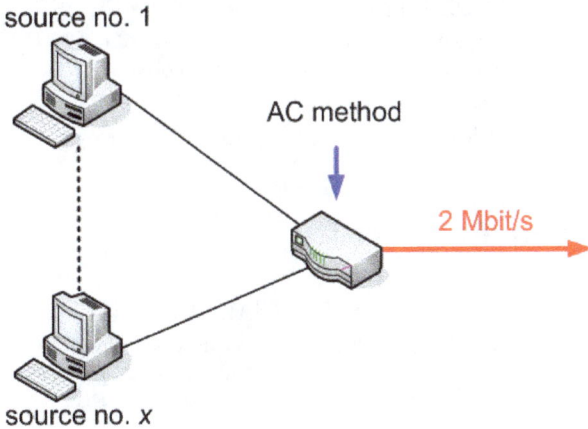

Fig. 1: Network topology.

Admission control method is applied in router. An assumption of this topology is that there is the number of users generating stochastic requests in time. These requests represent stochastic transmission rates from variable bit rate sources. Each source requires different demands on transmission rate in different time samples. Transmission rates vary from 0 to 128 kbit·s^{-1}. It is appropriate to use Gaussian distribution for simulation with a high number of users [19].

4.1. Simple Sum Simulations

The Simple sum simulations are shown in the Fig. 2.

Simple sum algorithm is the simplest admission control algorithm. The main goal of this algorithm is to admit or reject incoming flow on the basis of available bandwidth which can not be exceeded. The simulation results are shown in the Fig. 2. The black curve represents defined link capacity - 2 Mbit·s^{-1}. The red curve represents actual traffic of all users in the network before decision process. Based on the incoming admission request comparison with available bandwidth is done and connection is then accepted (green curve). This two courses show situation when the network is not loaded for maximum.

Comparison with opposed scenario (traffic load near to the link capacity 2 Mbit·s^{-1}) are made through blue and violet curves. Blue curve represents traffic on the border of bandwidth (i.e. 2 Mbit·s^{-1}, or 21 users). Any

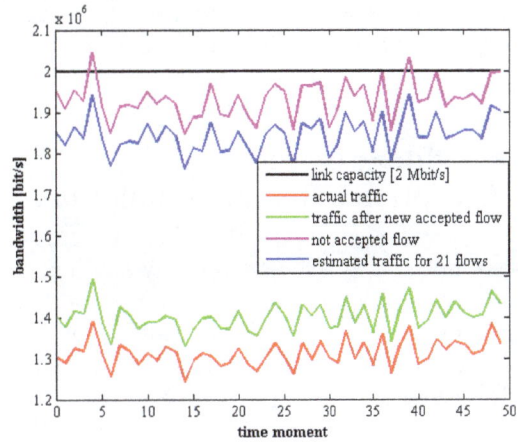

Fig. 2: Simple sum traffic simulation.

new connection can not be admitted due to lack of the bandwidth. Acceptance of new connection in such a situation will lead to QoS degradation (violet curve).

4.2. Measured Sum Simulations

Measured sum algorithm does not regard the total link capacity, but only its part (Fig. 3) compared to simple sum algorithm. The result of the decision process

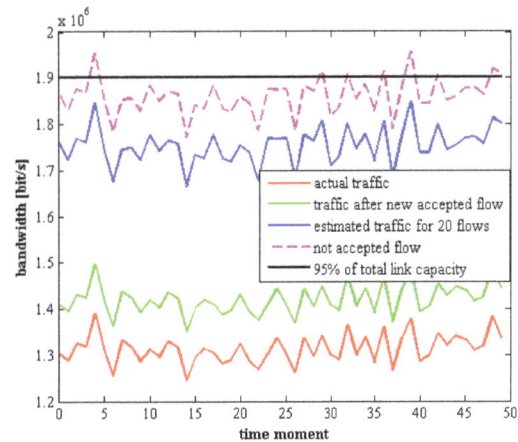

Fig. 3: Measured sum traffic simulation.

is shown in the Fig. 3. Red curve and green curve represent total traffic before and after connection admission. Blue and violet curves represent bound cases. Black curve represents 95 % of total link capacity (i.e. 1,9 Mbit·s^{-1}). Blue curve represents 20 accepted connections. Violet curve represents overrun of defined link capacity, therefore Measured sum algorithm will reject this connection.

Here we can see the difference between Simple sum and Measured sum algorithm. Another connection would be accepted by Simple sum algorithm because it not uses the transmission medium utilization parameter. Measured sum algorithm leaves reserve for the case of unexpected bandwidth increase of connections.

4.3. Hoeffding Bound and Acceptance Region Simulations

In the case of unchanged network topology and higher load the proper selection of AC algorithm is the key element in order to save some bandwidth. In the Fig. 4 the bound scenario with Hoeffding bound (violet curve) and Acceptance region (blue curve) algorithms are shown. Red curve represents actual traffic in the network.

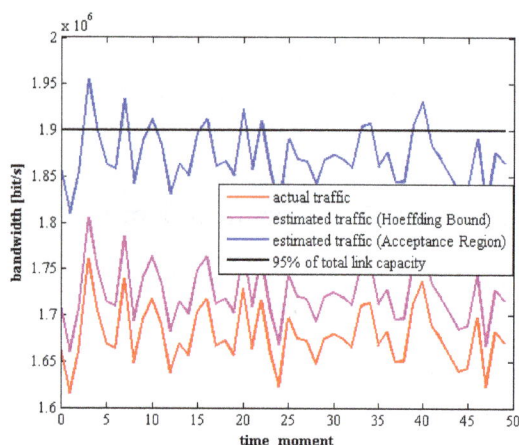

Fig. 4: Acceptance of connections in network with higher load.

We can see that Hoeffding bound algorithm is more suitable. At the moment 19 connections were in the network (actual traffic) and new connection is accepted only by Hoeffding bound algorithm. Acceptance region algorithm has rejected new connection despite of sufficient bandwidth available, therefore from effective bandwidth utilization, the Hoeffding bound algorithm is more preferable.

5. Conclusion

Realized simulations have shown the different allocation of bandwidth while each of simulated algorithms works on different measurement principles. In the case of wrong AC algorithm selection the waste of the bandwidth can occur, what is inefficient and economic unprofitable. Based on the simulations we have shown that in the case of higher traffic load the selection of Admission Control algorithm is very important. From the effective bandwidth utilization point of view the Hoeffding bound algorithm is suitable.

Acknowledgment

This work is a part of research activities conducted at Slovak University of Technology Bratislava, Faculty of Electrical Engineering and Information Technology, Institute of Telecommunications, within the scope of the projects "Grant programme to support young researchers of STU - Modelling of Traffic and Traffic Parameters in IPTV Networks", "Support of Center of Excellence for SMART Technologies, Systems and Services II, ITMS 26240120029, cofunded by the ERDF" and project VEGA no. 1/0106/11 "Analysis and Proposal for Advanced Optical Access Networks in the NGN Converged Infrastructure Utilizing Fixed Transmission Media for Supporting Multimedia Services".

References

[1] LAI, Y.-Ch. and S.-F. TSAI. Unfairness of measurement-based admission controls in a heterogeneous environment. In: *Proceedings. Eighth International Conference on Parallel and Distributed Systems. ICPADS 2001.* Kyongju City: IEEE, 2001, pp. 667–674. ISBN 0-7695-1153-8. DOI: 10.1109/ICPADS.2001.934882.

[2] VOZNAK, M. and J. ROZHON. Approach to stress tests in SIP environment based on marginal analysis. *Telecommunication Systems.* 2013, vol. 52, iss. 3, pp. 1583–1593. ISSN 1018-4864. DOI: 10.1007/s11235-011-9525-1.

[3] VOZNAK, M. and F. REZAC. Web-based IP telephony penetration system evaluating level of protection from attacks and threats. *WSEAS Transactions on Communications.* 2011, vol. 10, iss. 2, pp. 66–76. ISSN 1109-2742.

[4] VOZNAK, M. and F. REZAC. Threats to voice over IP communications systems. *WSEAS Transactions on Computers.* 2010, vol. 9, iss. 11, pp. 1348–1358. ISSN 1109-2750.

[5] ROKA, R. and F. CERTIK. Modeling of environmental influences at the signal transmission in the optical transmission medium. *International Journal of Electrical Communication Networks and Information Security.* 2012, vol. 4, no. 3, pp. 144–162. ISSN 2076-0930.

[6] KYRBASHOV, B., I. BARONAK, M. KOVACIK and V. JANATA. Evaluation and Investigation of the Delay in VoIP Networks. *Radioengineering.* 2011, vol. 20, iss. 2, pp. 540–547. ISSN 1210-2512.

[7] KOCKOVIC, L. and I. BARONAK. Alternatives of Providing IPTV Using IP Multimedia Subsystem. *International Journal of Computers and*

Technology. 2012, vol. 3, no. 2, pp. 188–192. ISSN 2277-3061.

[8] MISUTH, T. and I. BARONAK. Application of M/G/1/K Model for Aggregated VoIP Traffic Packet Loss Estimation. In: *2012 35th International Conference on Telecommunications and Signal Processing (TSP).* Prague: IEEE, 2012, pp. 42–46. ISBN 978-1-4673-1118-2. DOI: 10.1109/TSP.2012.6256249.

[9] ETSI ES 282 003. *Telecommunications and Internet converged Services and Protocols for Advanced Networking (TISPAN); Resource and Admission Control Sub-system (RACS); Functional Architecture.* V1.1.1. 650 Route des Lucioles: F-06921 Sophia Antipolis Cedex - FRANCE, 2006. Available at: http://www.etsi.org/deliver/etsi_es/282000_282099/282003/01.01.01_60/es_282003v010101p.pdf

[10] WUTHNOW, M., M. STAFFORD and J. SHIH. *IMS: A New Model for Blending Applications.* Boca Raton: CRC Press, 2010. ISBN 978-1-4200-9285-1.

[11] YERIMA, S. Implementation and Evaluation of Measurement-Based Admission Control Schemes Within a Converged Networks QoS Management Framework. *International Journal of Computer Networks & Communications (IJCNC).* 2011, vol. 3, no. 4, pp. 137–152. ISSN 0975-2293.

[12] ALIPOUR, E. and K. MOHAMMADI. Adaptive Admission Control for QoS Guarantee in Differentiated Services Networks. *International Journal of Computer Science and Network Security (IJCSNS).* 2008, vol. 8, no. 6, pp. 93–98. ISSN 1738-7906.

[13] JAMIN, S. and S. SHENKER. *Measurement-based Admission Control Algorithms for Controlled-load Service: A Structural Examination.* Michigan: University of Michigan, 1997, Technical Report. CSE-TR-333-97.

[14] JAMIN, S., P. DANZIG, S. SHENKER and L. ZHANG. A Measurement-Based Admission Control Algorithm for Integrated Services Packet Networks. *IEEE/ACM Transactions on Networking.* 1997, vol. 5, iss. 1, pp. 56–70. ISSN 1063-6692. DOI: 10.1109/90.554722.

[15] EGYHAZY, M. and Y. LIANG. Predicted Sum: A Robust Measurement-Based Admission Control with Online Traffic Prediction. *IEEE Communications Letters.* 2007, vol. 11, iss. 2, pp. 204–206. ISSN 1089-7798. DOI: 10.1109/LCOMM.2007.061127.

[16] FLOYD, S. Comments on Measurement-based Admission Control for Controlled-Load Services. *Computer Communications Review.* 1996, vol. 26, iss. 3, pp. 1–16. ISSN 0146-4833.

[17] GIBBENS, R., F. KELLY and P. KEY. A Decision–Theoretic Approach to Call Admission Control in ATM Networks. *IEEE Journal on Selected Areas in Communications.* 1995, vol. 13, iss. 6, pp. 1101–1114. ISSN 0733-8716. DOI: 10.1109/49.400665.

[18] ASHOUR, M. and T. LE-NGOC. Performance of Weighted Fair Queuing Systems with Long Range Dependent Traffic Inputs. In: *Canadian Conference on Electrical and Computer Engineering, 2005.* Saskatoon: IEEE, 2005, pp. 1002–1005. ISBN 0-7803-8885-2. DOI: 10.1109/CCECE.2005.1557145.

[19] BRICHET, F. and A. SIMONIAN. Conservative Gaussian models applied to Measurement-based Admission Control. In: *Sixth International Workshop on Quality of Service (IWQoS'98).* San Francisco: IEEE, 1998, pp. 68–71. ISBN 0-7803-4482-0. DOI: 10.1109/IWQOS.1998.675222.

About Authors

Erik CHROMY was born in Velky Krtis, Slovakia, in 1981. He received the M.Sc. degree in telecommunications in 2005 from Faculty of Electrical Engineering and Information Technology of Slovak University of Technology (FEEIT STU) in Bratislava. In 2007 he submitted Ph.D. work from the field of Observation of statistical properties of input flow of traffic sources on virtual paths dimensioning. Nowadays he works as assistant professor at the Institute of Telecommunications of FEEIT STU Bratislava.

Marcel JADRON was born in Levice, Slovakia, in 1990. He is a student at the Institute of Telecommunications, FEEIT STU in Bratislava.

Matej KAVACKY was born in Nitra, Slovakia, in 1979. He received the M.Sc. degree in telecommunications in 2004 from FEEIT STU in Bratislava. In 2006 he submitted Ph.D. work "Quality of Service in Broadband Networks". Nowadays he works as assistant professor at the Institute of Telecommunications of FEEIT STU Bratislava.

Stanislav KLUCIK was born in Bratislava, Slovakia, in 1984. He received the M.Sc. degree in telecommunications in 2009 from FEEIT STU in Bratislava. In 2012 he submitted Ph.D. work from the field of Source and QoS traffic parameters modeling

in NGN networks. Nowadays he works as researech at
the Institute of Telecommunications of FEEIT STU
Bratislava.

5

Transition Based Synthesis with Modular Encoding of Petri Nets into FPGAs

Arkadiusz BUKOWIEC, Jacek TKACZ, Marian ADAMSKI

Institute of Computer Engineering and Electronics,
Faculty of Electrical Engineering, Computer Science and Telecommunications,
University of Zielona Gora,
Podgorna 50, 65-246 Zielona Gora, Poland

a.bukowiec@iie.uz.zgora.pl, j.tkacz@iie.uz.zgora.pl, m.adamski@iie.uz.zgora.pl

Abstract. *The paper describes a new method for the synthesis of the application specific logic controllers, targeted into the FPGA. The initial steps of the proposed control algorithm rely on the notion of a Petri net, which is an easy way to describe parallel processes. The algorithm is oriented on transition based logic description. It allows easy analysis of dynamics and functioning of the circuit. The logic circuit is also decomposed into logic blocks responsible for particular functions. It leads to the compact implementation with usage of different kind of logic elements like. Additionally such decomposition allows easy analysis of circuit.*

Keywords

Application specific logic controller, FPGA, logic synthesis, Petri net, structural decomposition.

1. Introduction

A Petri net (PN) [10], [9] is one of the most popular models used in formal design and synthesis of the application specific logic controllers (ASLCs) [13], [15], [6]. The digital design of such controllers is very often implemented using field programmable gate arrays (FPGAs) [1], [15], [2], [16]. The most typical implementation of Petri nets in the FPGA devices uses the one-hot local state encoding method, where each place is represented by a flip-flop [11]. Such approaches are oriented towards places based logic description. Additionally, this approach requires hardware implementation of a large number of logic functions and flip-flops included in logic blocks.

In this paper we propose a new method for the synthesis of a Petri net. To allow its effective synthesis, the Petri net is initially converted into Petri macronet [9], [14]. The proposed algorithm is oriented towards transition based logic description. It means, that combinational equations describe transitions [5] in opposite to classical algorithms where they describe places [4]. It easy allows to analyze the dynamics of logic controller by exporting variables that describes transition in Boolean algebra. Additionally, the operations are encoded with a minimal-length binary vector. This encoding allows the realization of logic circuit in compact way. A microoperation decoder can be implemented with the use of embedded memory blocks of an FPGA [12]. It permits the stable work of whole controller.

2. Petri Net

A Petri net [10], [9] is defined as a triple:

$$PN = (P, T, F), \tag{1}$$

where:

- P is a finite non-empty set of places,
  $P = \{p_1, \ldots, p_M\}$,

- T is a finite non-empty set of transitions,
  $T = \{t_1, \ldots, t_S\}$,

- F is a set of arcs (from places to transitions and from transitions to places):

$$\begin{aligned} F &\subseteq (P \times T) \cup (T \times P), \\ P \cap T &= \varnothing. \end{aligned}$$

The sets of input and output transitions of a place $p_m \in P$ are defined respectively as follows:

$$\begin{aligned} \bullet p_m &= \{t_s \in T : (t_s, p_m) \in F\}, \\ p_m \bullet &= \{t_s \in T : (p_m, t_s) \in F\}. \end{aligned}$$

Sets of input and output places of a transition $t_s \in T$ are defined respectively as follows:

$$\bullet t_s = \{p_m \in P : (p_m, t_s) \in F\},$$
$$t_s \bullet = \{p_m \in P : (t_s, p_m) \in F\}.$$

A marking of a Petri net is defined as a function:

$$M : P \rightarrow \mathrm{N}.$$

For a given place p_m the function $M(p_m)$ returns the number of tokens in p_m. A place or a set of places is marked if it contains a token. A transition t_s can be fired if all its input places are marked. Firing a transition removes one token from each input place and puts one token in each output place. When the initial marking M_0 is additionally specified, the Petri net can be represented as a tuple:

$$PN = (P, T, F, M_0). \tag{2}$$

2.1. Interpreted Petri Net

A Petri net enhanced with an additional feature for information exchange is called an interpreted Petri net [9]. This exchange is made by use of binary signals. Interpreted Petri nets are used as models of concurrent logic controllers.

The Boolean variables occurring in the interpreted Petri net can be divided into three sets:

- X is the set of input variables, $X = \{x_1, \ldots, x_L\}$,

- Y is the set of output variables (microoperations - μO), $Y = \{y_1, \ldots, y_N\}$,

- Z is the set of internal communication variables (usually not used, with $Z = \varnothing$).

An interpreted Petri net has a guard condition φ_s associated with every transition t_s. The guard condition φ_s is defined to be a Boolean function of a subset of variables from the sets X and Z. In a special case, the condition φ_s can be defined as 1 (always true). Now, a transition t_s can be fired if all its input places are marked and the current value of the corresponding Boolean function φ_s is equal to 1.

The conjunction ψ_m associated with a place p_m is an elementary conjunction of positive literals formed from output variables from the set Y. If the place p_m is marked, the output variables from corresponding conjunction ψ_m are set and other variables are reset. The conjunction ψ_m correspond to microinstruction (μI).

2.2. Macro Petri Net

Macro Petri net is a Petri net where part of the net (subnet) is replaced by one macroplace [9]. It allow to enhance Petri nets with hierarchy [7] and it simplifies algorithms of coloring and verification of Petri net. There are many classes of subnets that could be replaced by macroplace, for e.g.:

- State machine subnets [10],

- Two-pole blocks [9],

- Parallel places [10],

- P-blocks [9].

These classes create to many possibilities of merging Petri net into macro Petri net. For the synthesis purpose, the best solution is application of mono-active macroplaces [9]. This is macroplaces that have one input and one output and consist of only sequential places. Only Petri macronets with such macroplaces will be used in this article.

3. Idea of Synthesis Method

The idea of proposed synthesis method is based on the modular encoding of places together with functional parallel decomposition of the Petri net-based logic circuit [4]. The novelty of this approach is that places are encoded with use of minimal length code separately inside each macroplace and macroplaces are encoded with use of one-hot encoding. The state of Petri net is determined by concatenation of these codes. Combinational circuit is oriented towards transition generation, and output variables (names of particular microoperations) are placed in configured memories of FPGA. It leads to realization of a logic circuit in double-level architecture (Fig. 1), where the transition coder (TC) of first level is responsible for activation of the transitions:

$$T = \mathrm{TC}(X, Q), \tag{3}$$

The register block (REG) holds a current state of Petri net in the register (RG). It also has additional custom combinational logic (LOGIC) connected to its inputs. This logic is responsible for generation of the next state based on active transitions and current state:

$$Q^* = \mathrm{RG}(\mathrm{LOGIC}(T, Q)), \tag{4}$$

where Q is the set of variables used to store the codes of currently marked places and macroplaces. The internal custom combinational logic of the register also generates the code of microoperation:

$$Z = \mathrm{LOGIC}(T, Q), \tag{5}$$

where Z is the set of variables used to store the codes of currently executed microinstruction. The second level decoder (D) is responsible for generation of microoperations based on microinstruction code and it is implemented using memory blocks. Their functionality can be described by function:

$$Y = D(Z). \qquad (6)$$

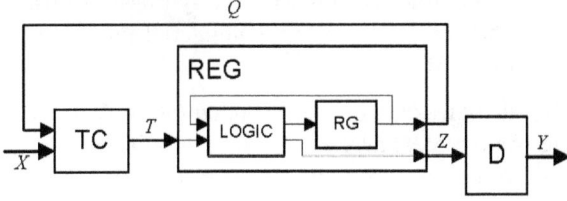

Fig. 1: Logic circuit of Petri net.

Such approach allows to use logic elements and embedded memory blocks available in modern FPGA devices.

The entry point to the synthesis method is the interpreted Petri macronet. The outline of synthesis process includes following steps:

- Modular encoding of places. The purpose of this step is to assign the shortest binary local code $K_o(p_m)$ to each place p_m inside each macroplace mp_o, where $o = 1, \ldots, O$ and it is an number of macroplace. Macroplaces are encoded by assigning the one-hot code $K(mp_o)$ to each macroplace mp_o. The global code $C(p_m)$ of global place p_m is determined as concatenation of these codes:

$$C(p_m) = K(mp_o) * K_o(p_m). \qquad (7)$$

The total required number of variables for encoding is equal to:

$$R = O + \sum_{o=1}^{O} r_o, \qquad (8)$$

where r_o is a required number of variables for o-th macroplace:

$$r_o = \lceil log_2(|P_o|) \rceil, \qquad (9)$$

where $P_o \subseteq P$ is a set of places that are placed inside macroplace mp_o.

To store the macroplace code we use $Q^0 = \{q_1, \ldots, q_O\}$ variables and to store the local place codes we use $Q^o = \{q_{O+r+1}, \ldots, q_{O+r+r_o}\}$ ($r = \sum_{j=1}^{o-1} r_j$). Now, these variables create set $Q = \bigcup_{o=0}^{O} Q^o$ and $|Q| = R$.

The process of encoding begins from assigning the one-hot codes to macroplaces. Then, places receive minimal length codes inside each macroplace independently.

- Formation of microinstructions. Let all microoperations create U different microinstructions $Y_u \subseteq Y$, $\Upsilon = \{Y_1, \ldots, Y_U\}$ in the Petri net. Let create O subsets $\Upsilon_o \subseteq \Upsilon$, where Υ_o consists only of microinstructions associated with places from set P_o.

- Encoding of microinstructions. All microinstructions are encoded by binary code $C_o(Y_u)$ separately in each subset Υ_o. The number of variables used is

$$\rho_o = \lceil log_2(|\Upsilon_o| + 1) \rceil. \qquad (10)$$

To store this code we use $Z^o = \{z_{\rho+1}, \ldots, z_{\rho+\rho_o}\}$ ($\rho = \sum_{j=1}^{o-1} \rho_j$) variables and $Z = \bigcup_{i=1}^{O} Z^o$. The process of encoding is trivial, and it required to assign binary code $C_o(Y_u)$ to each microinstruction $Y_u \in \Upsilon_o$ starting from value 1. The value 0 is reserved for situation where considered place do not generate any microinstruction. Let assume that particular one microinstruction Y_u can belong to several subsets. In such situation it will receive several codes.

- Formation of conjunctions. Conjunctions describe macroplaces, places and global places. They are needed for easier form of Eq. (3) and Eq. (5) that describe digital circuit. The conjunction describing the macroplace mp_o equals to the affirmation of variable $q_o \in Q^0$. This variable is equal to 1 in the code $K(mp_o)$. The conjunction describing the place p_m consists of affirmation or negation of variables $q_r \in Q^o$ that are used to store the code $K_o(p_m)$ of this place. If variable q_r is equal to 1 in the code $K_o(p_m)$ then affirmation of this variable is used otherwise its negation is used. The conjunction describing the global place p_m consists of the macroplace mp_o and the place p_m conjunction. It corresponds to the code $C(p_m)$.

- Formation of logic equations. Logic equations describe Eq. (3) and Eq. (5) of combinational circuit TC and custom combinational logic LOGIC of register REG. The characteristic function of transition is defined as conjunction of conjunctions of all its input global places and guard condition:

$$t_s = \bigwedge(\bullet t_s) \wedge \varphi_s. \qquad (11)$$

The function to generate the code of next place calculated by the custom combinational logic LOGIC is defined as:

$$q_r = q_r \oplus \bigvee(\bullet q_r) \oplus \bigvee(q_r \bullet), \qquad (12)$$

and the function to generate code microoperation is defined as:

$$z_\rho = \bigvee(P_{z_\rho}), \qquad (13)$$

where P_{z_ρ} is a set of global places conjunctions that generate microoperations Y_u represented by the $C_o(Y_u)$ that has variable z_ρ set to 1.

- Formation of memory contents. The memory content can be described as tables or as equations according to the function Eq. (6). In case of tabular description there is required to create O tables. The table consists of two columns. First column is an address and it is described by variables $z_\rho \in Z^o$. The second column is a binary value (vector) of operations. It is based on value of output variables form the set $Y^o = \bigcup_{u=1}^{u=|\Upsilon_o|} Y_u | Y_u \in \Upsilon_o$. In each line of the table, there should be placed a binary value with only these bits y_n set that are in microinstruction Y_u represented by code $C_o(Y_u)$ that equals to the address from the first column of this line.

- Formation of logic circuit and implementation. This step describes the rules of design of the Petri net HDL model and its implementation into FPGA device. Here is applied a bottom-up approach. Conjunctions of places can be described using standard bit-wise operators. Then logic equations can be described with the use of these conjunctions using continuous assignments or procedural assignments as well as bit-wise operators. There should be created a module for circuit TC with inputs X and Q and outputs T. The register REG should be described as R-bits register with an asynchronous set. The typical synthesis template can be used [3]. the decoder D can be described as processes with the **case** statement. As, the embedded memory blocks are synchronous, the sensitivity list of such processes includes only clock signal. The reset has to be realized as a synchronous one because typical memory blocks do not support any asynchronous control signal. To ensure that such a described module could be synthesized as a memory block it is required to set the value of the special synthesis directive. The syntax of this directive depends on FPGA vendor. The top-level module should describe connections of all components according to the block diagram presented in Fig. 1. Additionally the global reset and clock signals are connected to set and clock inputs of register and reset and clock inputs of decoder. The edge that trigs the decoder has to be opposite to the edge that trigs the register, and then operations are generated during only one clock cycle. The created model of logic circuit can be passed into third-party synthesis tool.

4.   Example of Method Application

The method of Petri net synthesis, described in the previous section, is illustrated by its application on Petri net PN_1 (Fig. 2a). This Petri net describes control process of an industrial mixer of aggregate content and water [8]. This Petri net is not complicated and it is a good example to illustrated a synthesis steps. For the synthesis purpose it was compacted into Petri macronet (Fig. 2b).

Firstly, the places have to be encoded (step Modular encoding of places). There is $O = 6$ macroplaces, so it is required to use $r_0 = 6$ variables $Q^0 = \{q_1, \ldots, q_6\}$ to encode macroplaces. Macroplaces contains respectively 2, 1, 3, 1, 2, and 2 places, so it is required to use $r_1 = 1$, $r_2 = 1$, $r_3 = 2$, $r_4 = 1$, $r_5 = 1$, and $r_6 = 1$ variables $Q^1 = \{q_7\}$, $Q^2 = \{q_8\}$, $Q^3 = \{q_9, q_{10}\}$, $Q^4 = \{q_{11}\}$, $Q^5 = \{q_{12}\}$, and $Q^6 = \{q_{13}\}$ to encode places inside each macroplace. In total, it is required to use $R = 13$ variables $Q = \{q_1, \ldots, q_{13}\}$ to encode all places. Macroplaces receive following one-hot codes $K(mp_o)$ using variables from Q^0 subset:

$$K(mp_1) = 1-----; \quad K(mp_2) = -1----;$$
$$K(mp_3) = --1---; \quad K(mp_4) = ---1--;$$
$$K(mp_5) = ----1-; \quad K(mp_6) = -----1;$$

and places receive following binary codes $K_o(p_m)$ inside each macroplace using variables from corresponding Q^o subset:

$$K_1(p_1) = 0; \; K_1(p_2) = 1; \; K_2(p_3) = 1;$$
$$K_3(p_4) = 00; \; K_3(p_5) = 01; \; K_3(p_6) = 10;$$
$$K_4(p_7) = 1; \; K_5(p_8) = 1; \; K_5(p_9) = 0;$$
$$K_6(p_{10}) = 0; \; K_6(p_{11}) = 1;$$

As an alternative, the Gray code can be applied also for places.

When encoding of places is finished the microinstructions can be formed (step Formation of microinstructions) and encoded (step Encoding of microinstructions). The set of microinstructions Υ is formed based on control algorithm and for the example Petri net PN_1 there is $U = 6$ different microinstructions: $Y_1 = \{YT1\}$, $Y_2 = \{YV1\}$, $Y_3 = \{YT2\}$, $Y_4 = \{YV2\}$, $Y_5 = \{YM\}$, $Y_6 = \{YV3\}$, and $\Upsilon = \{Y_1, Y_2, Y_3, Y_4, Y_5, Y_6\}$. This set is divided into $O = 6$ subsets: $\Upsilon_1 = \{Y_1\}$, $\Upsilon_2 = \{\varnothing\}$, $\Upsilon_3 = \{Y_2, Y_3\}$, $\Upsilon_4 = \{\varnothing\}$, $\Upsilon_5 = \{Y_4\}$, and $\Upsilon_6 = \{Y_5, Y_6\}$. Empty sets are omitted in the encoding process and now, microinstructions can be encoded with $\rho_1 = 1$, $\rho_3 = 2$, $\rho_5 = 1$, and $\rho_6 = 2$ variables $Z^1 = \{z_1\}$, $Z^3 = \{z_2, z_3\}$, $Z^5 = \{z_4\}$, and $Z^6 = \{z_5, z_6\}$. The sample encoding can be as follows:

$$C_1(Y_1) = 1; \; C_3(Y_2) = 01; \; C_3(Y_3) = 10;$$
$$C_5(Y_4) = 1; \; C_6(Y_5) = 01; \; C_6(Y_6) = 10;$$

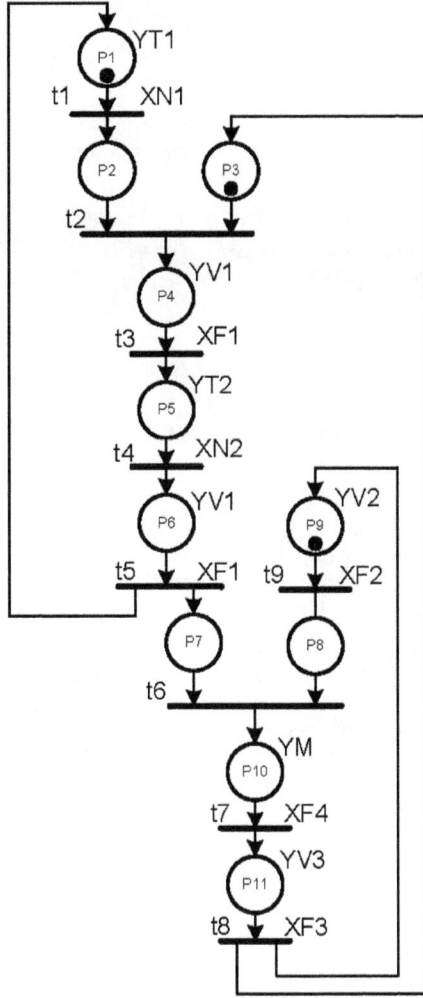

(a)

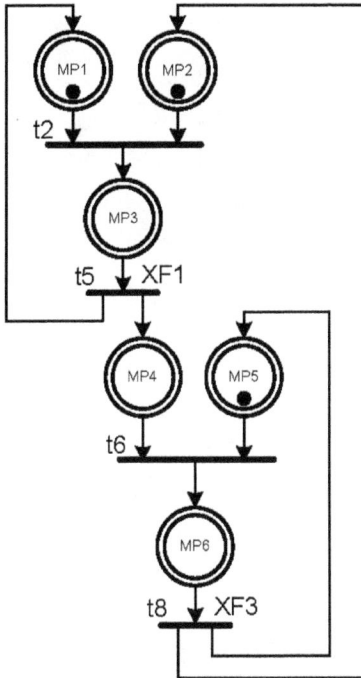

(b)

Fig. 2: (a) Petri net PN_1 and (b) its Petri macronet.

When all encodings are finished place conjunctions (step Formation of conjunctions) and logic equations (step Formation of logig equations) can be formed. Place conjunctions are created on the base of the place codes. First place conjunctions of macroplaces have to be denoted and for the Petri net PN_1 that is as follows:

$$mp_1 = q_1;\ mp_2 = q_2;\ mp_3 = q_3;$$
$$mp_4 = q_4;\ mp_5 = q_5;\ mp_6 = q_6;$$

Then, conjunctions of local place inside each macroplace can be formed:

$$lp_1 = \overline{q_7};\ lp_2 = q_7;\ lp_3 = q_8;$$
$$lp_4 = \overline{q_9}\ \overline{q_{10}};\ lp_5 = \overline{q_9}\ q_{10};\ lp_6 = q_9\ \overline{q_{10}};$$
$$lp_7 = q_{11};\ lp_8 = q_{12};\ lp_9 = \overline{q_{12}};$$
$$lp_{10} = \overline{q_{13}};\ lp_{11} = q_{13};$$

Finally, conjunctions of global places can be created based on Eq. (7):

$$p_1 = mp_1\ lp_1;\ p_2 = mp_1\ lp_2;\ p_3 = mp_2\ lp_3;$$
$$p_4 = mp_3\ lp_4;\ p_5 = mp_3\ lp_5;\ p_6 = mp_3\ lp_6;$$
$$p_7 = mp_4\ lp_7;\ p_8 = mp_5\ lp_8;\ p_9 = mp_5\ lp_9;$$
$$p_{10} = mp_6\ p_{10};\ p_{11} = mp_6\ p_{11};$$

After that, logic equations for each transition t_s can be formed:

$$t_1 = p_1\ xn_1;\ t_2 = p_2\ p_3;\ t_3 = p_4\ xf_1;$$
$$t_4 = p_5\ xn_2;\ t_5 = p_6\ xf_1;\ t_6 = p_7\ p_8;$$
$$t_7 = p_{10}\ xf_4;\ t_8 = p_{11}\ xf_3;\ t_9 = p_9\ xf_2$$

and logic equation for each variable q_r can be denoted:

$$q_1 = q_1 \oplus t_5 \oplus t_2;\ q_2 = q_2 \oplus t_8 \oplus t_2;$$
$$q_3 = q_3 \oplus t_2 \oplus t_5;\ q_4 = q_4 \oplus t_5 \oplus t_6;$$
$$q_5 = q_5 \oplus t_8 \oplus t_6;\ q_6 = q_6 \oplus t_6 \oplus t_8;$$
$$q_7 = q_7 \oplus t_1 \oplus t_2;\ q_8 = q_8 \oplus t_8 \oplus t_2;$$
$$q_9 = q_9 \oplus t_4 \oplus t_5;\ q_{10} = q_{10} \oplus t_3 \oplus t_4;$$
$$q_{11} = q_{11} \oplus t_5 \oplus t_6;\ q_{12} = q_{12} \oplus t_9 \oplus t_6;$$
$$q_{13} = q_{13} \oplus t_7 \oplus t_8;$$

and logic equation for each variable z_ρ can be created:

$$z_1 = p_1;\ z_2 = p_5;\ z_3 = p_4 \vee p_6$$
$$z_4 = p_9;\ z_5 = p_{11};\ z_6 = p_{10}$$

Then, the content of operation memory can be formed (step Formation of memory contents). In case of Petri net PN_1 there have to be created four such tables which are shown in Tab. 1. Two macroplaces do not generate any microinstructions so they are omitted.

Finally, the logic circuit can be described (step Formation of logic circuit and implementation) In our approach the VHDL was used. But in similar way it can be also described with the use of Verilog. The module for circuit TC (Fig. 3) uses input variables and

```
entity TC is
  port(XN1, XF1, XN2, XF2, XF3, XF4 : in
       STD_LOGIC;
       Q : in STD_LOGIC_VECTOR(1 to 13);
       T : out STD_LOGIC_VECTOR(1 to 9));
end TC;
architecture TC of TC is
  signal mp : STD_LOGIC_VECTOR(1 to 6);
  signal lp : STD_LOGIC_VECTOR(1 to 11);
  signal p : STD_LOGIC_VECTOR(1 to 11);
begin
  mp(1) <= Q(1);
  mp(2) <= Q(2);
  ...
  mp(6) <= Q(6);
  lp(1) <= not Q(7);
  lp(2) <= Q(7);
  ...
  lp(11) <= Q(13);
  p(1) <= mp(1) and lp(1);
  p(2) <= mp(1) and lp(2);
  ...
  p(11) <= mp(6) and lp(11);
  T(1) <= p(1) and XN1;
  T(2) <= p(2) and p(3);
  ...
  T(9) <= p(9) and XF2;
end TC;
```

Fig. 3: VHDL description of transition coder.

```
entity REG is
  port(clk, res : in STD_LOGIC;
       T : in STD_LOGIC_VECTOR(1 to 9);
       Q : out STD_LOGIC_VECTOR(1 to 13);
       Z : out STD_LOGIC_VECTOR(1 to 6));
end REG;
architecture REG of regREGis
  signal intQ : STD_LOGIC_VECTOR(1 to 13);
  signal mp : STD_LOGIC_VECTOR(1 to 6);
  signal lp : STD_LOGIC_VECTOR(1 to 11);
  signal p : STD_LOGIC_VECTOR(1 to 11);
begin
  RG: process (clk, res) begin
    if res='1' then
      intQ <= "1100100100000";
    elsif (RISING_EDGE(clk)) then
      intQ(1) <= intQ(1) xor T(5) xor T(2);
      ...
      intQ(12) <= intQ(12) xor T(9) xor T(6);
      intQ(13) <= intQ(13) xor T(7) xor T(8);
    end if;
  end process;
  Q <= intQ;
  mp(1) <= ...
  lp(1) <= ...
  p(1) <= ...
  Z(1) <= p(1);
  ...
  Z(6) <= p(10);
end reg;
```

Fig. 4: VHDL description of register.

conjunctions in continuous assignments for transition. Conjunctions are defined as internal signals and they are described as continuous assignments.

```
entity D is
  port(clk, res : in STD_LOGIC;
       Z : in STD_LOGIC_VECTOR(1 to 6);
       YT1, YV1, YT2, YV2, YM, TV3 : out
            STD_LOGIC);
  attribute bram_map: string;
  attribute bram_map of D: entity is "yes";
end D;
architecture D of D is
begin
  O1: process (clk) begin
    if FALLING_EDGE(clk) then
      if res='1' then
        YT1 <= '0';
      else case Z(1) is
        when '0' => YT1 <= '0';
        when '1' => YT1 <= '1';
        when others => YT1 <= '0';
      end case;
      end if;
    end if;
  end process;
  O3: process (clk) begin
    if FALLING_EDGE(clk) then
      if res='1' then
        YV1 <= '0'; YT2 <= '0';
        else case Z(2 to 3) is
        when "00" => YV1 <= '0'; YT2 <= '0';
        when "01" => YV1 <= '1'; YT2 <= '0';
        when "10" => YV1 <= '0'; YT2 <= '1';
        when others => YV1 <= '0'; YT2 <= '0';
      end case;
      end if;
    end if;
  end process;
  O5: process ...
  O6: process ...
end D;
```

Fig. 5: VHDL description of operations decoder.

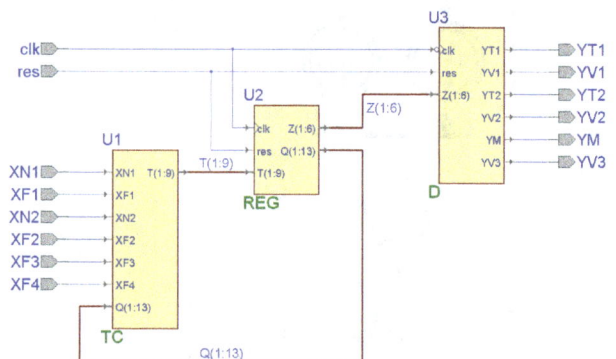

Fig. 6: Block diagram of top-level module.

The module of register REG (Fig. 4) describes the logic of code changes and generate code of microinstruction. It also requires definition of conjunctions as internal signals.

The decoder D (Fig. 5) generates outputs signals.

It can be synthesized as embedded memory block if there is added special synthesis directive. In this file, there is such directive for Xilnix devices that

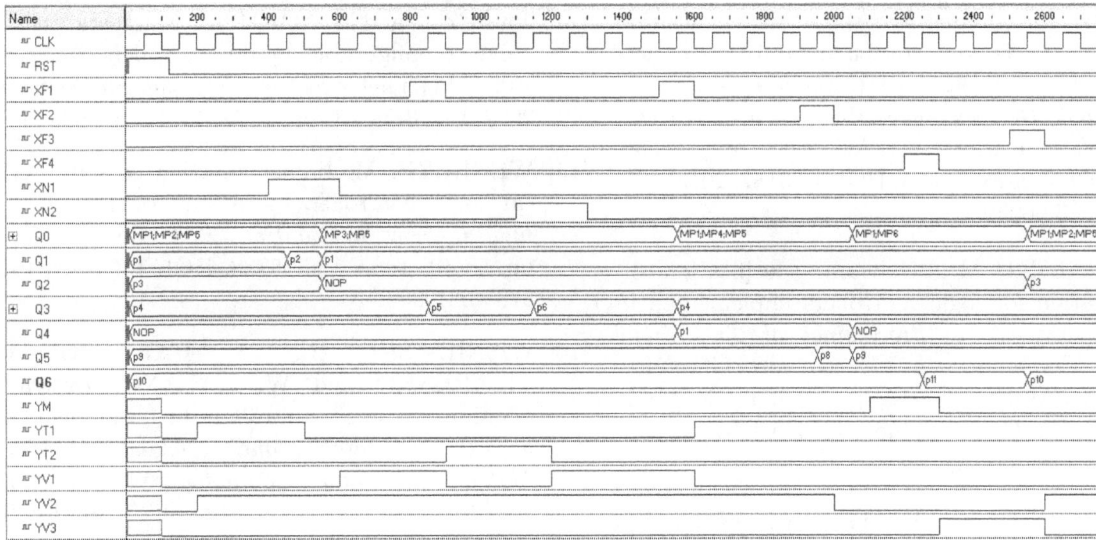

Fig. 7: Simulation results of logic circuit.

sets `bram_map` attribute to `yes`. The top-level module (Fig. 6) describes connections of all modules. In our case it is cerated in graphical editor of Active-HDL environment.

The designed circuit is verified in Active-HDL environment with use of a test-bench. The test-bench is described in VHDL and it emulate one cycle of work of an industrial mixer. The simulation results are sown in the Fig. 7.

Tab. 1: Operation memories tables of PN_1.

Addr. z_1	μO $YT1$	Addr. $z_2 z_3$	μO $YV1\ YT2$
0	0	00	00
1	1	01	10
		10	01

Addr. z_4	μO $YV2$	Addr. $z_5 z_6$	μO $YM\ YV3$
0	0	00	00
1	1	01	10
		10	01

5. Summary

The paper presents a method of realization of application specific logic controller. A formal description of the method is then accompanied with a simple example. The specification of the control algorithm uses the notion of a Petri net, which allows an easy description of parallel processes. We note that it is possible to apply formal verification methods to test the algorithm. The proposed method of synthesis is based on transition based logic description of the logic circuit and modular encoding of places. It allows to extend formal verification methods by additional analysis the dynamics of the circuit. Additionally the logic circuit is decomposed into three logic blocks responsible for particular functions: dynamic generation of transitions, store the state of the controller and generate output control signals. It allows the compact implementation of logic circuit into FPGA device with usage of different kind of logic elements like: LUTs, flip-flops and embedded memories. Additionally such decomposition allows easy analysis of circuit functioning.

References

[1] BOMAR, B. W. Implementation of microprogrammed control in FPGAs. *IEEE Transactions on Industrial Electronics.* 2002, vol. 49, iss. 2, pp. 415–422. ISSN 0278-0046. DOI: 10.1109/41.993275.

[2] BOROWIK, G., M. RAWSKI, G. LABIAK, A. BUKOWIEC and H. SELVARAJ. Efficient logic controller design. In: *2010 Fifth International Conference on Broadband and Biomedical Communications.* Malaga: IEEE, 2010, pp. 1–6. ISBN 978-1-4244-6952-9. DOI: 10.1109/IB2COM.2010.5723633.

[3] BROWN, S. and Z. VERNESIC. *Fundamentals of digital logic with VHDL design.* New York: McGraw-Hill Higher Education. 2005. ISBN 978-0077221430.

[4] BUKOWIEC, A. and M. ADAMSKI. Synthesis of Petri nets into FPGA with operation flexible memories. In: *2012 IEEE 15th International Symposium on Design and Diagnostics of Electronic Circuits.* Tallin: IEEE, 2012, pp. 16–21. ISBN 978-1-4673-1186-1. DOI: 10.1109/DDECS.2012.6219016.

[5] BUKOWIEC, A. and M. ADAMSKI. Transition based synthesis with code markers of Petri nets into GPGAs. In: *12th IFAC Conference on Programmable Devices and Embedded Systems PDeS 2013*. Velke Karlovice: IFAC, 2013. pp. 181–86. ISBN 978-3-902823-53-3. DOI: 10.3182/20130925-3-CZ-3023.00030.

[6] DOLIGALSKI, M. Behavioral specification diversification for logic controllers implemented in FPGA devices. In: *Proceedings of the Annual FPGA Conference on - FPGAworld '12*. New York: ACM Press, 2012, pp. 1–5. ISBN 978-1-4503-1645-3. DOI: 10.1145/2451636.2451642.

[7] ESPARZA, J. and M. SILVA. On the analysis and synthesis of free choice systems. *Advances in Petri Nets 1990*. Berlin: Springer, 1991, pp. 243–286. ISBN 978-3-540-53863-9. DOI: 10.1007/3-540-53863-1_28.

[8] GNIEWEK, L. and J. KLUSKA. Hardware Implementation of Fuzzy Petri Net as a Controller. *IEEE Transactions on Systems, Man and Cybernetics, Part B (Cybernetics)*. 2004, vol. 34, iss. 3, pp. 1315–1324. ISSN 1083-4419. DOI: 10.1109/TSMCB.2003.822956.

[9] KARATKEVICH, A. *Dynamic Analysis of Petri Net-Based Discrete Systems*. Berlin: Springer, 2007. ISBN 978-3-540-71464-4.

[10] MURATA, T. Petri nets: Properties, analysis and applications. *Proceedings of the IEEE*. 1989, vol. 77, iss. 4, pp. 541–580. ISSN 0018-9219. DOI: 10.1109/5.24143.

[11] PASTOR, E. and J. CORTADELLA. Efficient encoding schemes for symbolic analysis of Petri nets: Properties, analysis and applications. In: *Proceedings Design, Automation and Test in Europe*. Paris: IEEE, 1998, pp. 790–795. ISBN 0-8186-8359-7. DOI: 10.1109/DATE.1998.655948.

[12] RAWSKI, M., G. BOROWIK, T. LUBA, P. TOMASZEWSKI and B. FALKOWSKI. Logic synthesis strategy for FPGAs with embedded memory blocks. In: *Mixed Design of Integrated Circuits & Systems, 2009. MIXDES '09. MIXDES-16th International Conference*. Lodz: IEEE, 2009, pp. 296–301. ISBN 978-1-4244-4798-5.

[13] ROKYTA, P., W. FENGLER and T. HUMMEL. Electronic System Design Automation Using High Level Petri Nets. In: *Hardware Design and Petri Nets*. Boston: Springer, 2000, pp. 193–204. ISBN 978-1-4757-3143-9. DOI: 10.1007/978-1-4757-3143-9_10.

[14] TKACZ, J. and M. ADAMSKI. Macrostate encoding of reconfigurable digital controllers from topological Petri net structure. *Przeglad Elektroniczny*. 2012, vol. 8, no. 8, pp. 137–140. ISSN 0033-2097.

[15] WEGRZYN, M. Implementation of safety critical logic controller by means of FPGA. *Annual Reviews in Control*. 2003, vol. 27, iss. 1, pp. 55–61. ISSN 1367-5788. DOI: 10.1016/S1367-5788(03)00007-5.

[16] WISNIEWSKI, R., A. BARKALOV, L. TITARENKO and W. HALANG. Design of microprogrammed controllers to be implemented in FPGAs. *International Journal of Applied Mathematics and Computer Science*. 2011, vol. 21, iss. 2, pp.401-412. ISSN 1641-876X. DOI: 10.2478/v10006-011-0030-1.

About Authors

Arkadiusz BUKOWIEC received a Bachelor degree in computer engineering from Technical University of Zielona Gora. During these studies he completed industrial practice at Aldec Inc. in Henderson, NV, USA. Then, he received Master degree and a Ph.D. degree in computer science from the University of Zielona Gora. During the master thesis he was working for Aldec Poland. During the Ph.D. studies he spent one semester at Universidade Nova de Lisboa. Since 2003, he has been working at the University of Zielona Gora. His research interests include methods of design, synthesis and verification of digital circuits.

Jacek TKACZ graduated from the University of Zielona Gora and since 2009 works in the Chair of Computer Engineering. Dr. Tkacz's research is devoted to symbolic methods of theorem proving and their application to computer science and electronics. He is also interested in novel design and development technologies for application software, including mobile applications. During the years 1997-2005 he was involved in design and development of the PROLIB software, used by many Polish libraries.

Marian ADAMSKI is a retired head of the Institute of Computer Science and Electronics at the University of Zielona Gora. His research interests include the design of digital systems, understood as digital microsystems, and formal methods in programming of logical controllers. A member of IEEE, IEE, ACM, PTEiTS (Polish Society for Theoretical and Applied Electrical Engineering) and PTI (Polish Computer Science Society).

6

THE RESPONSE OF POLARIZATION MAINTAINING FIBERS UPON TEMPERATURE FIELD DISTURBANCE

Filip DVORAK[1], Jan MASCHKE[2], Cestmir VLCEK[2]

[1]Department of radar technology, Faculty of Military Technology, University of Defence, Kounicova 65, Brno, 602 00, Czech Republic
[2]Department of electrical engineering, Faculty of Military Technology, University of Defence, Kounicova 65, Brno, 602 00, Czech Republic

filip.dvorak@unob.cz, jan.maschke@email.cz, cestmir.vlcek@unob.cz

Abstract. *The paper deals with the response of polarization maintaining (PM) fibers upon the variation of the temperature field, on condition both polarization axes are excited. Proper use can be applied in the area of optical fiber thermal field disturbance sensor. For a description of polarization properties the coherent Jones and Muller matrices were used. The Poincare sphere was applied to depict the development of the output of the polarization state of PM fibers for wavelengths 633 nm, 1310 nm and 1550 nm were arranged in a proposed sensor setup and were studied in the laboratory. A wide file of results containing dependencies of phase shift variations upon different configurations of measured PM fiber was obtained. The thermal field disturbance of the PM fiber was applied by an object with defined proportions and a defined range of temperatures. Dependencies of phase shift upon the object's temperature, its distance from the fiber and exposed length of PM fiber were measured.*

Keywords

Beat length, coherency matrix, fiber response, Poincare sphere, sensor, state of polarization.

1. Introduction

From the beginning of fiber realization in the telecommunications area it got to the big development of the theory and practical applications. Almost in the same time there were ideas on the application of fibers also in sensors. Isotropic fibers [1], polarization maintaining fibers (PMF) and rare earth doped fibers [2] have been used in the number of sensor applications. One example of an application is the interferometric sensor of gyroscope [3], [4]. Polarization maintaining fibers were established for exciting of optical radiation to one polarization axis only and maintenance polarization state along the whole length of the fiber. According to the induced birefringence, fibers were applied in sensors, e.g. for measurement of mechanical pressure [5].

The artificial birefringence caused by the different thermal expansibility between strength elements and the fiber cladding reaches high sensitivity to the ambient temperature of surrounding environment [6]. For uniform excitation of both fiber axes there is a polarization state change of optical radiation in the fiber output caused by the ambient thermal field disturbance. The change of the polarization state of output optical radiation immediately detects any thermal field disturbance caused by an ambient thermal source. High fiber sensitivity to the outer thermal field initiated the idea to use the PMF application as a sensor of thermal field disturbance.

Infant of the research work, time development of output polarization fluctuation in PANDA PMF and bowtie for equal excitation of both polarization axes were studied. Expressive fluctuation led to the construction of a fiber sensor realized and evaluated for $\lambda = 633$ nm, utilizing good polarization properties of He-Ne laser. The measured results validated high sensor sensitivity upon the external thermal field. The disadvantage of this sensor is its construction arrangement using gas He-Ne laser as a source of optical radiation. To eliminate this disadvantage and to compare different properties of the components used, the experiment was extended for PANDA PM fibers and semiconductor lasers with wavelengths of 1550 nm and 1310 nm wavelengths.

Results of theoretical analysis and partial experimental tests for $\lambda = 633$ nm and 1550 nm were published in previous papers [7], [8], [9], [10]. Two partial experiments for $\lambda = 1550$ nm have been made there. The rate of fiber response depending upon the distance between external thermal source and the fiber was studied in the first one. Effects of PMF response to its exposed length were studied in the second one. Presented work is an extension of previous experiments for a wavelength of 1310 nm. This work focuses mainly on the determination of fiber response in both dependences during an excitation by an external thermal field with variable initial temperature. All the results obtained for the three particular PM fibers will be compared.

2.   Principle of Sensor Function

The part of PMF can be, in principle, studied as a multiple linear retarder containing no partial polarizers and can be described by means of a unitary Jones matrix. Output optical radiation from the PMF can be described by coherency matrix $\mathbf{C}'$, that is determined by the unitary Jones matrix of the given component $\mathbf{L}$ and also by a coherency matrix of input optical radiation $\mathbf{C}$ according to: [11]

$$
\begin{aligned}
\mathbf{C}' &= (\mathbf{LE}) \otimes (\mathbf{LE})^+ = (\mathbf{LE}) \otimes (\mathbf{L}^+\mathbf{E}^+) = \\
&= \mathbf{L}(\mathbf{E}) \otimes (\mathbf{E}^+)\mathbf{L}^+ = \mathbf{LCL}^+,
\end{aligned} \tag{1}
$$

where $+$ is labelling of the Hermitian conjugate matrix. By decomposition of the coherency matrix of the output optical radiation elements of Stokes vector can be found, enabling description of the fiber sensor features for thermal field disturbance from the point of view optical intensity I.

To determine a coherency matrix of output optical radiation a coherency matrix of input optical radiation and Jones unitary matrix of PMF were needed. For excitation of both polarization axes at any angle of the fiber a clockwise polarized optical radiation was introduced, described by Jones matrix $\mathbf{J}_{\mathrm{RC}}$:

$$
\mathbf{J}_{\mathrm{RC}} = \begin{pmatrix} -i \\ 1 \end{pmatrix}. \tag{2}
$$

The suffix RC expresses right circular polarization. The coherency matrix of an input optical radiation provides as:

$$
\begin{aligned}
\mathbf{C} &= \langle \mathbf{E} \otimes \mathbf{E}^+ \rangle = \left\langle \begin{pmatrix} E_x \\ E_y \end{pmatrix} \otimes \begin{pmatrix} E_x^* & E_y^* \end{pmatrix} \right\rangle = \\
&= \begin{pmatrix} \langle E_x E_x^* \rangle & \langle E_x E_y^* \rangle \\ \langle E_y E_x^* \rangle & \langle E_y E_y^* \rangle \end{pmatrix} = \begin{pmatrix} C_{xx} & C_{xy} \\ C_{yx} & C_{yy} \end{pmatrix},
\end{aligned} \tag{3}
$$

where $\mathbf{E}$ is the Jones column matrix (Jones vector) and $\mathbf{E}^+$ is its Hermitian conjugate matrix . Equation (2)

is expressed by matrix $\mathbf{C}_{ij}$ to define Eq. (10). After substitution of Eq. (2) we obtain the coherency matrix of circular optical radiation:

$$
\mathbf{C}_{\mathrm{RC}} = \begin{pmatrix} -i \\ 1 \end{pmatrix} \otimes \begin{pmatrix} i & 1 \end{pmatrix} = \begin{pmatrix} 1 & -i \\ i & 1 \end{pmatrix}. \tag{4}
$$

The phase shift of the linear retarder ϕ is put into the Jones matrix of a linear retarder with azimuth 0 °. By this common relation of linear retarder is obtained as:

$$
\mathbf{L} = \begin{pmatrix} e^{i\frac{\phi}{2}} & 0 \\ 0 & e^{-i\frac{\phi}{2}} \end{pmatrix}. \tag{5}
$$

Hermitian conjugate matrix to the matrix Eq. (5) can then be expressed as:

$$
\mathbf{L}^+ = \begin{pmatrix} e^{-i\frac{\phi}{2}} & 0 \\ 0 & e^{i\frac{\phi}{2}} \end{pmatrix}. \tag{6}
$$

By the substitution of matrices Eq. (4), Eq. (5) and Eq. (6) into Eq. (1) we obtain coherency matrix of output optical radiation:

$$
\begin{aligned}
\mathbf{C}'_{\mathrm{RC}} &= \mathbf{L}\mathbf{C}_{\mathrm{RC}}\mathbf{L}^+ = \\
&= \begin{pmatrix} e^{i\frac{\phi}{2}} & 0 \\ 0 & e^{-i\frac{\phi}{2}} \end{pmatrix} \begin{pmatrix} 1 & -i \\ i & 1 \end{pmatrix} \begin{pmatrix} e^{-i\frac{\phi}{2}} & 0 \\ 0 & e^{i\frac{\phi}{2}} \end{pmatrix}.
\end{aligned} \tag{7}
$$

By calculating the matrix product in Eq. (7), a resultant relation of coherency matrix for output optical radiation can be obtained:

$$
\mathbf{C}' = \begin{pmatrix} 1 & -ie^{i\phi} \\ ie^{-i\phi} & 1 \end{pmatrix}. \tag{8}
$$

By decomposition of Eq. (8) into the spin matrices using the following equation,

$$
\begin{aligned}
\mathbf{C} &= \frac{C_{xx} + C_{yy}}{2} \begin{pmatrix} 1 & 0 \\ 0 & 1 \end{pmatrix} + \frac{C_{xx} - C_{yy}}{2} \begin{pmatrix} 1 & 0 \\ 0 & -1 \end{pmatrix} + \\
&\quad + \frac{C_{xy} + C_{yx}}{2} \begin{pmatrix} 0 & 1 \\ 1 & 0 \end{pmatrix} + \\
&\quad + \frac{C_{xy} - C_{yx}}{2} \begin{pmatrix} 0 & -i \\ i & 0 \end{pmatrix},
\end{aligned} \tag{9}
$$

we obtain:

$$\mathbf{C'} = \begin{pmatrix} 1 & 0 \\ 0 & 1 \end{pmatrix} + \sin\phi \begin{pmatrix} 0 & 1 \\ 1 & 0 \end{pmatrix} + \cos\phi \begin{pmatrix} 0 & -i \\ i & 0 \end{pmatrix}. \quad (10)$$

From the relation Eq. (10) it is obvious an assignment of components on the right side of the equation to the corresponding components of Stokes vector. Unit matrix corresponds to the component S_0, the component with multiple $\sin\phi$ corresponds to S_2 and the component with multiple $\cos\phi$ corresponds to S_3. Distribution of optical intensity I in the vertical preference is zero according to the definition $S_1 = C_{xx} - C_{yy}$.

For the length of the optical fiber equaling to the multiple of its beat length and for excitation of both polarization axes by means of circular polarized optical radiation with no disturbance of external thermal field, the phase shift ϕ will be equal to zero and the corresponding Stokes component $S_2 = 0$. For optical fiber length different from a multiple of its beat length and without effect of the external thermal field disturbance the phase shift will be none zero and will be determining the particular state of polarization. In the case of ambient thermal field disturbance by means of an external thermal source the phase shift will be excited and it will result as a change of the output polarization state. This variation of the polarization state is directly proportional to the change of ϕ according to the relation Eq. (11). Determination of this change is realized by means of polarizer - analyzer placed in the output of fiber. For obtaining the maximum sensor sensitivity the orientation of the output polarizer – analyzer is 45 ° towards polarization axes. By this arrangement we can measure the whole range of the phase shift, expressed by the minimum and the maximum value of the measured optical intensity I.

Due to the character of the measured quantity, the optical intensity I, is preferable for next description to use a Mueller matrix. Behavior of the input circular polarized optical radiation propagating through the sensor under consideration, the linear retarder, is expressed by the Mueller matrix as follows:

$$\begin{pmatrix} S_0 \\ S_1 \\ S_2 \\ S_3 \end{pmatrix} = \begin{pmatrix} 1 & 0 & 0 & 0 \\ 0 & 1 & 0 & 0 \\ 0 & 0 & \cos\phi & \sin\phi \\ 0 & 0 & -\sin\phi & \cos\phi \end{pmatrix} \begin{pmatrix} 1 \\ 0 \\ 0 \\ 1 \end{pmatrix} = \begin{pmatrix} 1 \\ 0 \\ \sin\phi \\ \cos\phi \end{pmatrix}. \quad (11)$$

The output optical radiation Eq. (11) hits the linear polarizer – analyzer in the required orientation towards the polarization axes. In the output of polarizer we obtain the optical radiation expressed by the Stokes vector:

$$\begin{pmatrix} S'_0 \\ S'_1 \\ S'_2 \\ S'_3 \end{pmatrix} = \frac{1}{2} \begin{pmatrix} 1 & 0 & 1 & 0 \\ 0 & 0 & 0 & 0 \\ 1 & 0 & 1 & 0 \\ 0 & 0 & 0 & 0 \end{pmatrix} \begin{pmatrix} 1 \\ 0 \\ \sin\phi \\ \cos\phi \end{pmatrix} =$$

$$= \frac{1}{2} \begin{pmatrix} 1 + \sin\phi \\ 0 \\ 1 + \sin\phi \\ 0 \end{pmatrix}. \quad (12)$$

From this resultant equation it is clear that the output optical radiation of the polarizer – analyzer will be linearly polarized with an orientation of 45 ° and the change of optical intensity is directly proportional to the change of phase shift in fiber. From Eq. (12) it is evident that for given arrangement we measured by the photodetector directly the value of Stokes element S_2. Since the variations of phase shift in fiber and development of the polarization state of propagating optical radiation through the fiber, these variations are proportional to the S_2 and S_3 according to Eq. (10). To evaluate these variations we need know only one measured Stokes element S'_2 or S'_3. If S'_2 is known, S'_3 can be calculated. The measured Stokes element S'_2 is determined from Eq. (12) as:

$$S'_2 = \frac{1}{2}(1 + \sin\phi) = \left(\sin\frac{\phi}{2} + \cos\frac{\phi}{2} \right)^2. \quad (13)$$

As we measure optical intensity I of Stokes element S'_2 and not a phase shift, we determine ϕ from Eq. (13) as:

$$\phi = \arcsin(2S'_2 - 1). \quad (14)$$

Such analysis deals with the description of sensor from the point of view of the outer optical intensity I. In order to describe internal phenomenon in more detail theoretical analysis focusing on the change of the beat length depending upon the temperature variation excited by an external subject was conducted.

3.   Phase Shift Variation Depending on Temperature

When studying PM fiber temperature dependency, it can be presumed that the output polarization state variation is dependent on the refractive indices change of fast and slow polarization axes, i.e. the refractive indices change invoked by a temperature change is proportional to the beat length L_B variation. This is the length where 2π phase shift is introduced between the two polarization axes. The variation of beat length depending upon temperature can be described by the following method. The beat length can be expressed as: [11]

$$L_B = \frac{\lambda}{\Delta n_e}, \tag{15}$$

where λ is the wavelength and Δn_e is the differential index. The differential index is defined as:

$$\Delta n_e = \Delta n_s - \Delta n_f, \tag{16}$$

where Δn_s and Δn_f are differences between refractive indices of particular axes n_s, n_f and the cladding refractive index n_c:

$$\begin{aligned} \Delta n_s &= n_s - n_c, \\ \Delta n_f &= n_f - n_c. \end{aligned} \tag{17}$$

The dependency of beat length variation L_B on Δn_e can be written as:

$$\frac{dL_B}{d\Delta n_e} = -\frac{\lambda}{\Delta n_e^2} = -L_B \frac{1}{\Delta n_e}. \tag{18}$$

Under the condition that phase shift variation δ is linearly dependent on the beat length variation L_B, it can be expressed as:

$$-\frac{d\delta}{2\pi} = \frac{dL_B}{L_B}, \tag{19}$$

where the sign $-$ (minus) indicates the state for which the beat length L_B decreases and the phase shift variation δ increases.

By integration of equation Eq. (19) the following relation is obtained:

$$\delta = -2\pi \int \frac{dL_B}{L_B} = -2\pi(\ln L_B + C). \tag{20}$$

For the ambient temperature the beat length is L_{B0} and the phase shift variation $\delta = 0$. For these conditions the following formula is valid:

$$C = -\ln L_{B0}. \tag{21}$$

By substituting the constant C Eq. (21) into Eq. (20), the phase shift variation shall be:

$$\delta = -2\pi \ln \frac{L_B}{L_{B0}}. \tag{22}$$

By substituting Eq. (15) into Eq. (22), the beat length depending on a differential index Δn_e, phase shift variation can be obtained:

$$\delta = 2\pi \ln \frac{\Delta n_e}{\Delta n_{e0}}, \tag{23}$$

where Δn_{e0} is corresponding to L_{B0}. Under the condition that Δn_e is linearly changing with temperature ϑ, the following formula applies:

$$\Delta n_e(\vartheta - \vartheta_0) = K(\vartheta - \vartheta_0) + \Delta n_{e0}, \tag{24}$$

where K is a proportion constant, ϑ is the excitation source temperature and ϑ_0 is the ambient temperature. The phase shift variation dependence on temperature is then expressed by substituting Eq. (24) into Eq. (23):

$$\delta(\vartheta - \vartheta_0) = 2\pi \ln \left(\frac{K(\vartheta - \vartheta_0)}{\Delta n_{e0}} + 1 \right). \tag{25}$$

The phase shift variation dependence on temperature expressed in Eq. (25) is valid on condition that the difference of refractive indices of fast and slow axes is proportional to the temperature. The resultant PM fiber response is dependent on the number of exposed beat lengths the heat source is affecting. The response is different due to the constant source length and different beat lengths of particular PM fibers. Therefore it is suitable to consider PM fiber responses related only to one beat length. The total phase shift variation δ dependence on the total number of exposed beat lengths and the phase shift variation δ_{LB} dependence on one beat length is expressed by the following formula, where N is the number of exposed beat lengths L_B effected by the heat source:

$$N = \frac{l}{L_B}, \tag{26}$$

where l is the exposed length of PM fiber. As the heat source effect at the ends is not defined exactly the number N can be rounded to an integer number. The phase shift variation of one beat length δ_{LB} is expressed as:

$$\delta_{LB} = \frac{\delta}{N} = \delta \frac{L_B}{l}. \tag{27}$$

For a differential index the following formula is valid:

$$\Delta n_e = \frac{\lambda}{L_B}. \tag{28}$$

Tab. 1: Features of PM fibers.

Type of OF	Wavelength [nm]	Beat length L_B [mm]	Δn_e [$\times 10^{-3}$]	Factor L_B/l [$\times 10^{-3}$]
PM630-HP	633	2	0.316	7.2
PM1300-HP	1310	4	0.327	14.5
PM1550-HP	1550	5	0.31	18

Tab. 2: Phase shift variation per one beat length for fiber PM630-HP.

		Phase shift variation per one beatlength			
		Temperature source distance from PMF			
		5 [cm]	8 [cm]	11 [cm]	
	50 °C	13.00	12.00	8.90	
	45 °C	10.90	8.20	7.10	$\pi \times 10^{-3}$ [rad]
Temp.	40 °C	7.70	6.80	5.30	
	35 °C	5.70	4.70	2.90	

Factors L_B/l and differential refractive index values Δn_e of particular PM fibers are presented in Tab. 1. In the case there is m mumber of exposed lengths of PMF the value of the factor shall be L_B/ml. The differential indices Δn_e of studied PM fibers are practically the same and that the difference of refractive indices between fast and slow axes and mechanism of birefringence generation is very similar for all the selected PM fibers. This conclusion is supported by the production technology of identical producer during industrial process of PANDA structures of particular PM fibers. The temperature response related to the one beat length should be literally the same.

A coefficient for determining sensitivity per beat length for length $l = 27.5$ cm is presented in Tab. 1. The length l is corresponding to the length of the applied heat source.

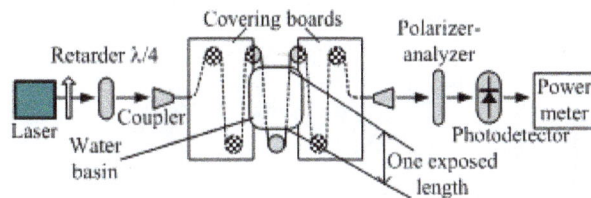

Fig. 1: Test setup for measuring two exposed lengths [9].

4. Experimental Results

The aim of our experimental tests is to determine PM fiber PANDA type response during its thermal field disturbance when an external thermal source with different initial temperatures was applied. The temperature was changed from 50 °C to 35 °C in 5 °C steps. Measurements were taken for two different exposed fiber lengths. The fiber response was investigated for three different distances (5 cm, 8 cm and 11 cm) between the thermal source and the optical fiber. Disturbance of the thermal PM fiber field was done using a plastic basin with a constant amount of water heated to the desired temperature. It was placed on a gap between 2 polystyrene boards Fig. 1 that cover the unexposed part of the PMF.

The above layout in the sensor block enables the exposition of the required length of investigated fiber and also enables to arrange the required distance of ambient thermal field from the fiber. In this experiment, a fiber response for 1 and 3 exposed lengths were tested between cork cylinders. One exposed length was 27.5 cm. The source of optical radiation used for $\lambda = 633$ nm was He-Ne laser, for $\lambda = 1310$ nm and 1550 nm laser diodes ML 725 B8F and ML 925 B45F, respectively, were used.

The beat length catalogue value of fiber PM630-HP is $L_B \leq 2$ mm, for PM1300-HP is $L_B \leq 4$ mm and for PM1550-HP is $L_B \leq 5$ mm. Optical power for $\lambda = 1310$ nm and 1550 nm was measured in intervals of 0.5 s using a GENTEC-EO P-LINK power meter with a PH78-Ge photodiode. For $\lambda = 633$ nm, an analog power meter in combination with a HP 34401A digital multimeter was used controlled by MATLAB. The interval for taking these readings was 0.64 s.

Figure 2 shows an example of measured polarization state in the above described given configuration.

For easier illustration and computer modeling of development of the output polarization state, the Poincare sphere [12], [13] was created in the MATLAB environment. The polarization state in Fig. 2 for 1550 nm is depicted on Poincaré sphere in Fig. 3.

The measured results were processed in two graphs versions of phase shift, as a dependency on temperature for different distances between the thermal source and

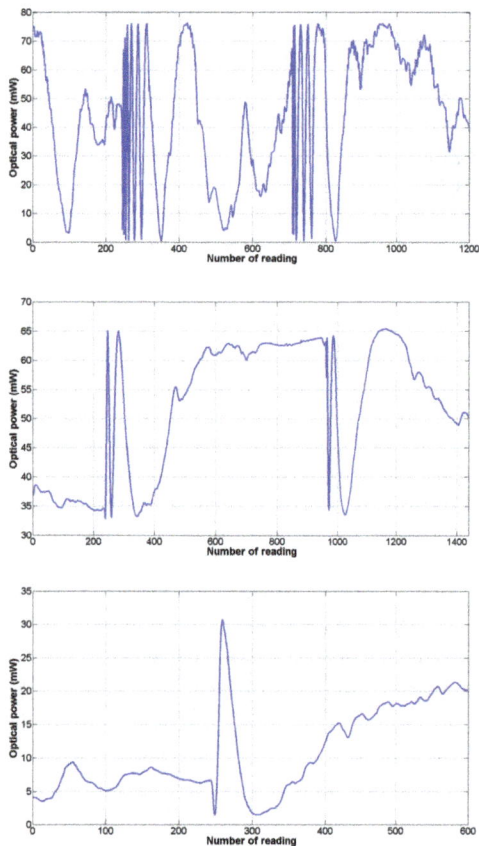

Fig. 2: Example of output optical power for $\lambda = 1310$ nm, 1310 nm and 1550 nm, 3 exposed lengths of fiber, initial temperature 50 °C and 5 cm distance between the external thermal source and the PM fiber.

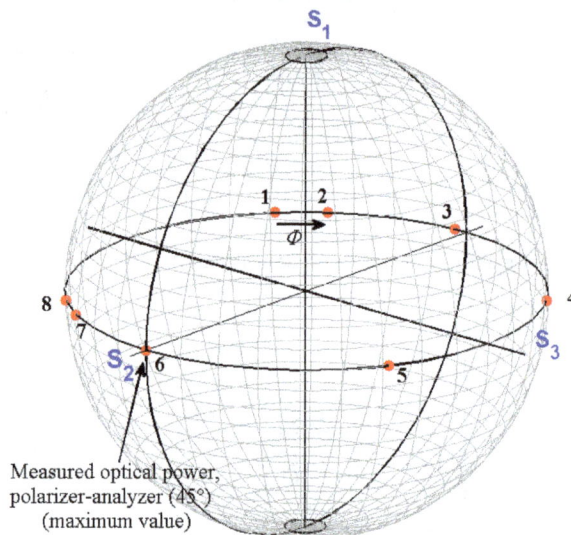

Fig. 3: Polarization state readings of the output optical radiation corresponding to the Fig. 2 at 1550 nm.

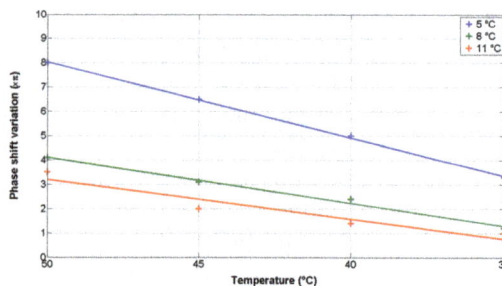

Fig. 4: Phase shift variation dependency on temperature for one exposed length of winding for PM630-HP.

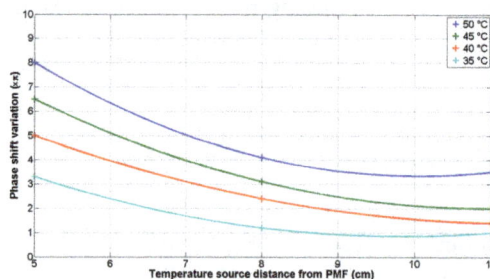

Fig. 5: Phase shift variation dependency on the distance of thermal source from PM fiber for one exposed length of winding for PM630-HP.

the fiber and as a dependency on the distance for the selected temperatures. The presented approximations come from readings at temperatures of 50 °, 45 °, 40 ° and 35 ° and at distances of 5 cm, 8 cm and 11 cm. Examples of these distances for one exposed length are shown in Fig. 4, Fig. 5, Fig. 8, Fig. 9, Fig. 12 and Fig. 13, results for three exposed lengths are in Fig. 6, Fig. 7, Fig. 10, Fig. 11, Fig. 14 and Fig. 15.

Fig. 4 and Fig. 6 show that dependencies of phase shift variation on the temperature are approximately linear. Similarly with increasing exposed length the phase shift variation increases too, but slowly. There was only one anomaly presented at the distance of 5 cm, where a big value of phase shift variation was caused by the thermal source affecting also the neighboring lengths of the fiber that were not supposed to be exposed.

Examples of the phase shift variation dependence on the distance for selected temperatures are shown in Fig. 5 and Fig. 7. Nonlinearity of the dependency for one exposed length validates the fact that for shorter distances and greater temperature emission of thermal source there are undesired interferences, effecting other

than the required exposed length. This heat effect is substantially smaller in the case of three exposed lengths, as the undesired coverage of effected lengths is smaller. The presented examples are a part of a large set of measurements for sensitivity tests depending on the fiber configuration and the ambient temperature.

One of the aims of this study is to compare thermal field disturbance for three PM fibers working atn different wavelengths of 633 nm, 1310 nm and 1550 nm. As

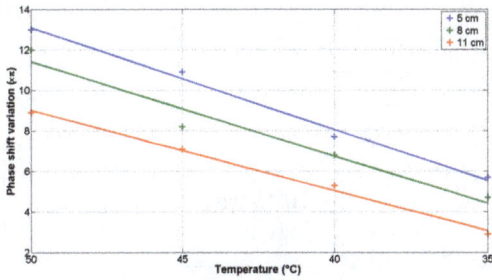

Fig. 6: Phase shift variation dependence on temperature for three exposed lengths of winding for PM630-HP.

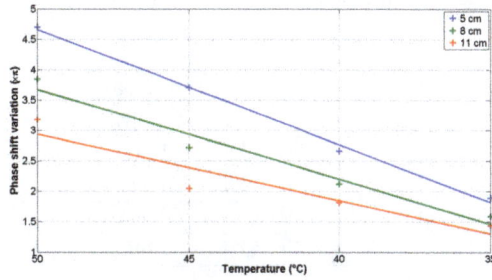

Fig. 10: Phase shift variation dependence on temperature for three exposed lengths of winding for PM1300-HP.

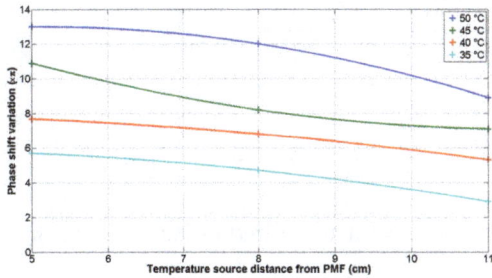

Fig. 7: Phase shift variation dependence on distance between thermal source and PM fiber for three exposed lengths of winding for PM630-HP.

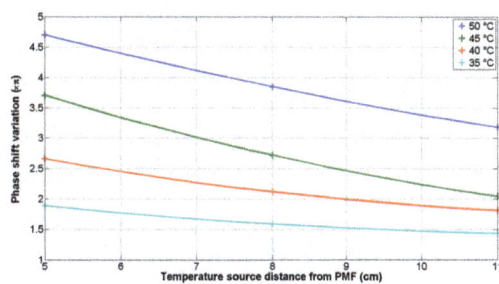

Fig. 11: Phase shift variation dependence on distance between thermal source and PM fiber for three exposed lengths of winding for PM1300-HP.

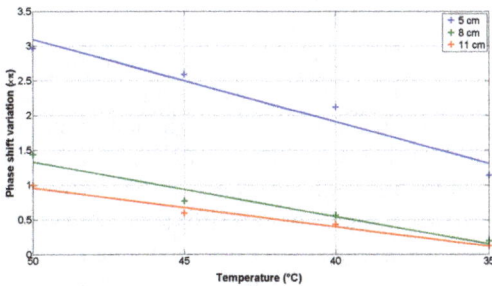

Fig. 8: Phase shift variation dependence on temperature for one exposed length of winding for PM1300-HP.

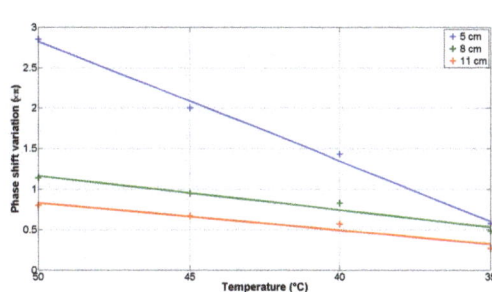

Fig. 12: Phase shift variation dependence on temperature for one exposed length of winding for PM1550-HP.

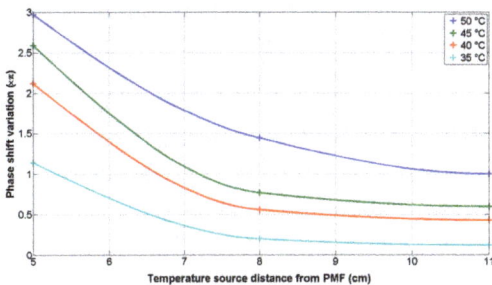

Fig. 9: Phase shift variation dependence on distance between thermal source and PM fiber for one exposed length of winding for PM1300-HP.

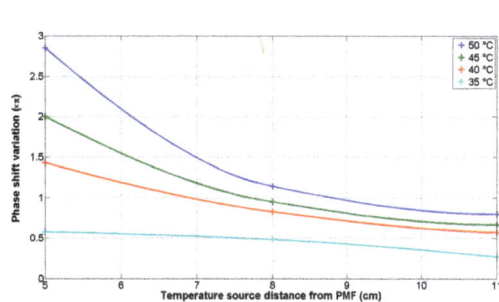

Fig. 13: Phase shift variation dependence on distance between thermal source and PM fiber for one exposed length of winding for PM1550-HP.

the phase shift variation δ is dependent on the number of beat lengths, the measured responses were recalculated according to Eq. (27). Obtained results for par-

ticular PM fibers are presented in Tab. 2, Tab. 3 and Tab. 4.

Tab. 3: Phase shift variation per one beat length for fiber PM1330-HP.

		Phase shift variation per one beatlength			
		Temperature source distance from PMF			
		5 [cm]	8 [cm]	11 [cm]	
Temp.	50 °C	4.70	3.85	3.18	$\pi \times 10^{-3} \ [rad]$
	45 °C	3.71	2.72	2.05	
	40 °C	2.66	2.12	1.82	
	35 °C	1.89	1.56	1.44	

Tab. 4: Phase shift variation per one beat length for fiber PM1550-HP.

		Phase shift variation per one beatlength			
		Temperature source distance from PMF			
		5 [cm]	8 [cm]	11 [cm]	
Temp.	50 °C	2.85	1.14	0.80	$\pi \times 10^{-3} \ [rad]$
	45 °C	2.00	0.95	0.67	
	40 °C	1.43	0.83	0.57	
	35 °C	0.58	0.48	0.27	

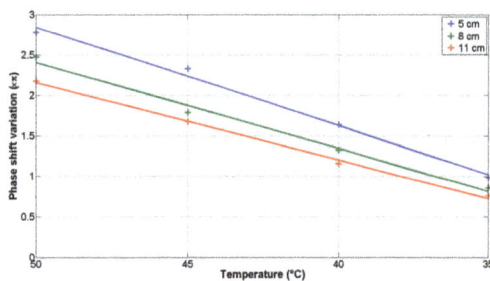

Fig. 14: Phase shift variation dependence on temperature for three exposed lengths of winding for PM1550-HP.

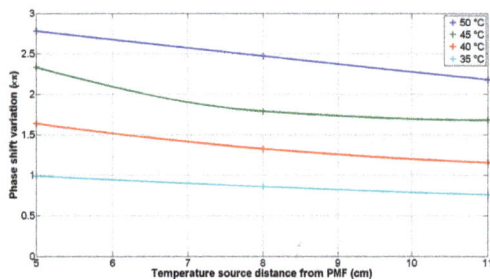

Fig. 15: Phase shift variation dependence on distance between thermal source and PM fiber for three exposed lengths of winding for PM1550-HP.

All measurements show relatively high instability and sensitivity to other surrounding effects such as random temperature changes, mechanical vibrations etc. Therefore this application is suitable as a sensor of thermal field disturbance, but not as a precision temperature sensor. High sensitivity to temperature changes could be combined with suitable fiber configuration to get the sensitivity to mechanical vibrations, Such sensor applications could be used for territory monitoring, to prevent undesired objects approaching, etc.

5. Conclusion

The presented results complete the tests from a previous study for $\lambda = 633$ nm. By comparing these results, the expected conclusion can be stated, that there is a higher sensitivity for $\lambda = 633$ nm, mainly due to the short beat length and high coherency of the He-Ne laser source. To reach comparable sensitivity at 1550 nm a longer segment of the exposed fiber should be used. As advantage of LD for 1550 nm is the compact set up of the source-fiber-polarizer-detector.

The next study should focus on a detailed analysis of the effect of source coherency on the fiber behavior in respect to a phase shift variation during thermal field disturbance. Generally, also analysis of phase shift variation in time during all the transformations stages could be very interesting, first the initial absorption of the thermal radiation, then successive temperature stabilizing and then transition into a new stable state. Based on this, a compact solutions could be built to be utilized in other practical applications.

Acknowledgment

This work has been supported by Project for the development of K217 Department, Brno University of Defense – Modern electrical elements and systems.

References

[1] PERLICKI, K. Identification of Polarization configuration based on torsion and Curvature Calculation. *WSEAS Transactions of Communications.* 2006, vol. 5, iss. 4, pp. 631–633. ISSN 1109-2742.

[2] VLCEK, C. Study of external effects on the polarization properties of selected fiber segments. *WSEAS Transactions on Communications.* 2006, vol 5, iss. 4, pp. 611–616. ISSN 1109-2742.

[3] MARTELLUCCI, S., A. N. CHESTER and A. G. MIGNANI. *Optical sensors and microsystems, new concepts, materials, technologies.* New York: Kluwer Academic Publishers, 2000. ISBN 978-0306463808.

[4] FRADEN, Jacob. *Handbook of modern sensors: physics, designs, and applications.* New York: Springer, 2010. ISBN 978-1-4419-6465-6.

[5] HOTATE, K. and S. O. S. LENG. Transversal force sensor using polarization-maintaining fiber independent of direction of applied force: proposal and experiment. In: *15th Optical Fiber Sensors Conference Technical Digest. OFS 2002.* Portland: IEEE, 2002, pp. 363–366. ISBN 0-7803-7289-1. DOI: 10.1109/OFS.2002.1000586.

[6] ZHANG, F. and J. W. Y. LIT. Temperature sensitivity measurements of high-birefringent polarization-maintaining fibers. *Applied Optics.* 1993, vol. 32, iss. 13, pp. 2213–2218. ISSN 1559-128X.

[7] DVORAK, F., J. MASCHKE and C. VLCEK. Fiber sensor of temperature field disturbance. In: *4th International Conference on Circuits, Systems and Signal (CSS'10).* Corfu: World Scientific and Engineering Academy and Society, 2010, pp. 134–139. ISBN 978-960-474-208-0.

[8] DVORAK, F., J. MASCHKE and C. VLCEK. Utilization of birefringent fiber as sensor of temperature field disturbance. *Radioengineering.* 2009, vol. 18, no. 4, pp. 639–643. ISSN 1210-2512.

[9] DVORAK, F., J. MASCHKE and C. VLCEK. Response analysis of thermal field disturbance sensor. In: *Electro-Optical and Infrared Systems: Technology and Applications VIII.* Prague: SPIE, 2011, pp. 501–508. ISBN: 978-081948813-8. DOI: 10.1117/12.897707.

[10] DVORAK, F., J. MASCHKE and C. VLCEK. Study of fiber PM1550-HP response in the set of thermal field disturbance sensor. In: *Proc. of NAUN International Multi-conference, Recent Researchers in Circuits, Systems, Communications & Computers.* Puerto De La Cruz: WSEAS, 2011. pp. 137–141. ISBN 978-1-61804-056-5.

[11] COLLETT, Edward. *Polarized light in fiber optics.* Lincroft: The PolaWave Group, 2003. ISBN 978-081-9457-615.

[12] COLLETT, E. and B. SCHAEFER. Visualization and calculation of polarized light. I. The polarization ellipse, the Poincaré sphere and the hybrid polarization sphere. *Applied optics.* 2008, vol. 47, iss. 22. ISSN 1559-128X. DOI: 10.1364/AO.47.004009.

[13] COLLETT, E. and B. SCHAEFER. Visualization and calculation of polarized light. II. Applications of the hybrid polarization sphere. *Applied optics.* 2008, vol. 47, iss. 22. ISSN 1559-128X. DOI: 10.1364/AO.47.004017.

About Authors

Filip DVORAK was born in 1977. He received his M.Sc. degree from the Brno Military Academy in 2004 and Ph.D. degree in 2011. He is currently lecturer with the Department of Radar Technology, Brno University of Defense. His work is focused on the modeling of fibers and optical components by means of matrix methods in the MATLAB environment and analysis of fiber sensors.

Jan MASCHKE was born in 1942. He received his M.Sc. degree in 1965 and Ph.D. degree in 1978. He was a teacher at the Technical school at Liptovský Mikulas and at the Military Academy since 1968, and associate professor of the Department of electrical engineering since 1985. He retired in 2005. His research work focused on the problems of fiber optics and fiber sensors.

Cestmir VLCEK was born in 1946. He received his M.Sc. degree in 1969 and Ph.D. degree in 1980. He was a teacher at the Military Academy since 1969, associate professor since 1985, head of the Electrical Engineering and Electronics Department since 1997 and professor since 2000. His research work during the last years was aimed at problems of optoelectronic signals and systems, single mode fiber components modeling, atmospheric optical communication systems and analysis of sensors for military applications.

7

Investigation on Superior Performance by Fractional Controller for Cart-Servo Laboratory Set-Up

Ameya Anil KESARKAR, Selvaganesan NARAYANASAMY

Department of Avionics, Indian Institute of Space Science and Technology, Department of Space, Government of India, Valiamala P.O., Thiruvananthapuram – 695 547 Kerala, India.

ameyakesarkar@iist.ac.in, selvag@iist.ac.in

Abstract. *In this paper, an investigation is made on the superiority of fractional PID controller ($PI^\alpha D^\beta$) over conventional PID for the cart-servo laboratory set-up. The designed controllers are optimum in the sense of Integral Absolute Error (IAE) and Integral Square Error (ISE). The paper contributes in three aspects: 1) Acquiring nonlinear mathematical model for the cart-servo laboratory set-up, 2) Designing fractional and integer order PID for minimizing IAE, ISE, 3) Analyzing the performance of designed controllers for simulated plant model as well as real plant. The results show a significantly superior performance by $PI^\alpha D^\beta$ as compared to the conventional PID controller.*

Keywords

Cart-servo, fractional PID, IAE, ISE.

1. Introduction

Fractional calculus [1] has recently found new applications in control engineering resulting in an area popularly known as 'Fractional Order Control (FOC)'. FOC is nothing but designing the controllers which are governed by fractional order the differential equations. The compact form expressions of these controllers possess easily tunable characteristics for meeting stringent loop performance [2], [3], [4], [5], [6].

The tuning of three-parameter fractional order controllers such as PI^α, $[PI]^\alpha$, PD^β, $[PD]^\beta$ has been addressed in the literature [7], [8], [9], [10], [11]. The formulation in these works consists of a set of three equations which are solved analytically or graphically.

Literature also covers tuning of five-parameter $PI^\alpha D^\beta$ controller to minimize certain performance indices such as Integral Absolute Error (IAE), Integral Square Error (ISE), etc. [12], [13]. This is an unconstrained, five-dimensional and multi-modal optimization problem in which the objective function is optimized with respect to five parameters. The works in [12], [13] have considered linear plants. However, one can also design IAE, ISE minimizing fractional controllers for the given nonlinear plant model. A few early works in this regard are seen in the literature [14], [15], [16], [17] in which the superiority of fractional order controllers over integer ones is investigated.

In the present paper, we explore further in this direction to examine the fractional superiority for cart-servo lab set-up which contains a few nonlinear elements.

The major contributions of this paper are as follows:

- The mathematical model of the cart-servo lab setup is obtained which is further validated by performing a model-matching test.

- For the acquired model, optimum fractional and integer order PID controllers are designed so as to minimize performance indices such as IAE and ISE.

- The designed controllers are analyzed in detail for their performance with the plant model as well as real plant to examine the superiority of fractional controllers.

Organization of the paper: Section 2 presents mathematical modelling of the cart-servo lab setup and its validation using model-matching test. In section 3, preliminaries of the fractional order control are discussed and also the controller design problem for minimizing performance indices such as IAE, ISE is explained. Section 4 demonstrates the design of integer and fractional order PID controllers to meet the control requirements. Also, the performance of designed controllers is discussed and compared in section 4. Finally, section 5 provides the concluding remarks.

2. Mathematical Modelling of Cart-Servo Lab Set-Up

2.1. Plant Description

The original cart-pendulum lab set-up designed by Feedback Instruments, UK consists of a cart moving along a 1 metre long track [18]. The cart has a shaft to which the pendulum is attached. The cart can move back and forth causing the pendulum to swing.

For the cart-servo control purpose intended for the current paper, the pendulum is detached from the above set-up as shown in Fig. 1. The movement of the cart is caused by pulling the belt in two directions by the DC motor attached at one end of the rail. The control task is to attain the desired cart position on the rail which is realized by controlling the input voltage to the motor.

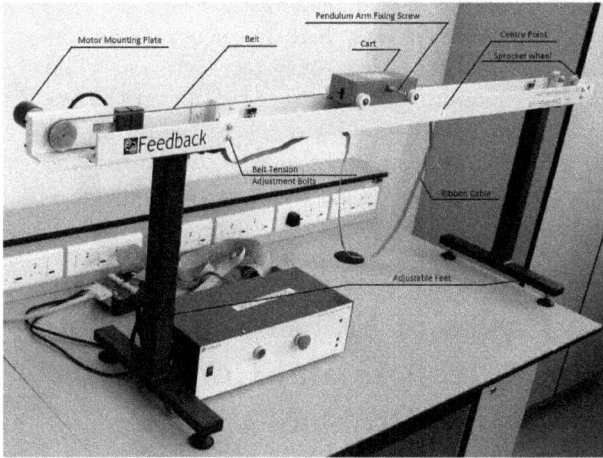

Fig. 1: Cart-servo plant experimental set-up.

2.2. Model Identification

In model-based controller tuning approach, it is essential to have the sufficiently captured mathematical model for the real plant dynamics. It is carried out as explained below.

- The equations governing cart-servo plant dynamics are [18]:

$$V_a - K_b \omega_m = R_m I_a + L_m \frac{dI_a}{dt}, \tag{1}$$

$$T_m = K_t I_a, \tag{2}$$

$$T_{load} = T_m - J_m \alpha_m - B_m \omega_m, \tag{3}$$

$$F_{load} = T_{load} g_r, \tag{4}$$

$$F_{load} = M_c a_c + B_c v_c + F_{frict}. \tag{5}$$

- Incorporating Eq. (1), Eq. (2), Eq. (3), Eq. (4), Eq. (5) along with the current-loop type of power amplifier dynamics [18], we construct the complete mathematical model of the cart-servo set-up as shown in Fig. 2. In Fig. 2, the gray shaded blocks are the nonlinear elements present in the system. Also, the cart auxiliary velocity v_{cd} has been obtained from cart velocity v_c to eliminate the 'algebraic loop' while simulating the mathematical model. (Note: Table of parameters and the table of variables have been given in Tab. 1 and Tab. 2.)

- To define the nonlinear relation between $[F_{frict}, \text{Reset}]$ and $[F_{load}, v_{cd}]$, we propose the following embedded MATLAB code (Refer the block, 'Cart Friction Model' in Fig. 2):

$function[F_{frict}, Reset] = cartfrict(F_{load}, v_{cd})$
$fr = 1.68; Mc = 2.3; Ts = 0.0001;$
$Fc = fr * sign(v_{cd}); \%fr = frictioncoefficient,$
$Ts = SamplingInterval$
if $v_{cd} == 0$
$Fc = fr * sign(F_{load});$
end
if $v_{cd} == 0$
if $abs(F_{load}) < fr$
$Reset = 1; F_{frict} = F_{load};$
else
$Reset = 0; F_{frict} = fr * sign(F_{load});$
end
else
$vc2 = v_{cd} + (F_{load} - F_c)/Mc * Ts;$
if$((sign(vc2 * v_{cd}) < 0)\&\&(sign(F_{load} * vc2) < 0))$
$||((sign(vc2 * v_{cd}) < 0)\&\&$
$((sign(F_{load} * vc2) > 0)\&\&(abs(F_{load}) < fr)))$
$Reset = 1; F_{frict} = F_{load};$
else
$Reset = 1; F_{frict} = F_{load};$
end
end

2.3. Model-Matching

In order to validate the acquired model as shown in Fig. 2, we perform a model-match test. For this purpose, a sweep signal of amplitude 0.2 is generated [1]. The sweep signal is given as an input to a closed loop

---

[1] The sweep signal is a composite signal which is constructed by the sine-waves of different frequencies such that the time instance at which one sine-wave ends, the other one begins.

Tab. 1: Table of parameters.

Description	Symbol	Unit	Value
Power amplifier input voltage saturation limit	V_c	V	2.5
Power amplifier supply voltage saturation limit	V_s	V	24
Power amplifier current loop gain	K_1	-	$\frac{1}{35}$
Power amplifier forward gain	K_2	-	200
Motor armature resistance	R_m	Ω	2.5
Motor armature inductance	L_m	H	0.0025
Motor back-emf constant	K_b	V/(rad/sec)	0.05
Motor torque constant	K_t	N-m/A	0.05
MI of motor rotating assembly	J_m	Kg-m^2	0.000014
Viscous damping coefficient of motor shaft	B_m	N-m/(rad/sec)	0.000001
Rotary to linear motion conversion ratio	g_r	m^{-1}	600
Viscous damping coefficient of cart	B_c	N-m/(rad/sec)	0.00005
Mass of cart	M_c	Kg	2.3
Second order filter natural frequency	ω_f	rad/sec	2215.7
Second order filter damping factor	ζ_f	-	0.7

Tab. 2: Table of variables.

Description	Symbol
Armature voltage	V_a
Armature current	I_a
Motor torque	T_m
Load torque	T_{load}
Load force	F_{load}
Friction force	F_{frict}
Motor shaft angular velocity	ω_m
Motor shaft angular acceleration	α_m
Cart position	x_c
Cart velocity	v_c
Cart auxillary velocity	v_{cd}
Cart acceleration	a_c

system which contains simulated/real plant with unity gain controller.

The response is as shown in Fig. 3. It is seen from Fig. 3 that the responses for simulated as well as real plant are close enough to confirm sufficient capture of plant dynamics in its mathematical model.

The response to the sweep signal is composed of responses to each (frequency) sine-wave in the corresponding time-intervals. By considering the fundamental harmonic in such an output response corresponding to each sine-wave, one can construct the frequency response of the closed loop system. We obtain the closed loop frequency responses with real and simulated plant as presented in Fig. 4.

The mathematical model (refer Fig. 2) consisting of elements such as Cart-Friction Model, current loop power amplifier, etc. sufficiently captures the lower mode dynamics of cart-servo which is in the required control-passband frequency range. High frequency

phenomena such as mechanical vibrations, switching in power amplifier circuits, high frequency noise signal etc. have not been taken into consideration while modelling. Therefore, one can see in Fig. 4 that the frequency responses with simulated and real plant match closely for the lower range of frequency while there is a mismatch between these responses at the higher frequency side.

3. Basics of Fractional Controllers and Controller Design Problem

3.1. Preliminaries of Fractional Controllers

1) Fractional Calculus

Conventional calculus deals with integer order differentiation and integration. Generalization of conventional calculus so as to consider differentiation and integration of any order (not necessarily integer) leads to 'Fractional Calculus (FC)' [1]. In FC, the fundamental differ-integration operator $_aD_t^\alpha$ (where a and t are the limits of the operation) is defined as [2]:

$$_a\mathrm{D}_t^\alpha = \begin{cases} \frac{d^\alpha}{dt^\alpha} & \alpha > 0 \\ 1 & \alpha = 0 \\ \int_a^t (d\tau)^{-\alpha} & \alpha < 0 \end{cases}, \qquad (6)$$

where α is the order of the operation, generally $\alpha \in \mathbb{R}$ but α could also be a complex number.

Out of many definitions of fractional differ-integration in FC, the popular ones are [2]:

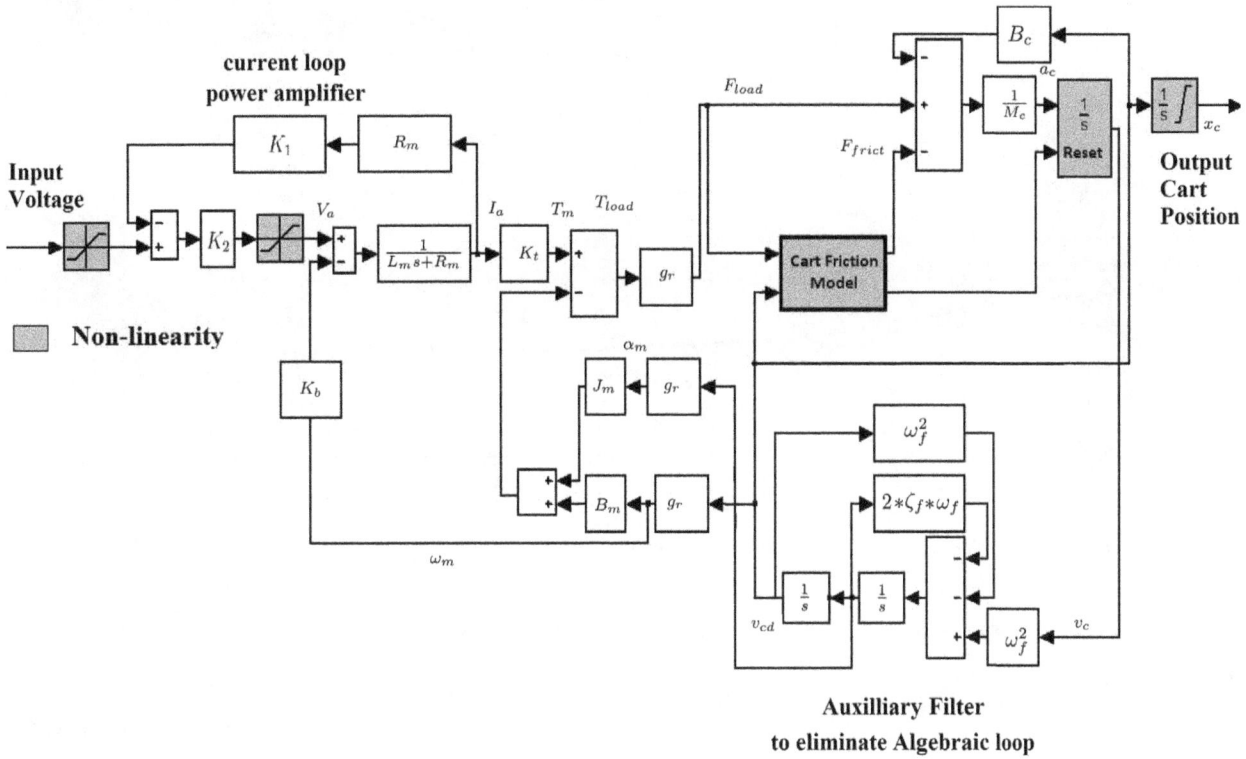

Fig. 2: Mathematical model of cart-servo plant.

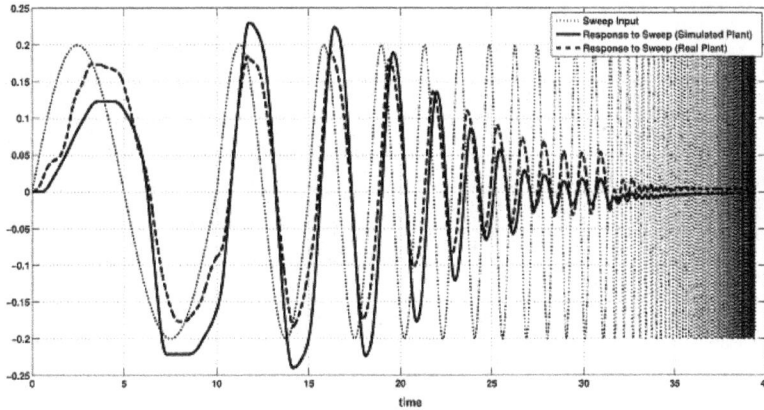

Fig. 3: Closed loop response with real and simulated plant to sweep signal.

- Grunwald-Letnikov Definition:

$$_aD_t^\alpha f(t) = \lim_{h \to 0} \frac{1}{h^\alpha} \sum_{j=0}^{\left[\frac{t-a}{h}\right]} (-1)^j \binom{\alpha}{j} f(t-jh), \quad (7)$$

where, $\left[\frac{t-a}{h}\right]$ truncates $\frac{t-a}{h}$ to an integer.

- Riemann-Liouville (R-L) Definition:

$$_aD_t^\alpha f(t) = \frac{1}{\Gamma(n-\alpha)} \left(\frac{d}{dt}\right)^n \int_a^t \frac{f(\tau)}{(t-\tau)^{\alpha-n+1}} d\tau, \quad (8)$$

where n is an integer, a is a real number, and α satisfies $(n-1) \le \alpha < n$.

- Caputo Definition:

$$_aD_t^\alpha f(t) = \frac{1}{\Gamma(n-\alpha)} \int_a^t \frac{f^{(n)}(\tau)}{(t-\tau)^{\alpha-n+1}} d\tau. \quad (9)$$

2) Fractional Order Transfer Function Model

The equation of Laplace transform for the defined fractional-order operator is [2]:

$$L(_aD_t^\alpha f(t)) = s^\alpha F(s), \quad (10)$$

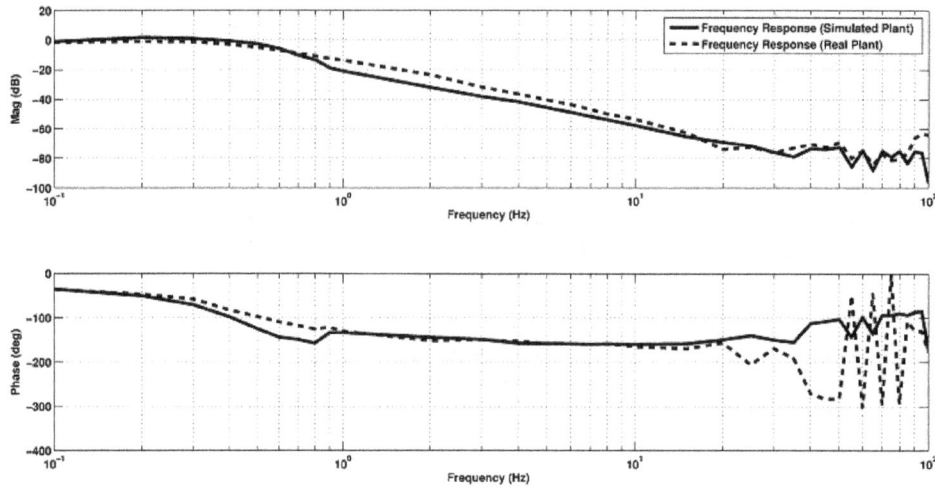

Fig. 4: Closed loop frequency response with real and simulated plant.

with zero initial conditions.

Linear time invariant fractional model of a system with input u, and output y takes the following form [2]:

$$a_n D^{\alpha_n} y(t) + a_{n-1} D^{\alpha_{n-1}} y(t) + \ldots + a_0 D^{\alpha_0} y(t)$$
$$= b_m D^{\beta_m} u(t) + b_{m-1} D^{\beta_{m-1}} u(t) + \ldots + b_0 D^{\beta_0} u(t), \tag{11}$$

where, a_i, α_i $(i = 0, 1, \ldots, n)$, b_k, β_k $(k = 0, 1, \ldots, m)$ are real constants. n and m are positive integers.

Therefore, Laplace transform on both sides (assuming zero initial conditions) results into the following transfer function:

$$\frac{Y(s)}{U(s)} = \frac{b_m s^{\beta_m} + b_{m-1} s^{\beta_{m-1}} + \ldots + b_0 s^{\beta_0}}{a_n s^{\alpha_n} + a_{n-1} s^{\alpha_{n-1}} + \ldots + a_0 s^{\alpha_0}}. \tag{12}$$

3) Fractional Order Controller

From control engineering point of view, the application of FC can be in either system modelling or controller design. The typical fractional order controllers $C(s)$ found in the literature are as follows:

Fractional order proportional-integral controller, which is of two types [9]:

- PI^α

$$C(s) = K_p \left(1 + \frac{K_i}{s^\alpha}\right). \tag{13}$$

- $[\text{PI}]^\alpha$

$$C(s) = K_p \left(1 + \frac{K_i}{s}\right)^\alpha, \tag{14}$$

with $\alpha = 1$, we get Integer PI of the form: $C(s) = K_p \left(1 + \frac{K_i}{s}\right)$.

Fractional order proportional-derivative controller, which is of two types [7], [8]:

- PD^β

$$C(s) = K_p \left(1 + K_d s^\beta\right). \tag{15}$$

- $[\text{PD}]^\beta$

$$C(s) = K_p \left(1 + K_d s\right)^\beta, \tag{16}$$

with $\beta = 1$, we get Integer PD of the form: $C(s) = K_p \left(1 + K_d s\right)$.

Fractional order proportional-integral-derivative controller [2]:

- $\text{PI}^\alpha \text{D}^\beta$

$$C(s) = K_p \left(1 + \frac{K_i}{s^\alpha} + K_d s^\beta\right), \tag{17}$$

with $\alpha = 1, \beta = 1$, we get Integer PID of the form: $C(s) = K_p \left(1 + \frac{K_i}{s} + K_d s\right)$.

3.2.   Design of Optimum Controller to Minimize IAE, ISE

The typical unity feedback control system is shown in Fig. 5. $r(t)$, $e(t)$, $u(t)$, and $y(t)$ denote reference input, error, controller output, and plant output respectively.

The following performance indices are considered:

- Integral Absolute Error (IAE):

$$J = \int_0^\infty |e(t)| dt. \tag{18}$$

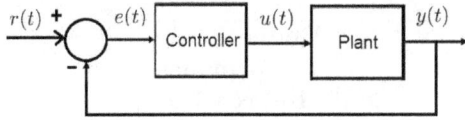

Fig. 5: Typical unity feedback control system.

- Integral Square Error (ISE):

$$J = \int_0^\infty |e(t)|^2 dt. \qquad (19)$$

Let $e(kT)$ be the sampled value of the error $e(t)$ at an instant (kT), where T is the sampling interval. $k=0,1,...,N$. For the given T, N is an integer which depends on the time span considered for computing $e(t)$. The following cost functions are considered corresponding to the performance indices defined in Eq. (18), Eq. (19):

- IAE cost function:

$$J_c = \sum_{k=0}^N |e(kT)|T. \qquad (20)$$

- ISE cost function:

$$J_c = \sum_{k=0}^N |e(kT)|^2 T. \qquad (21)$$

Each performance index emphasizes different aspects of the system response [19]. Large errors contribute more in ISE than IAE. Consequently, the controller tuned for minimizing ISE ensures lower overshoot in the transient response than IAE minimizing controller. The ISE, however tends to give larger settling time.

For the optimum performance of control system, controller parameters are tuned by minimizing the selected performance index.

4. Design and Performance Analysis of Integer and Fractional PID for Cart-Servo

Mathematical model of the cart-servo plant (as developed in Section 2) is considered for designing $PI^\alpha D^\beta$ and PID controllers (refer Eq. (17) for the controller structure). The controller design problem for minimizing IAE and ISE (as discussed in Section 3) is solved numerically with MATLAB using fminsearch() function.

While tuning, a step input of 0.2 m is given to the closed loop system for 10 sec. The sampling interval is taken as 0.001 sec. The search space for the controller is limited by assigning certain bounds to its parameters. Fractional order integration and differentiation are ensured by choosing $\alpha \in (0,1)$ and $\beta \in (0,1)$ respectively. In $PI^\alpha D^\beta$ controller, if $\alpha = 1$ and $\beta = 1$, it becomes the PID controller. Therefore, to eliminate the case of PID while tuning $PI^\alpha D^\beta$, the values $\alpha = 1$ and $\beta = 1$ are not included in the bounds for α and β. The bounds for K_p, K_i and K_d ($K_p \in (0,15]$, $K_i \in (0,5]$, $K_d \in (0,1]$) have been suitably chosen by referring their typical values for the design examples given in the manual [18]. However, one is free to choose any other valid bounds. For the selected bounds, the emphasis is on the investigation of possible superior performance by $PI^\alpha D^\beta$ over PID.

Oustaloup [20] approximated transfer function model of the $PI^\alpha D^\beta$ controller is considered for the simulation. The order of Oustaloup approximation is taken as 7 and the approximation is valid over the frequency range $[0.001, 1000]$ rad·s^{-1}.

After the design, controllers are tested with simulated as well as real plant. The results are presented in Tab. 3 and Tab. 4. The following conclusions are derived based on Tab. 3 and Tab. 4:

- For simulated plant, fractional PID reduces J_{IAE} by nearly 65 % and J_{ISE} by nearly 62 % as compared to integer PID.

- For the real plant case, fractional PID reduces J_{IAE} by nearly 75 % and J_{ISE} by nearly 65 % as compared to the integer PID.

- The differences in the performance index values obtained for simulated and real plants are due to slight imperfections in the captured plant dynamics.

Figure 6 presents cart-position and error signals for the closed loop system with simulated plant and fractional/integer PID controller tuned for IAE minimization. A step signal of amplitude 0.2 m is given for 10 sec duration as input. The corresponding response with real plant is shown in Fig. 7. The responses for ISE minimization case are shown in Fig. 8 and Fig. 9.

From Fig. 6, Fig. 7, Fig. 8, Fig. 9, we observe that the cart-position response with fractional PID shows significantly smaller rise time, settling time, peak overshoot as compared to the integer PID. This means that the performance with fractional PID is far superior over its integer counter part for cart-servo lab set-up.

Tab. 3: IAE for designed controllers with simulated and real plants.

Performance Index		Controller		Reduction in Performance Index with $\mathbf{PI^{\alpha}D^{\beta}}$ over PID (%)
		PID	$\mathbf{PI^{\alpha}D^{\beta}}$	
		$K_p = 14.9815,$ $K_i = 2.7509,$ $K_d = 0.5$	$K_p = 1.3124, K_i = 4.9906,$ $\alpha = 0.1874, K_d = 2.9322,$ $\beta = 0.5452$	
IAE	$J_{simulated}$	0.1562	0.0551	64.725
	J_{real}	0.1636	0.0393	75.978

Tab. 4: ISE for designed controllers with simulated and real plants.

Performance Index		Controller		Reduction in Performance Index with $\mathbf{PI^{\alpha}D^{\beta}}$ over PID (%)
		PID	$\mathbf{PI^{\alpha}D^{\beta}}$	
		$K_p = 14.9762, K_i = 3.2929,$ $K_d = 0.5001$	$K_p = 4.999, K_i = 4.7835,$ $\alpha = 0.0535, K_d = 0.8685,$ $\beta = 0.7384$	
ISE	$J_{simulated}$	0.0120	0.0046	61.667
	J_{real}	0.0128	0.0045	64.844

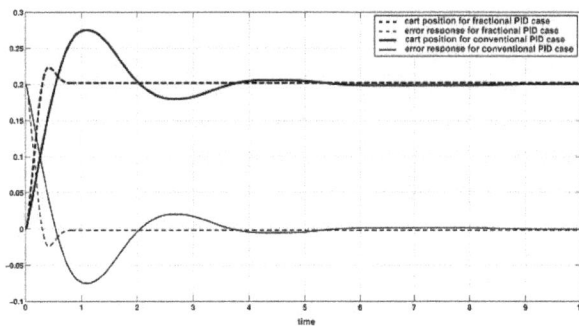

Fig. 6: Step and error response with simulated plant (IAE minimization).

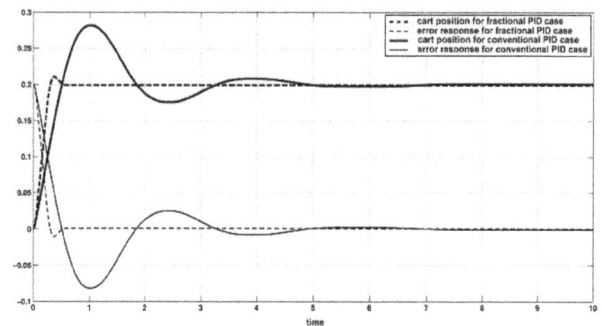

Fig. 8: Step and error response with simulated plant (ISE minimization).

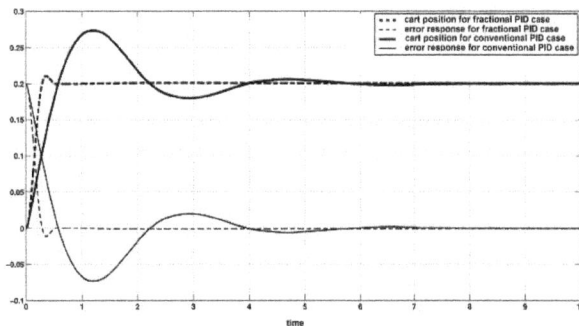

Fig. 7: Step and error response with real plant (IAE minimization).

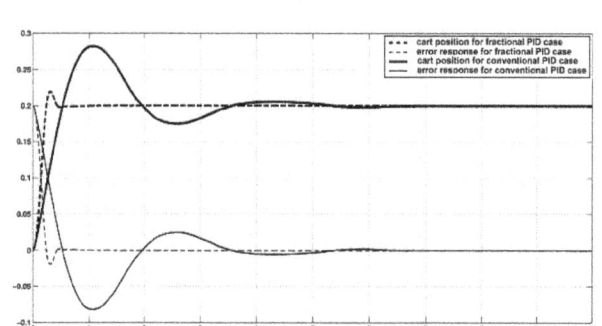

Fig. 9: Step and error response with real plant (ISE minimization).

5.   Conclusion

The paper presented the ability of fractional PID controller ($PI^{\alpha}D^{\beta}$) to produce superior performance over conventional PID for cart-servo lab set-up. For this purpose, these controllers were designed for the acquired mathematical model of the plant so as to minimize performance indices, IAE and ISE. The designed controllers were tested with the simulated as well as real plant. It was observed that the fractional PID outperformed integer order PID by significantly reducing the performance indices (more than 60 % in each case).

References

[1] OLDHAM, K. B. and J. SPANIER. *The Fractional Calculus.* New York: Academic Press, 1974. ISBN 9780125355506.

[2] CHEN, Y. Q., I. PETRAS and D. XUE. Fractional order control - A tutorial. In: *American Control Conference, 2009. ACC '09.* St. Louis: IEEE, 2009, pp. 1397–1411. ISBN 978-1-4244-4523-3. DOI: 10.1109/ACC.2009.5160719.

[3] PADULA, F. and A. VISIOLI. Tuning rules for optimal PID and fractional-order PID controllers. *Journal of Process Control.* 2011, vol. 21, iss. 1, pp. 69–81. ISSN 0959-1524. DOI: 10.1016/j.jprocont.2010.10.006.

[4] PODLUBNY, I. Fractional-order systems and $PI^\lambda D^\mu$-controllers. *IEEE Transactions on Automatic Control.* 1999, vol. 44, iss. 1, pp. 208–214. ISSN 0018-9286. DOI: 10.1109/9.739144.

[5] VALERIO D. and J. SA DA COSTA. Introduction to single-input, single-output fractional control. *Control Theory & Applications, IET.* 2011, vol. 5, iss. 8, pp. 1033–1057. ISSN 1751-8644. DOI: 10.1049/iet-cta.2010.0332.

[6] LI, H., Y. LUO and Y. CHEN. A Fractional Order Proportional and Derivative (FOPD) Motion Controller: Tuning Rule and Experiments. *IEEE Transactions on Control Systems Technology.* 2010, vol. 18, iss. 2, pp. 516–520. ISSN 1063-6536. DOI: 10.1109/TCST.2009.2019120.

[7] LI, H. and Y. CHEN. A fractional order proportional and derivative (FOPD) controller tuning algorithm. In: *Control and Decision Conference, 2008. CCDC 2008.* Yantai, Shandong: IEEE, 2008, pp. 4059–4063. ISBN 978-1-4244-1733-9. DOI: 10.1109/CCDC.2008.4598094.

[8] LUO, Y. and Y. CHEN. Fractional-order [proportional derivative] controller for robust motion control: Tuning procedure and validation. In: *American Control Conference, 2009. ACC '09.* St. Louis: IEEE, 2009, pp. 1412–1417. ISBN 978-1-4244-4523-3. DOI: 10.1109/ACC.2009.5160284.

[9] WANG, CH., Y. LUO and Y. CHEN. An Analytical Design of Fractional Order Proportional Integral and [Proportional Integral] Controllers for Robust Velocity Servo. In: *4th IEEE Conference on Industrial Electronics and Applications, 2009. ICIEA 2009.* Xi'an: IEEE, 2009, pp. 3448–3453. ISBN 978-1-4244-2799-4. DOI: 10.1109/ICIEA.2009.5138847.

[10] WANG, CH., Y. LUO and Y. CHEN. Fractional Order Proportional Integral (FOPI) and [Proportional Integral] (FO[PI]) Controller Designs for First Order Plus Time Delay (FOPTD) Systems. In: *Control and Decision Conference, 2009. CCDC '09. Chinese.* Guilin: IEEE, 2009, pp. 329–334. ISBN 978-1-4244-2722-2. DOI: 10.1109/CCDC.2009.5195105.

[11] LUO, Y., Y. CHEN, CH. WANG and Y. PI. Tuning fractional order proportional integral controllers for fractional order systems. *Journal of Process Control.* 2010, vol. 20, iss. 7, pp. 823–831. ISSN 0959-1524.

[12] VALERIO, D. and J. SA DA COSTA. Tuning-rules for fractional PID controllers. In: *2nd IFAC Workshop on Fractional Differentiation and its Applications, 2006.* Porto: IFAC, 2010, pp. 28–33. ISBN 978-3-902661-12-8. DOI: 10.3182/20060719-3-PT-4902.00004.

[13] KHALFA B. and CH. ABDELFATAH. Parameter Tuning of Fractional PI^α-PD^β Controllers with Integral Performance Criterion. In: *1st National Conference on Fractional Order Systems and Applications.* Skikda: SOFA, 2010, pp. 1–6.

[14] BARBOSA, S. R., J. A. TENREIRO MACHADO and A. M. GALHANO. Performance of Fractional PID Algorithms Controlling Nonlinear Systems with Saturation and Backlash Phenomena. *Journal of Vibration and Control.* 2007, vol. 13, iss. 9, pp. 1407–1418. ISSN 1741-2986.

[15] IKEDA, F. and S. TOYAMA. Fractional Derivative Control Designs by Inhomogeneous Sampling for Systems with Nonlinear Elements. In: *SICE Annual Conference.* Takamatsu: IEEE, 2007, pp. 1224–1227. ISBN 978-4-907764-27-2. DOI: 10.1109/SICE.2007.4421171.

[16] ALDAIR, A. A. and W. J. WANG. Design of Fractional Order Controller Based on Evolutionary Algorithm for a Full Vehicle Nonlinear Active Suspension Systems. *International Journal of Control and Automation.* 2010, vol. 3, iss. 4, pp. 33–46. ISSN 2005-4297.

[17] KESARKAR, A. A. and N. SELVAGANESA. Superiority of Fractional Order Controllers in Limit Cycle Suppression. *International Journal of Control and Automation.* 2013, vol. 7, iss. 3, pp. 166–182. ISSN 2005-4297.

[18] FEEDBACK INSTRUMENTS LTD. *Digital Pendulum Control Experiments, Manual: 33-936S Ed01 122006.* Feedback Part No. 1160-33936S.

[19] UMEZ-ERONINI, E. *System dynamics and control.* Singapore: Thomson Asia Pte. Ltd., 2002. ISBN 978-0534944513.

[20] OUSTALOUP, A., F. LEVRON, B. MATHIEU and F. M. NANOT. Frequency-Band Complex Noninteger Differentiator: Characterization and Synthesis. *IEEE Transactions On Circuits And Systems I: Fundamental Theory And Applications.* 2000, vol. 47, iss. 1, pp. 25–39. ISSN 1057-7122. DOI: 10.1109/81.817385.

About Authors

Ameya Anil KESARKAR Ameya Anil Kesarkar received his Bachelor of Engineering (BE) in Electronics from Mumbai University, Maharashtra, India in 2007, and Master of Technology (MTech) in Control and Automation from Electrical Department, Indian Institute of Technology Delhi (IITD), Delhi, India in 2009. He is currently pursuing his PhD from Department of Avionics, Indian Institute of Space Science and Technology (IIST), Kerala, India. His research interests include classical control, fractional calculus, fractional order control, numerical optimization, and non-linear dynamics.

Selvaganesan NARAYANASAMY Selvaganesan Narayanasamy received his BE in Electrical and Electronics, ME in Control Systems and PhD in System Identification and Adaptive Control from Mepco Schlenk Engineering College-Sivakasi, PSG College of Technology-Coimbatore and MIT Campus, Anna University, India in the year 1997, 2000, and 2005 respectively. He has more than 13 years of research and teaching experience. Currently, he is working as an Associate Professor and Head of Department of Avionics in Indian Institute of Space Science and Technology, Trivandrum, India. He has 19 international journal papers and 35 conference papers to his credit. His areas of interest include system identification, fault diagnosis, model reduction, converters, power filter control design and fractional order control. He is involved in many editorial activities.

8

An Application of Graph Theory in Markov Chains Reliability Analysis

Pavel SKALNY

Department of Applied Mathematics, Faculty of Electrical Engineering and Computer Science,
VSB–Technical University of Ostrava, 17. listopadu 15/2172, 708 33 Ostrava-Poruba, Czech Republic

pavel.skalny@vsb.cz

Abstract. *The paper presents reliability analysis which was realized for an industrial company. The aim of the paper is to present the usage of discrete time Markov chains and the flow in network approach. Discrete Markov chains a well-known method of stochastic modelling describes the issue. The method is suitable for many systems occurring in practice where we can easily distinguish various amount of states. Markov chains are used to describe transitions between the states of the process. The industrial process is described as a graph network. The maximal flow in the network corresponds to the production. The Ford-Fulkerson algorithm is used to quantify the production for each state. The combination of both methods are utilized to quantify the expected value of the amount of manufactured products for the given time period.*

Keywords

Discrete time Markov chain, flow in network, reliability.

1. Introduction

The reliability of production plays the fundamental role in an industrial sphere. Nowadays the reliability of industry process is on a high level. It is increased by improving the quality of each component or by redundancy of the production process. Even though it is the top reliability process, there is still a chance that system fails. In our case we analyse the process which has no redundancy. Thus the information about the probability of the systems failures is very valuable.

In the previous work an [1] a reliability analysis of the part of an industry process was realized. In the former application there was analysed a part of the industry process distributed in parallel. The main aim of the cited paper was to estimate the probability, that

the production will be equal or greater than the industrial partner demand. To deal with the task a Monte Carlo simulation of Discrete time Markov chains was used.

The goal of this paper is to present the Discrete time Markov chain analysis realized on the system, which is not distributed in parallel. Since the analysed process is more complicated the graph theory approach seems to be an appropriate tool.

Graph theory has previously been applied to reliability, but for different purposes than we intend. One of example of application of graph theory in reliability is a reliability polynomial [3] or network reliability. In this approach a graph describes a network where each component (edge) has the same probability of fail. The most related work to our paper is probably the research of Christopher Dabrowski [4], [5]. As well as in our paper Dabrowski uses graph theory as a tool for counting with discrete time Markov chains. In contrast to our work the Markov chains described in Dabrowski one are used for different purposes. Dabrowski analyses a large scale grid. His study describe the application for finding states of the grid, which could lead to the system degradation. Dabrowski uses the graph theory to describe transitions of the Markov chain model between the initial state and the absorbing state. We use the graph theory to describe the analysed process on which the discrete time Markov chain is be applied.

2. Markov Chains

Markov chain is a random process with a discrete time set $T \subset \mathbb{N} \cup \{0\}$, which satisfies the so called "Markov property". The Markov property means that the future evolution of the system depends only on the current state of the system and not on its past history.

$$\mathbf{P}\{X_{n+1} = x_{n+1} | X_0 = x_0, \dots, X_n = x_n\} =$$
$$= \mathbf{P}\{X_{n+1} = x_{n+1} | X_n = x_n\}, \qquad (1)$$

where $X_1, \ldots, X_n$ is a sequence of random variables. The index denotes certain time $t \in T$ $x_1, \ldots x_n$ is a sequence of states in time $t \in T$. As a transition probability p_{ij} we regard probability, that the system changes from the state i to the state j:

$$p_{ij} = \mathbf{P}\{X_{n+1} = x_j | X_n = x_i\}. \qquad (2)$$

Matrix $\mathbf{P}$, where p_{ij} is placed in row i and column j, is for all admissible i and j called transition probability matrix:

$$\mathbf{P} = \begin{pmatrix} p_{11} & p_{12} & \cdots & p_{1n} \\ p_{21} & p_{22} & \cdots & p_{2n} \\ \vdots & \vdots & \vdots & \vdots \\ p_{n1} & p_{n2} & \cdots & p_{nn} \end{pmatrix}. \qquad (3)$$

Clearly all elements of the matrix $\mathbf{P}$ satisfy the following property:

$$\forall i \in \{1, 2, \cdots, n\} : \sum_{j=1}^{n} p_{ij} = 1. \qquad (4)$$

As a probability vector we will understand a vector $\vec{v}$ where, an amount of its elements v_i are equal to the amount of states in Markov chain and the following equations holds:

$$\forall v_i \in \vec{v} \; : \; v_i \leq 1, \; and \sum_{all i} = 1. \qquad (5)$$

By $\vec{v}(t)$ we denote vector of probabilities of all the states in time $t \in T$. By $\vec{v}(0)$ we denote an initial probability vector. Usually, all its coordinates are equal to zero except the first, which is equal to 1. The vector $\vec{v}(0)$ denotes the probability in which state the system occurs in time $t = 0$. It is easy to proof that [6]:

$$\vec{v}(t) = \vec{v}(0) \cdot \mathbf{P}^t. \qquad (6)$$

As a stationary distribution we will understand a vector $\vec{z}$ which satisfies the property:

$$z = \vec{z}\mathbf{P}. \qquad (7)$$

Suppose that the limit $\pi = \lim_{n \to \infty} \vec{\pi}(n)$ exists, then the probability vector $\vec{\pi}$ is called limiting distribution (limiting vector):

$$\pi = \lim_{n \to \infty} \vec{\pi}(0)\mathbf{P}^n. \qquad (8)$$

In some literature there is written that, the stationary distribution is equal to the limiting distribution. Actually it is true only if the discrete time Markov chain (further as DTMC) satisfies the condition of ergodicity (irreducibility, aperiodicity). To be able to

calculate the limiting distribution π we will describe further properties of DTMC.

A Markov chain j is periodic with period p, if on leaving state j a return is possible only in a number of transitions that is a multiple of the integer $p > 1$. For example Markov chain with following transition probability matrix:

$$\begin{pmatrix} 0 & 1 \\ 1 & 0 \end{pmatrix}$$

is periodic with period $p = 2$. The DTMC is said to be irreducible if every state can be rached from every other state.

The probability of ever returning to state j is denoted by f_{jj} and is given by:

$$f_{jj} = \sum_{n=1}^{\infty} f_{jj}^n. \qquad (9)$$

If $f_{jj} < 1$, then state j is said to be transient. If $f_{jj} = 1$, the state is recurrent and we can define the mean recurrent time.

As mean recurrent time M_{jj} we will understand:

$$M_{jj} = \sum_{n=1}^{\infty} n f_{jj}^n. \qquad (10)$$

If $M_{jj} < \infty$ (or equal to ∞), we say that the state is positive recurrent (or null recurrent). In a finite Markov chain, each state is either positive-recurrent or transient, and furthermore, at least one state must be positive-recurrent [6]. Theorem Let a Markov chain C be irreducible. Then C is positive-recurrent or null recurrent or transient, i.e.:

- all the states are positive-recurrent, or

- all the states are null-recurrent, or

- all the states are transient.

If all the states of DTMC are positive-recurrent and aperiodic, then the probability distribution $\vec{v}(n)$ converges to a limiting distribution π, which is independent of initial distribution $\vec{v}(0)$. The π can be obtained be solving the following equations:

$$\pi = \vec{\pi}\mathbf{P}, \quad \pi > 0 \; and \sum_{for \; all \; i} \pi_i = 1. \qquad (11)$$

3.   Preliminaries

In this section we will establish fundamental terminology of graph theory which will be used further in this paper.

3.1. Definition of an Oriented Graph

An oriented graph is an ordered pair $G = (V; E)$, where V is the set of vertices and E is the set of edges. $E \subseteq V \times V$.

3.2. Definition of Network

Network is a four-tupple $S = (G; s; t; x)$, where:

- G is an oriented graph,

- vertices $s \in V(G)$, $t \in V(G)$ are the source and sink,

- $x : E(G) \to \mathbb{R}^+$ is a a positive labelling on edges, called edge capacities.

3.3. Definition of Flow in Network

Flow in a network $S = (G; s; t; x)$ is a function $f : E(G) \to \mathbb{R}_0^+$, where:

- no edge capacity is exceeded:
  $\forall e \in E(G) : 0 \leq\leq f(e) \leq x(e)$,

- the conservation of flow equation holds:
  $\forall v \in V(G), v \neq z, s : \sum_{e \to v} f(e) = \sum_{e \leftarrow v} f(e)$.

The value of a flow f is $\|f\| = \sum_{e \leftarrow z} f(e) - \sum_{e \to z} f(e)$.

3.4. Definition of Cut

Cut in the network $S = (G; s; t; x)$ is such a set of edges C, $C \subseteq E(G)$, such that in the factor of graph G, $G - C$, no oriented path remains. As minimal cut we will understand the cut where each proper subset of C is not a cut in the network.

4. Problem Formulation

The research presented here, was motivated by the practical problem. Analysed company was asked to quantify the probability that production fails was. Knowledge of risk, that the order won't be delivered in time, is important for the partner's firm to establish sufficient gods supplies. In the previous application see [1] a reliability analysis of the part of an industry process was realized. For each machine we could distinguish two modes – in order '1' or in fail '0'. Thus the whole system could occur in one of the 2^n states where n is an amount of machines of analysed industry process. Since machines were organized in parallel it was easy to calculate a whole production of each state. The production of the certain state was calculated as a sum of production of functional states. More complicated situation occurs when the system is not connected in parallel. The aim of this paper is to present a calculation of a maximal production of each of 2^n states by usage a graph flow in network theory. In the following application we will simplify the process that, even at the beginning the gods can be reached from the begin to end of the process. We assume that for each state the process will produce a maximum possible production w.

5. Application

In this section we will present an example how to use a flow in network theory in a reliability analysis. First, we will define states of the industry process and calculate a maximum production of each state by using the Ford Fulkerson algorithm. At the end we will demonstrate on the certain data how to calculate an expected value of production for a given time t.

The analyzed industry process consists of six machines. For each machine we distinguish 2 different states one - work, zero - in fail. Thus the industrial process consists of 2^6 different states. State denotes an ordered six -tuple of ones and zeroes.

Let us describe the industry process of a firm with an oriented network. Every machine is represented as a vertex. The begin and the end of the processes represented by a source and sink of the network. The begin of a process consists of acceptance of gods and division between machines of the process. The end of a process consists of product inspection and expediting to the customers. Oriented edges describes the direction of the production process. Labelling of vertices represents the maximum amount of gods processed by the certain machine. To be able to work with labelled edges each vertex V_1, V_2, ... V_6 from the structure of the process (Fig. 1) is replaced by two vertices connected by the edge labelled by the same value as former vertex.

For our purposes all edges except newly created edges (with original labelling of vertices) incident with source and sink are labelled by ∞. For each of 2^6 state we will find the maximum flow in a network. For each state the edges incident with the vertices representing the machines occurred in failure will be removed from the network. To calculate the maximal production of each state we will find a max flow in network. In our application we will assume that the maximal flow is achieved for any possible state. Without this simplification, there would be nearly impossible to estimate the production of certain state.

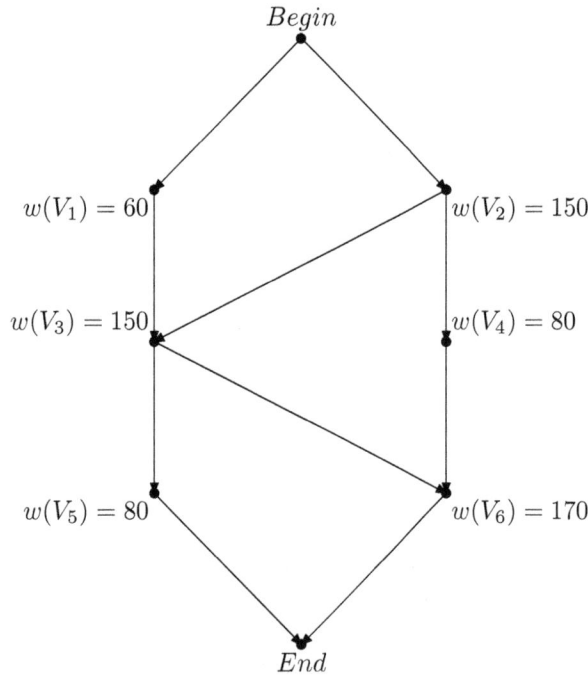

Fig. 1: Structure of process.

system changes from the state 8 to the state 9 in the Eq. 14, see the Tab. 1:

$$p_{89} = (1 - p_{r1}) \cdot p_{r2} \cdot p_{f3} \cdot (1 - p_{f4}) \cdot$$
$$(1 - p_{f5}) \cdot (1 - p_{f6}) = 0.64 \cdot 0.36 \cdot 0.005 \cdot$$
$$\cdot 0.997 \cdot 0.996 \cdot 0.997 = 0.0011. \qquad (14)$$

Tab. 1: Production of states.

state	V_1	V_2	V_3	V_4	V_5	V_6	prod. w
1	1	1	1	1	1	1	210
2	0	1	1	1	1	1	150
3	1	0	1	1	1	1	60
4	1	1	0	1	1	1	80
$\vdots$	$\vdots$	$\vdots$	$\vdots$	$\vdots$	$\vdots$	$\vdots$	
8	0	0	1	1	1	1	0
9	0	1	0	1	1	1	80
$\vdots$	$\vdots$	$\vdots$	$\vdots$	$\vdots$	$\vdots$	$\vdots$	
23	0	0	0	1	1	1	0
$\vdots$	$\vdots$	$\vdots$	$\vdots$	$\vdots$	$\vdots$	$\vdots$	
64	0	0	0	0	0	0	0

To find a maximal flow we will use a well known Ford-Fulkerson algorithm. For searching the graph edges a "Breadth-first" search was used. In the Tab. 1 there is a demonstration of several states and their production w_i.

To be able, to compute with DTMC, we need to calculate elements of transition probability matrix. To calculate the transition probability matrix $\mathbf{P}$ we need to calculate probabilities of failure p_f, and probability of repair p_r for each machine. To estimate probability p_f that the system fails during one hour we calculated the expected value as an average length of period fail Q:

$$p_f = \frac{1}{Q}. \qquad (12)$$

Using the maximum likelihood method we estimated the probability p_r which says that a machine will be repaired within one hour:

$$p_r = \frac{\Delta V}{V}, \qquad (13)$$

where V is an amount of all repairs that were realized and ΔV is an amount of all repairs that lasted less than one hour.

The calculated probabilities for given machines V_1, $V_2, \ldots V_6$ are presented in the Tab. 2. Since there are 2^6 different states the probability matrix will consist of 2^{6+6} elements. With presented probabilities p_f, p_r we can calculate the probability transition matrix. For example $p_{11} = \prod_{i=1}^{6} p_{fi}$ and p_{89} probability that the

Each element of transition probability matrix is a result of multiplication of non-zero probabilities. Since the probability matrix consists of non-zero elements (some of them re close to zero), every state can be reached from every other in time $t = 1$ the analyzed DTMC is irreducible and aperiodic. Because of the DTMC is finite and irreducible it is also positive-recurrent. Then we can use the Eq. 6 to calculate the limiting distribution π. The Eq. 11 were solved numerically by using the "backslash"(implemented solver of linear equations) from Matlab program. In Tab. 4 there are presented few Elements of the 1st row of transition probability matrix and limiting distribution π.

After calculating the transition probability matrix $\mathbf{P}$ and the production of each state we can quantify the reliability of the industry process. To describe the reliability of the process we will calculate the expected value of production. The expected value of production W within time T is a sum of all expected values for each time step $t \in \{1, 2, \cdots, T\}$. In our case, the expected value $E(W)(t)$ of production for the certain time t is equal to:

$$E(W)(t) = \sum_{i=1}^{t} \vec{v}(0) \mathbf{P}^t \vec{W}, \qquad (15)$$

where $\vec{W}$ is a vector of productions w of all 2^6 states. The results for calculating an expected value of production for time step 1 (time t) are presented in the Tab. 3 (maximal production for one time step is 170). According to the Tab. 3 we can conclude, that the expected value of production for the certain time quickly converges to the limit expected value ($\pi \cdot w$).

Tab. 2: Input probabilities.

prob.	V_1	V_2	V_3	V_4	V_5	V_6
p_f	0.100	0.006	0.005	0.003	0.004	0.003
p_r	0.360	0.360	0.400	0.375	0.240	0.190

6. Complexity of the Algorithm

Tab. 3: Calculation of expected value.

time	expected value	relative expected value
1	201.8095	0.9610
2	197.2069	0.9390
3	194.5720	0,9265
4	193.0340	0.9192
5	192.1160	0.9148
6	191.5550	0.9121
7	191.2030	0.9105
8	190.9750	0.9094
∞	190.4157	0.9067

In this section we aim to estimate the complexity of the Ford-Fulkerson (F-F) algorithm for calculating the production $\vec{W}$ of all states. As a $T(n)$ we will denote the worst-case time complexity of an algorithm with the input of size n. The complexity of F-F algorithm is $T(n) = O(n^3)$. In our case the size of input n is equal to the number of vertices $|V|$ of the network. The algorithm should run up to $2^{|V|}$. Thus $T(n) < O\left(\sum_{i=1}^{|V|} \binom{|V|}{i} (i)^3\right)$. The given formula is an upper bound of complexity of the algorithm for searching the vector of production $\vec{W}$ in a network. To be able to estimate a complexity of the whole application we have to add a complexity of matrix exponentiation which is less than $T(n) < \left(2^{|V|}\right)^3 t$ and multiplication of first row of transition probability matrix by the vector $\vec{W}$ which is equal tu $2^{|V|}$. The last form is not very significant In comparison with complexity of matrix exponentiation and calculating the vector $\vec{W}$ of production.

The complexity of calculating the production $\vec{W}$ will also depend on the amount of vertices of the network and on the character of the network. Another way how to decrease the amount of computation is to use the minimal cut. Let S_f fail state be a state with zero production w equal to zero. Let C denotes the set of vertices of the state S_f. Clearly the edges incident with C form the cut of the network. Any other state where the vertices C are in failure has the zero production.

For example in the Tab. 1 we can see that the production of the 8th (0,0,1,1,1,1) state is zero. Thus any state (for example the 23rd state), where the first and the second vertex is in fail has also the zero production. For those states, there is no need to use the Ford-Fulkerson algorithm to calculate the maximal flow. For the large time t there would be more suitable to to use limit probability distribution than to exponentiate the matrix.

Tab. 4: First row of transition probability matrix, limiting distribution π.

$p_{1,1}$	0.8810	π_1	0.7300
$p_{1,2}$	0.0980	π_2	0.2030
$p_{1,3}$	0.0050	π_3	0.0120
$p_{1,4}$	0.0040	π_4	0.0090
$p_{1,5}$	0.0030	π_5	0.0050
$p_{1,6}$	0.0040	π_6	0.0120
$p_{1,7}$	0.0010	π_7	0.0110
$p_{1,8}$	$5.1905e^{-5}$	π_8	0.0030
$\vdots$	$\vdots$	$\vdots$	$\vdots$
$p_{1,64}$	$1.0800e^{-13}$	π_{64}	$8.8980e^{-11}$

A problem could occur with saving the transition probability matrix in the computer memory. An amount of elements of the matrix grow exponentially (2^{2n}) with an amount of elements of analysed systems. The problem could by partially solved by rounding small values to zero and save the matrix as a sparse matrix. In this case there should be more complicated to proof the convergence of the DTMC. Another way how to work with large matrices is to dived the DTMC into more separate systems and calculate with their transition probability matrix separately. Other possibility how reduce the transition probability matrix is to shorter the time step and define only one possible change within the time step. It has the effect of simplifying the transition matrix.

7.   Conclusion

In the paper we have presented the application of graph flow in the network in the reliability analysis. The main worth of this work is an innovative usage of discrete time Markov chains and the flow in network theory in reliability analysis. The DTMC we used to describe the changes of the analysed process. For calculating for larger time step t, there is possible to use the limiting distribution instead of exponentiate the probability matrix. The Ford-Fulkerson algorithm was used to compute the industry production of analysed system. The main disadvantage of the presented approach is a high computational complexity. In further research there is possible to modify the Ford-Fulkerson algorithm to be more suitable for our purposes.

Acknowledgment

This work was supported by the Grant of SGS No. SP2013/116, VSB–Technical University of Ostrava, Czech Republic.

References

[1] SKALNY, P. and B. KRAJC. Discrete-Time Markov Chains in Reliability Analysis-Case Study. In: *International Joint Conference CISIS'12-ICEUTE'12-SOCO'12 Special Sessions*. Ostrava: Springer Berlin Heidelberg, 2012, pp. 421–427. ISBN 978-3-642-33017-9. DOI: 10.1007/978-3-642-33018-6_43.

[2] WEST, D. B. *Introduction to graph theory*. N.J.: Prentice Hall, 2001. ISBN 01-301-4400-2.

[3] PAGE, L. B. and J. E. PERRY. Reliability polynomials and link importance in networks. *IEEE Transactions on Reliability*. 1994, vol. 43, iss. 1, pp. 51–58. ISSN 0018–9529. DOI: 10.1109/24.285108.

[4] DABROWSKI, Ch., F. HUNT and Katherine MORRISON. *Improving the Efficiency of Markov Chain Analysis of Complex Distributed Systems*. National Institute of Standards and Technology. Gaithersburg MD, 2010. Available at: http://www.nist.gov/customcf/get_pdf.cfm?pub_id=906149.

[5] DABROWSKI, Ch. and F. HUNT. Using Markov Chain and Graph Theory Concepts to Analyze Behavior in Complex Distributed Systems. In: *The 23rd European Modeling and Simulation Symposium*. Rome: Curran Associates, Inc., 2011. Available at: http://www.nist.gov/itl/antd/upload/DabrowskiHunt-Paper129-updated.pdf.

[6] Stewart, W. *Introduction to the Numerical Solution of Markov Chains*. Princeton University Press, 1994. ISBN 0-691-03699-3.

[7] CORMEN, T. H. *Introduction to algorithms*. Cambridge: MIT Press, 2001. ISBN 02-620-3293-7.

About Authors

Pavel SKALNY was born in Ostrava. He received his M.Sc. from Mathematics and History in Silesian University in Opava in 2009. His research interests include Stochastic modelling in reliability and fracture mechanics modelling.

9

MPC-Based Path Following Control of an Omnidirectional Mobile Robot with Consideration of Robot Constraints

Kiattisin KANJANAWANISHKUL

Mechatronics Research Unit, Faculty of Engineering, Mahasarakham University, Kantarawichai District, Khamriang Sub-District, Maha Sarakham, 44150, Thailand

kiattisin_k@hotmail.com

Abstract. In this paper, the path following problem of an omnidirectional mobile robot (OMR) has been studied. Unlike nonholonomic mobile robots, translational and rotational movements of OMRs can be controlled simultaneously and independently. However the constraints of translational and rotational velocities are coupled through the OMR's orientation angle. Therefore, a combination of a virtual-vehicle concept and a model predictive control (MPC) strategy is proposed in this work to handle both robot constraints and the path following problem. Our proposed control scheme allows the OMR to follow the reference path successfully and safely, as illustrated in simulation experiments. The forward velocity is close to the desired one and the desired orientation angle is achieved at a given point on the path, while the robot's wheel velocities are maintained within boundaries.

Keywords

Model predictive control, omnidirectional mobile robots, path following control, robot constraints, virtual vehicle.

1. Introduction

In this paper, we address the path following problem where a mobile robot is forced to follow a desired spatial path without consideration in temporal specifications [1]. The solution of this problem offers several remarkable advantages over trajectory tracking [2]. For example, the time dependence of the trajectory tracking problem is eliminated, convergence to the path is achieved smoothly, and saturation of control signals is less likely reached. Original research on this area can be found in [3].

In general, the path following controller determines the robot's moving direction that can bring it to the path, while the robot's forward velocity tracks a desired velocity profile. In the literature, there are two control strategies for path parameterization [4], i.e., the Frenet frame with an orthogonal projection of a robot on the given path [3], [5], [6], [7] and the Frenet frame with a non-orthogonal projection of a robot on the given path [1], [8], [9], [10], [11], [12]. In the first method, the position of the virtual vehicle to be followed by a real one is defined by the orthogonal projection of the robot on the path. However, this method can be used only when the initial position of the robot is near the path. In the other method which can overcome the initial condition problem of the first method, a desired geometric path is parametrized by the curvilinear abscissa $s \in \mathbb{R}$ and the velocity of the virtual vehicle $(\dot{s}(t))$ can be controlled explicitly.

Although the path following problem has been solved for different types of robots over the past decade, omnidirectional mobile robots (OMRs) have been considered in this work since they have some distinct advantages over nonholonomic mobile robots. They can move instantly in any direction without reorientation [13]. They become increasingly popular in mobile robot applications as seen from a large number of publications dealing with OMRs. However, research study on path following control of OMRs is still rare. Some related work is as follows: Vazquez et al. [14] adapted computed torque control usually used in robot manipulators to the path following control problem of the OMR. Conceicao et al. [6] proposed a nonlinear model predictive control for an OMR. The cost function includes the robot pose error and the control effort. They followed the concept that the position of the virtual vehicle is defined by the orthogonal projection of the robot on the path. Huang and Tsai [15] applied an adaptive robust control method to the path following problem for an OMR with consideration of actuators' uncertainties in polar coordinates. Kanjanawanishkul and Zell [16] used model predictive control to generate an optimal velocity of the virtual vehicle. Recently, Oftadeh et al. [11] proposed a new solution to the path following problem where speed of the robot can be determined analytically to keep the steering and driving velocities of the wheels under predetermined values.

In this work, a model predictive control (MPC) approach for solving the path following problem of an OMR is designed. The proposed idea is that the velocity of the virtual vehicle can be controlled explicitly through the MPC scheme. Although the translational velocity and the rotational velocity of an OMR can be controlled separately, their boundaries are coupled via the robot's orientation angle. Thus, both velocities are included into the objective function, while the robot's wheel velocity constraints and other robot constraints are satisfied in the constrained minimization problem of the MPC scheme that is online solved at each sampling time.

The rest of the paper is structured as follows: in Section 2., the robot kinematics is derived. The path following control problem is described and the path following controller based on the MPC strategy with consideration of robot constraints are developed in Section 3. Then, simulation experiments are conducted in Section 4. to show the effectiveness of our proposed controller. Finally, our conclusions are given in Section 5.

2.   Kinematic Modeling

From Fig. 1, the kinematic model of an OMR can be given by:

$$
\begin{bmatrix} \dot{x} \\ \dot{y} \\ \dot{\theta} \end{bmatrix} = \mathbf{R}_z(\theta) \begin{bmatrix} u \\ v \\ \omega \end{bmatrix} = \begin{bmatrix} \cos\theta & -\sin\theta & 0 \\ \sin\theta & \cos\theta & 0 \\ 0 & 0 & 1 \end{bmatrix} \cdot \begin{bmatrix} u \\ v \\ \omega \end{bmatrix} , \quad (1)
$$

where $\mathbf{R}_z(\theta)$ is the rotation matrix that transforms the robot velocities with respect to the body frame (X_m, Y_m) to the world frame (X_w, Y_w). $\vec{x}(t) = [x, y, \theta]^T$ is the state vector of the robot in the world frame and θ denotes the angle of the robot's orientation, u and v are the translational velocities observed in the body frame, while ω is the rotational velocity.

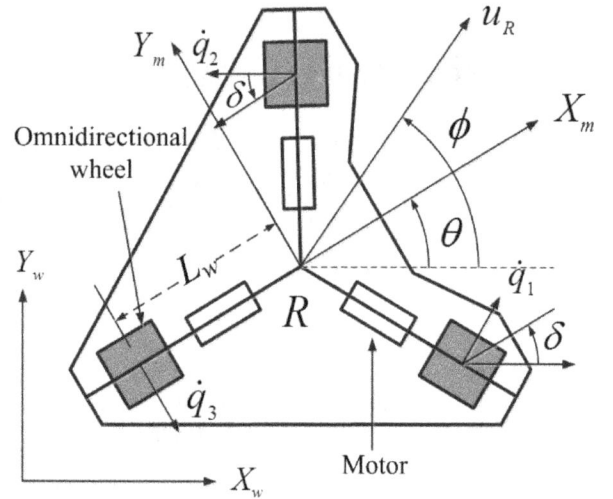

Fig. 1: Coordinate frames of an OMR.

Equation (1) can be rewritten by decoupling translation and rotation as follows:

$$
\begin{bmatrix} \dot{x} \\ \dot{y} \\ \dot{\phi} \\ \dot{\theta} \end{bmatrix} = \begin{bmatrix} u_R \cos\phi \\ u_R \sin\phi \\ \psi \\ \omega \end{bmatrix} , \quad (2)
$$

where ϕ denotes the angle of the robot's moving direction in the world frame and u_R is the forward linear velocity of the robot. Furthermore, the robot translational velocities can be determined by:

$$
\begin{bmatrix} u \\ v \end{bmatrix} = \begin{bmatrix} u_R \cos(\phi - \theta) \\ u_R \sin(\phi - \theta) \end{bmatrix} . \quad (3)
$$

When the wheel velocities are taken into account, the following lower level kinematic model with respect to the body frame is given:

$$
\begin{bmatrix} \dot{q}_1 \\ \dot{q}_2 \\ \dot{q}_3 \end{bmatrix} = \begin{bmatrix} \cos\delta & \sin\delta & L_w \\ -\cos\delta & \sin\delta & L_w \\ 0 & -1 & L_w \end{bmatrix} \cdot \begin{bmatrix} u \\ v \\ \omega \end{bmatrix} , \quad (4)
$$

where $\vec{q}(t) = [\dot{q}_1, \dot{q}_2, \dot{q}_3]^T$ is the vector of linear velocities of the wheel. It is equal to the wheel's radius multiplied by the wheel's angular velocity. L_w denotes the distance from the platform center to the wheel center (see Fig. 1) to the center of wheel. δ refers to the angle of the wheel orientation in the body frame. Since the translational and rotational velocities of an OMR can be separately controlled, the wheel velocities can be divided into two components:

$$
\vec{q}_t = \begin{bmatrix} \cos\delta & \sin\delta \\ -\cos\delta & \sin\delta \\ 0 & -1 \end{bmatrix} \cdot \begin{bmatrix} u \\ v \end{bmatrix} , \quad \vec{q}_r = \begin{bmatrix} L_w \\ L_w \\ L_w \end{bmatrix} \omega , \quad (5)
$$

where $\vec{q}_t$ and $\vec{q}_r$ are translational and rotational components for each wheel velocity, respectively.

As the motor's voltage and current are limited, the summation of these two components is bounded by $\dot{q}_{max}$, i.e., $\|\vec{q}_t + \vec{q}_r\|_\infty \leq \dot{q}_{max}$.

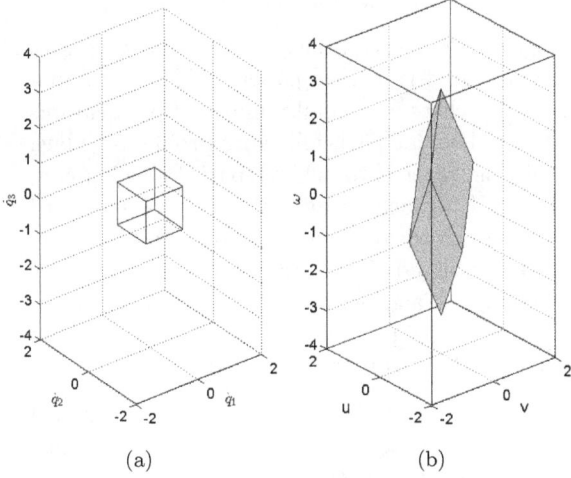

(a)   (b)

Fig. 2: The relationship between wheel velocities and robot velocities with respect to the body frame: (a) the cube defined by $\mathcal{Q}(t) = \{\vec{q}(t) \mid |\dot{q}_i(t)| \leq \dot{q}_{i,max}\}$. (b) the tilted cuboid $\mathbf{P}(0)\mathcal{Q}(t)$.

The relationship between wheel velocities and robot velocities with respect to the world frame can be calculated by substituting Eq. (4) into Eq. (1), resulting in:

$$\dot{\mathbf{x}}(t) = \mathbf{R}_z(\theta)\mathbf{P}(0)\vec{q}(t) \ , \qquad (6)$$

where:

$$\mathbf{P}(0) = \begin{bmatrix} \sqrt{3}/3 & -\sqrt{3}/3 & 0 \\ 1/3 & 1/3 & -2/3 \\ 1/(3L_w) & 1/(3L_w) & 1/(3L_w) \end{bmatrix}, \qquad (7)$$

with $\delta = \frac{\pi}{6}$ rad. As an example, the linear transformation $\mathbf{P}(0)$ maps the cube $\mathcal{Q}(t) = \{\vec{q}(t) \mid |\dot{q}_i(t)| \leq \dot{q}_{max}\}$, where $i = 1, 2, 3$, (see Fig. 2(a)) into the tilted cuboid $\mathbf{P}(0)\mathcal{Q}(t)$ (see Fig. 2(b)) with $L_w = 0.2$ m and $|\dot{q}_i(t)| \leq 0.6$ m·s^{-1}. The transformation $\mathbf{R}_z(\theta)$ then rotates this cuboid about the ω axis. As seen in Fig. 2(b), the boundary of translational and rotational velocities are coupled via θ. Furthermore, for a given θ, the translational velocity may decrease in order that the allowable rotational velocity can increase, while all wheel velocities are kept within the cube $\mathcal{Q}(t)$. This concept is useful when the large rotational velocity is required, e.g., during converging the robot to the path or moving along a sharp turning. One simple solution is to scale the translational and rotational velocities, as proposed in [17]. However, this solution does not utilize the full capacity of the wheel's maximum velocity. In this work, an MPC-based method with consideration of robot constraints is proposed to ensure that the path following control is attained and the robot constraints are within boundaries.

3. The Path Following Control Problem and Controller Design

The Frenet frame plays the role of the body frame of the virtual vehicle moving long the reference path. In general, the forward velocity u_R tracks a desired velocity profile, while the velocity of the virtual vehicle $\dot{s}$ converges to u_R. However, to utilize the full capacity of wheel velocities and to keep the wheel velocities within boundaries, the forward velocity cannot be fixed. Thus, an acceleration control input $a_m = \dot{u}_m$, where u_m is the robot's actual forward velocity, is introduced. Then, we obtain $\eta_e = u_m - u_R$ and $\dot{\eta}_e = a_m - \dot{u}_R$. From Fig. 3, the error state vector $\vec{x}_e = [x_e, y_e, \phi_e, \theta_e, \eta_e]^T$ between the state vector of the robot and that of the virtual vehicle can be expressed in the Frenet frame as follows:

$$\begin{bmatrix} x_e \\ y_e \\ \phi_e \\ \theta_e \\ \eta_e \end{bmatrix} =$$

$$= \begin{bmatrix} \cos\phi_d & \sin\phi_d & 0 & 0 & 0 \\ -\sin\phi_d & \cos\phi_d & 0 & 0 & 0 \\ 0 & 0 & 1 & 0 & 0 \\ 0 & 0 & 0 & 1 & 0 \\ 0 & 0 & 0 & 0 & 1 \end{bmatrix} \cdot \begin{bmatrix} x - x_d \\ y - y_d \\ \phi - \phi_d \\ \theta - \theta_d \\ u_m - u_R \end{bmatrix}, \qquad (8)$$

where θ_d is the desired orientation angle, and ϕ_d is the tangent angle to the path. Note that, in this work, the desired orientation angles of the robot are predetermined on the path. This setting is very useful when the robot is required to orient itself to a specific direction.

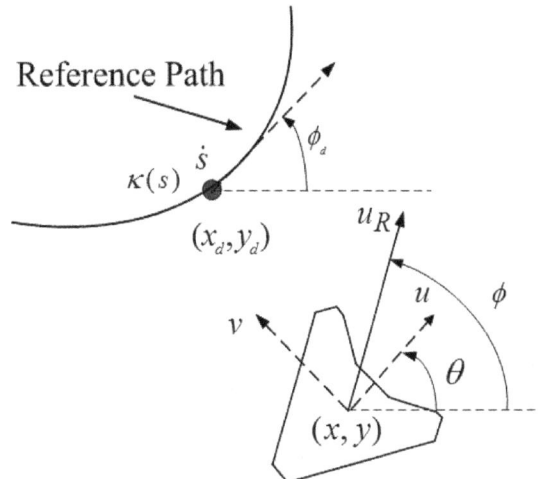

Fig. 3: The path following problem where a small black dot represents the location of the virtual vehicle.

The error state dynamic model chosen in the Frenet frame is derived using Eq. (1), Eq. (2) and Eq. (7), resulting in:

$$\dot{x}_e = y_e \kappa(s)\dot{s} - \dot{s} + u_m \cos\phi_e, \qquad (9)$$

$$\dot{y}_e = -x_e \kappa(s)\dot{s} + u_m \sin\phi_e, \qquad (10)$$

$$\dot{\phi}_e = \psi - \kappa(s)\dot{s}, \qquad (11)$$

$$\dot{\theta}_e = \omega - \frac{\partial\theta_d}{\partial s}\dot{s}, \qquad (12)$$

$$\dot{\eta}_e = a_m - \dot{u}_R, \qquad (13)$$

where $\kappa(s)$ is the path's curvature. From Eq. (9), the following system control inputs are redefined:

$$\vec{u}_e = \begin{bmatrix} u_1 \\ u_2 \\ u_3 \\ u_4 \end{bmatrix} = \begin{bmatrix} -\dot{s} + u_m \cos\phi_e \\ \psi - \kappa(s)\dot{s} \\ \omega - \frac{\partial\theta_d}{\partial s}\dot{s} \\ a_m - \dot{u}_R \end{bmatrix}. \qquad (14)$$

The error dynamic model defined in Eq. (9), Eq. (10), Eq. (11), Eq. (12), Eq. (13) and Eq. (14) is used in the MPC framework designed in the next subsection. Moreover, to show the effectiveness of our control scheme, a comparison with the feedback control laws proposed by Oftadeh et al. [11] has been conducted (see Subsection 3.2.).

3.1. MPC Design

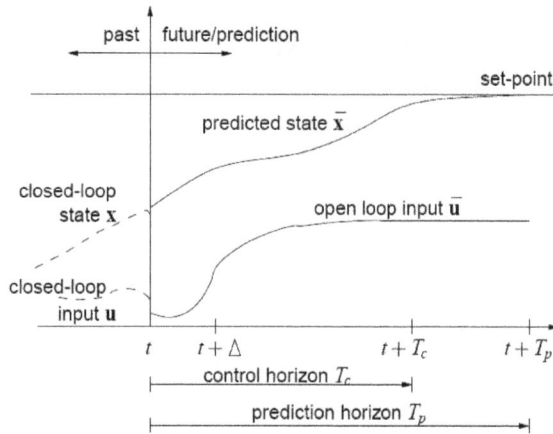

Fig. 4: Principle of model predictive control [18].

MPC is one of the most successful control techniques used in industry. It is based on a minimization of predicted tracking errors and control effort with constraints on the control inputs and the state variables over a finite horizon. At each sampling time, it generates an optimal control input sequence after the minimization problem is solved. The first element of this control input sequence

is applied to the system. The problem is then solved again at the next sampling time with the updated process measurements and a shifted horizon. The conceptual structure of MPC is depicted in Fig. 4. The reader is referred to [18] and [19] for more details.

Although MPC is obviously not a new control method according to a review paper on MPC for a mobile robot [20], there is a small number of publications dealing with MPC for path following problems. Thus, the aim of this paper is to achieve path following control for an OMR in such a way that the robot's forward velocity is close to the desired one and the path following control is attained. Furthermore, robot constraints can be handled straightforwardly as hard constraints in the optimization problem so that the robot can travel safely without constraint violations.

The control input applied to the system is obtained by solving the following finite horizon open-loop optimal control problem at each sampling time:

$$\min_{\bar{u}(\cdot)} \int_t^{t+T_p} F(\bar{x}(\tau), \bar{u}(\tau))\, d\tau, \qquad (15)$$

subject to:

$$\dot{\bar{x}}(\tau) = f(\bar{x}(\tau), \bar{u}(\tau)), \qquad (16)$$

$$\bar{u}(\tau) \in \mathcal{U} \quad \forall\tau \in [t, t+T_c], \qquad (17)$$

$$\bar{x}(\tau) \in \mathcal{X} \quad \forall\tau \in [t, t+T_p], \qquad (18)$$

$$\|\bar{x}(t+T_p)\|_P \le \beta\|\bar{x}(t)\|_P \quad \beta \in [0,1), \qquad (19)$$

where $F(\bar{x}, \bar{u}) = \bar{x}^T Q \bar{x} + \bar{u}^T R \bar{u}$. The bar denotes an internal controller variable. At the beginning of each sampling time, $\bar{x} = x_e, \bar{u} = u_e$. T_p represents the length of the prediction horizon, and T_c denotes the length of the control horizon ($T_c \le T_p$). The deviations from the desired values are weighted by the positive definite matrices Q, and R, where:

- $Q = \text{diag}\{q_{11}, q_{22}, q_{33}, q_{44}, q_{55}\}$,

- $R = \text{diag}\{r_{11}, r_{22}, r_{33}, r_{44}\}$.

Besides wheel velocities maintained within boundaries (i.e., $\dot{q}_i(t) \le \dot{q}_{max}$, where $i = 1, 2, 3$), the constraints in Eq. (17) denote bounded control inputs as follows:

$$\begin{bmatrix} 0 \\ \psi_{min} \\ \omega_{min} \\ a_{m,min} \\ \alpha_{min} \end{bmatrix} \le \begin{bmatrix} \dot{s} \\ \psi \\ \omega \\ a_m \\ \alpha \end{bmatrix} \le \begin{bmatrix} \dot{s}_{max} \\ \psi_{max} \\ \omega_{max} \\ a_{m,max} \\ \alpha_{max} \end{bmatrix}, \qquad (20)$$

where α is angular acceleration, i.e., $\alpha = \dot{\omega}$.

A so-called contractive constraint [21] is defined in the last inequality end constraint Eq. (19). It requires that, at time t, the system states at the end of the prediction horizon, i.e., $\bar{\mathbf{x}}(t + T_p)$ are contracted in norm with respect to the states at the beginning of the prediction, $\bar{\mathbf{x}}(t)$. $\beta \in [0, 1)$ and the positive definite matrix $\mathbf{P}$ are two parameters that determine how much contraction is required. The reader is referred to [21] for stability analysis.

3.2. Feedback Control Laws with Low-Level Constraint Handling

The following feedback control laws for $\dot{s}$, ω and ϕ taken from [11] with slight modification is given by:

$$\dot{s} = (k_1 x_e + \cos(\sigma(y_e)))u_R = k_s u_R, \qquad (21)$$

$$\omega = (-k_3 \theta_e + \frac{\partial \theta_d}{\partial s} k_s)u_R, \qquad (22)$$

$$\phi = \phi_d - \sigma(y_e), \qquad (23)$$

where:

$$\sigma(y_e) = \arcsin \frac{k_2 y_e}{|y_e| + \epsilon},$$

$$k_s = k_1 x_e + \cos(\sigma(y_e)), \qquad (24)$$

$$0 < k_1 \leq 1, \epsilon > 0,$$

$$k_1, k_3 > 0,$$

The proof for semi-global exponential stability can be found in [11] with $u_R > 0$. To handle the robot constraints, a conventional approach shown in Alg. 1 has been used. However, this solution does not utilize the full capacity of the wheel's maximum velocity as shown in the simulation results. In Alg. 1, each component of the wheel velocities is scaled down such that no components are out of acceptable bounds. Function max() returns the maximum value in the array.

Algorithm 1 Velocity scaling.

INPUT: $\vec{q}$ and $\dot{q}_{max}$
OUTPUT: $\vec{q}$
1: $factor \leftarrow 1$
2: $maxSpd \leftarrow \max(|\dot{q}_1|, |\dot{q}_2|, |\dot{q}_3|)$
3: **if** $maxSpd > \dot{q}_{max}$ **then**
4:     $factor \leftarrow \dot{q}_{max}/maxSpd$
5: **end if**
6: $\vec{q} \leftarrow \vec{q} \cdot factor$

---

4. Simulation Experiments

The control strategy proposed in this work was evaluated through simulation experiments. The following eight-shaped path was considered as a desired reference path:

$$x_d(t) = 1.8 \sin(t), \qquad (25)$$

$$y_d(t) = 1.2 \sin(2t). \qquad (26)$$

This reference path was numerically parameterized by the curvilinear abscissa s, while the robot's desired orientation angles were given as follows: $\theta_d(s) = \pi s \cdot 2^{-1}$, 0, and $\pi s \cdot 2^{-1}$ at $s = 0.0 - 4.0$ m, $s = 4.0 - 8.0$ m, and $s = 8.0 - 12.0$ m, respectively. All snapshots shown in the simulation results were taken at every 2 s (except for the last one).

Three simulation experiments were conducted and the results were compared. In each simulation, the initial conditions for the OMR were given as:

$$\begin{bmatrix} x \\ y \\ \phi \\ \theta \\ u_m \end{bmatrix} = \begin{bmatrix} -0.6 \\ -0.25 \\ -\pi/4 \\ 0 \\ 0 \end{bmatrix}. \qquad (27)$$

The forward velocity u_R was 0.6 m·s^{-1}, and a total traveling distance was 10 m. The feedback control laws without consideration of actuator constraints were first implemented. The control parameters were set as follows: $k_1 = 1.1$ m^{-1}, $k_2 = 0.5$, $k_3 = 10$ m^{-1}. As seen in Fig. 5(a) and Fig. 5(b), although the OMR followed almost exactly the reference path, this control scheme cannot be used in practice due to constraint violation.

In the second simulation experiment, the feedback control laws with low-level constraint handling described in Subsection 3.2. were implemented with the same values of control parameters. As seen in Fig. 6(a), there were some deviations from the desired reference path at sharp corners, which means poor performance for path following control. However, the wheels' velocity constraints were not violated, as required.

In the last simulation experiment, our proposed MPC-based control strategy was evaluated. It was carried out by using a set of the following parameters:

- $\mathbf{Q} = \text{diag}(200, 1000, 5, 0.1, 20)$,

- $\mathbf{R} = \text{diag}(1, 0.1, 0.01, 0.01)$,

- $\mathbf{P} = \text{diag}(1, 1, 0.1, 0.1, 0.1)$, $\beta = 0.99$,

- $\dot{s}_{max} = 1.2$ m·s^{-1},

- $\psi_{max} = 2$ rad·s^{-1}, $\psi_{min} = -2$ rad·s^{-1},

- $\omega_{max} = 2$ rad·s^{-1}, $\omega_{min} = -2$ rad·s^{-1},

- $a_{m,max} = 1$ m·s^{-2}, $a_{m,min} = -1$ m·s^{-2},

- $\alpha_{max} = 2$ rad·s^{-2}, $\alpha_{min} = -2$ rad·s^{-2},

- $T_p = T_c = 5\Delta$, $\Delta = 0.05$ s,

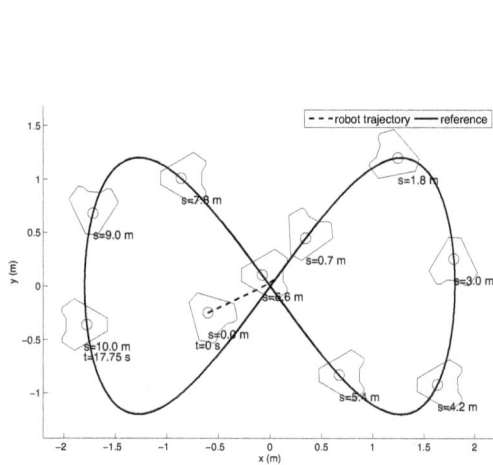

(a)

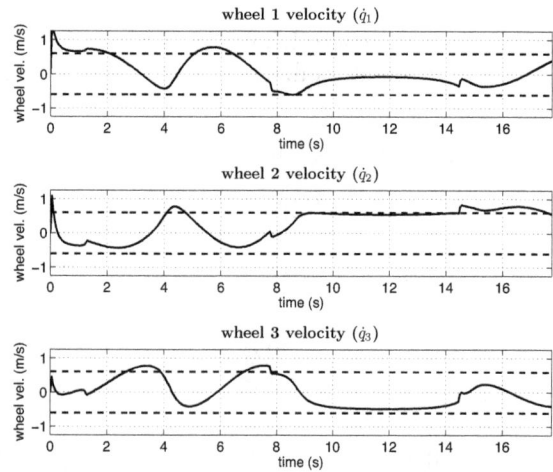

(b)

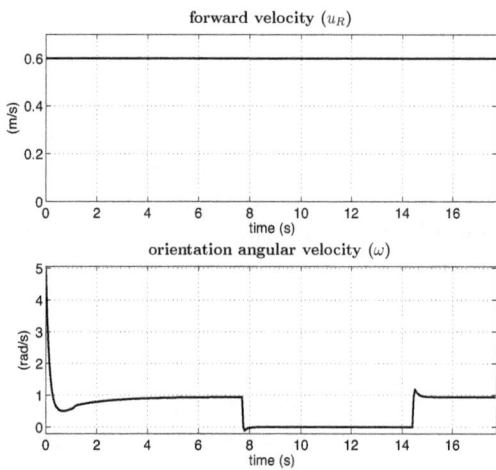

(c)

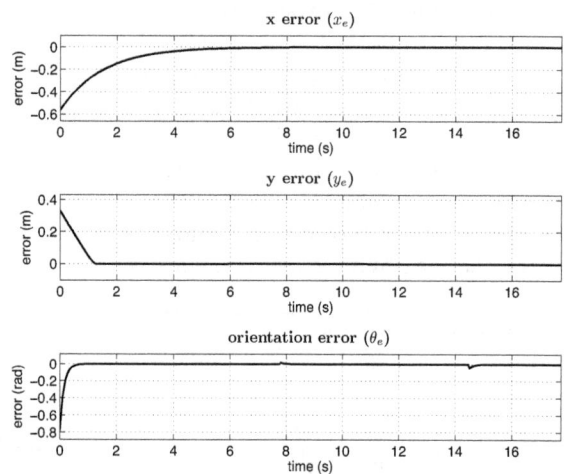

(d)

Fig. 5: The simulation results when the feedback control laws [11] without constraint handling were implemented: (a) superimposed snapshots, (b) wheel velocities, (c) robot velocities with respect to the body frame and (d) robot state errors with respect to the path coordinate.

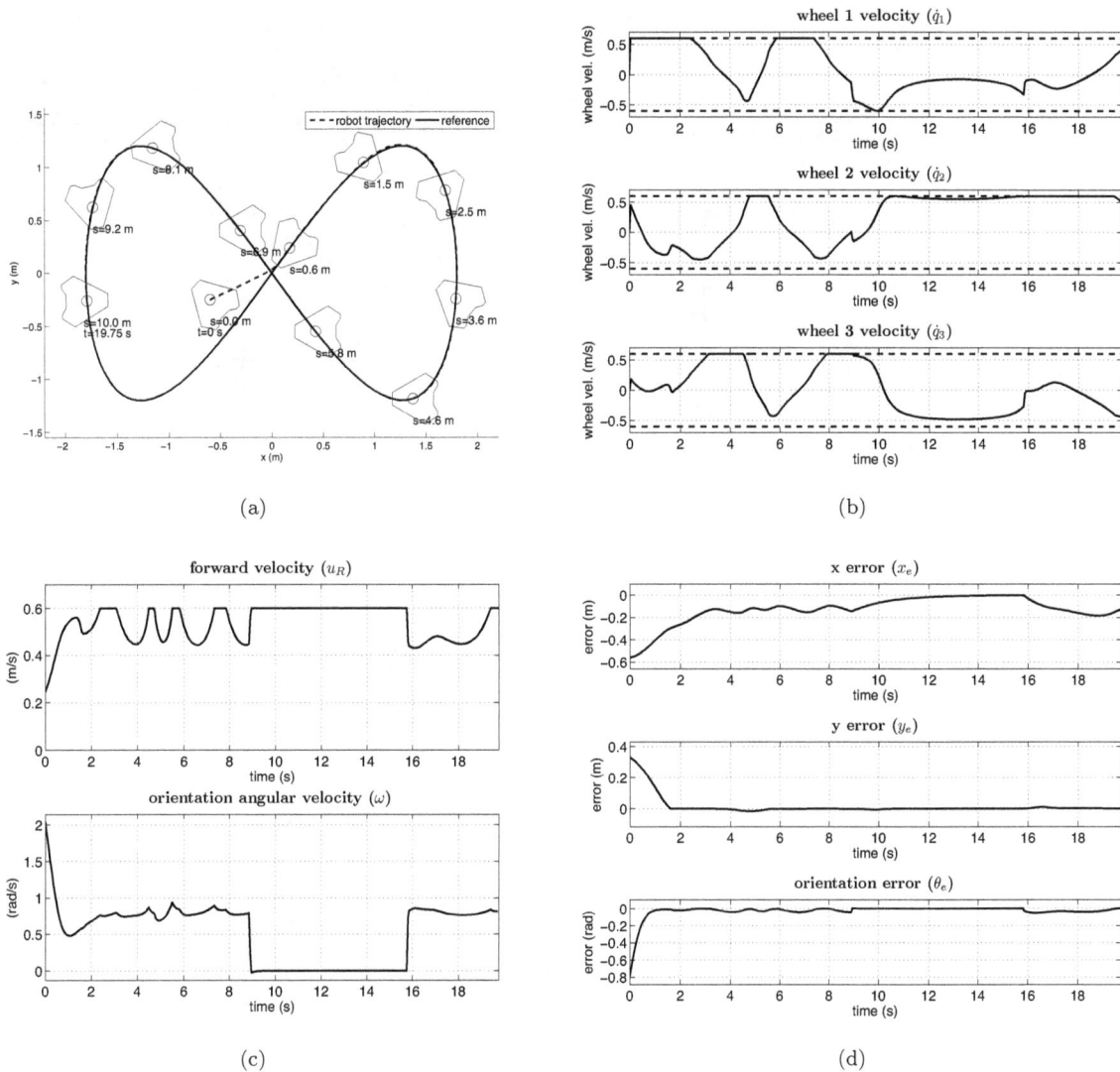

Fig. 6: The simulation results when the feedback control laws [11] with low-level constraint handling were employed: (a) superimposed snapshots, (b) wheel velocities, (c) robot velocities with respect to the body frame and (d) robot state errors with respect to the path coordinate.

(a)

(b)

(c)

(d)

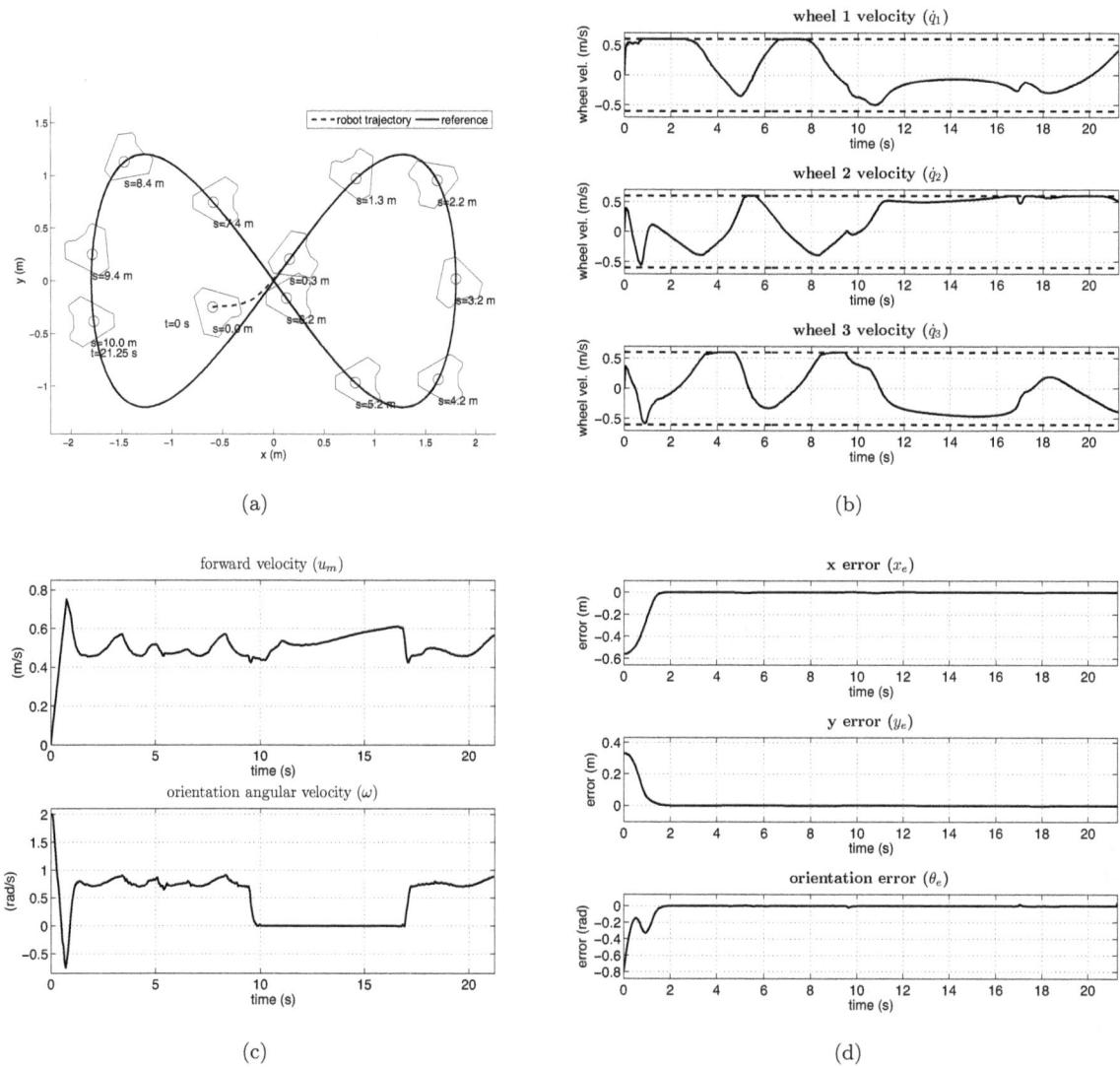

Fig. 7: The simulation results when the MPC-based control strategy was implemented: (a) superimposed snapshots, (b) wheel velocities, (c) robot velocities with respect to the body frame and (d) robot state errors with respect to the path coordinate.

where Δ is the sampling time.

Figure 7 shows the simulation results when our proposed control strategy was implemented. The OMR followed almost exactly the reference path (see Fig. 7(a)), i.e., the position errors and orientation errors defined in Eq. (8) converged to zero as depicted in Fig. 7(d). However, the forward velocity was less than 0.6 m·s^{-1} during $s = 0.0 - 4.0$ m since the large rotational velocity was required. When the requirement of the rotational velocity became lower, the forward velocity became closer to the desired one (see Fig. 7(c)), while each wheel's velocity was bounded (see Fig. 7(b)).

5.   Conclusions

An OMR with no nonholonomic constraints has remarkable advantages over more common design platforms like car-like robots and differential drive robots. In particular, it can move in any direction regardless of current pose and, at the same time, it can attain any desired orientation. Thus, this kind of maneuverability is especially preferred for congested applications.

In this paper, we presented an MPC scheme to solve the path following control problem of an OMR. The proposed MPC controller can handle robot constraints straightforwardly. Thus, the OMR can follow a reference path safely. Moreover, the forward velocity was optimized in the sense that it decreased when the large rotational velocity was required and it was close to the desired one when the capacity of the wheel's velocity was available.

Acknowledgment

The author gratefully acknowledges the financial support from National Science and Technology Development Agency (NSTDA), Thailand.

References

[1] SOETANTO, D., L. LAPIERRE and A. PASCOAL. Adaptive, non-singular path-following control of dynamic wheeled robots. In: *42nd IEEE International Conference on Decision and Control*. Piscataway: IEEE, 2003, pp. 1765–1770. ISBN 0-7803-7924-1. DOI: 10.1109/CDC.2003.1272868.

[2] AGUIAR, A. P., D. B. DACIC, J. P. HESPANHA and P. KOKOTOVIC. Path-following or reference-tracking? An answer relaxing the limits to performance. In: *Proceedings of the IFAC/EURON Symposium on Intelligent Autonomous Vehicles*. Lisbon: Elsevier, 2004, pp. 1–6. ISBN 008-044237-4.

[3] MICAELLI A. and C. SAMSON. Trajectory-tracking for unicycle-type and two-steering-wheels mobile robots. In: *HAL-Inria* [online]. 1993. Available at: https://hal.inria.fr/inria-00074575/document.

[4] PLASKONKA, J. Different Kinematic Path Following Controllers for a Wheeled Mobile Robot of (2,0) Type. *Journal of Intelligent*. 2013, vol. 71, iss. 3, pp. 1–18. ISSN 0921-0296. DOI: 10.1007/s10846-013-9879-6.

[5] ALTAFINI, C. Following a path of varying curvature as an output regulation problem. *IEEE Transactions on Automatic Control*. 2002, vol 47, iss. 9. pp. 1551–1556. ISSN 0018-9286. DOI: 10.1109/TAC.2002.802750.

[6] CONCEICAO, A., H. OLIVEIRA, A. SILVA, D. OLIVEIRA and A. MOREIRA. A nonlinear model predictive control of an omni-directional mobile robot. In: *Proceedings of the IEEE International Symposium on Industrial Electronics*. Vigo: IEEE, 2007, pp. 2161–2166. ISBN 978-1-4244-0754-5. DOI: 10.1109/ISIE.2007.4374943.

[7] PETROV, P. and L. DIMITROV. Nonlinear path control for a differential-drive mobile robot. *RECENT Journal*. 2010, vol. 11, iss. 1. pp. 41–45. ISSN 1582-0246.

[8] AICARDI, M., G. CASALINO, A. BICCHI and A. BALESTRINO. Closed loop steering of unicycle like vehicles via Lyapunov techniques. *IEEE Robotics and Automation Magazine*. 1995, vol. 2, iss. 1, pp. 27–35. ISSN 1070-9932. DOI: 10.1109/100.388294.

[9] FAULWASSER, T., B. KERM and R. FINDEISEN. Model predictive path-following for constrained nonlinear systems. In: *Proceedings of the 48th IEEE Conference on Decision and Control*. Shanghai: IEEE, 2009, pp. 8642–8647. ISBN 978-1-4244-3871-6. DOI: 10.1109/CDC.2009.5399744.

[10] KANJANAWANISHKUL, K. Path following control of a mobile robot using contractive model predictive control. *Applied Mechanics and Materials*. 2013, vol. 397–400, iss. 1, pp. 1366–1372. ISSN 1662-7482. DOI: 10.4028/www.scientific.net/AMM.397-400.1366.

[11] OFTADEH, R., R. GHABCHELOO and J. MATTILA. A novel time optimal path following controller with bounded velocities for mobile robots

with independently steerable wheels. In: *Proceedings of the 2013 IEEE/RSJ International Conference on Intelligent Robots and Systems.* Tokyo: IEEE, 2013, pp. 4845–4851. ISBN 978-146736358-7. DOI: 10.1109/IROS.2013.6697055.

[12] YU, S., X. LI, H. CHEN and F. ALLGOWER. Nonlinear model predictive control for path following problems. In: *Proceedings of the 4th IFAC Nonlinear Model Predictive Control Conference.* Noordwikkerhout: IFAC, 2012, pp. 145–150. ISBN 978-390282307-6. DOI: 10.3182/20120823-5-NL-3013.00003.

[13] CAMPION, G., G. BASTIN, B. DANDREA-NOVEL and Y. SHUYOU. Structural properties and classification of kinematic and dynamic models of wheeled mobile robots. *IEEE Transactions on Robotics and Automation.* 2013, vol. 12, iss. 1. pp. 47–62. ISSN 1042-296X. DOI: 10.1109/70.481750.

[14] VAZQUEZ, A. and M. VELASCO-VILLA. Path-tracking dynamic model based control of an omnidirectional mobile robot. In: *Proceedings of the World Congress.* Seoul: IFAC, 2008, pp. 5365–5370. ISBN 978-1-1234-7890-2. DOI: 10.3182/20080706-5-KR-1001.00904.

[15] HUANG, H. and C. TSAI. Adaptive robust control of an omnidirectional mobile platform for autonomous service robots in polar coordinates. *Journal of Intelligent and Robotic Systems.* 2008, vol. 51, iss. 4. pp. 439–460. ISSN 0921-0296. DOI: 10.1007/s10846-007-9196-z.

[16] KANJANAWANISHKUL, K. and A. ZELL. Path following for an omnidirectional mobile robot based on model predictive control. In: *Proceedings of the 2009 IEEE International Conference on Robotics and Automation (ICRA 2009).* Kobe: IEEE, 2009, pp. 3341–3346. ISBN 978-1-4244-2788-8. DOI: 10.1109/ROBOT.2009.5152217.

[17] ORIOLO, G., A. DE LUCA and M. VENDITTELLI. WMR control via dynamic feedback linearization: design, implementation and experimental validation. *IEEE Transactions on Control Systems Technology.*2002, vol. 10, iss. 6, pp. 835–852. ISSN 1063-6536. DOI: 10.1109/TCST.2002.804116.

[18] ALLGOWER, F., R. FINDEISEN and Z. K. NAGY. Nonlinear model predictive control: from theory to application. *Journal of the Chinese Institute of Chemical Engineers.* 2004, vol. 35, no. 3, pp. 299–315. ISSN 0368-1653.

[19] MAYNE, D. Q., J. B. RAWLINGS, C. V. RAO and P. O. M. SCOKAERT. Constrained model predictive control: Stability and optimality. *Automatica.* 2000, vol. 36, iss. 6. pp. 789–814. ISSN 0005-1098. DOI: 10.1016/S0005-1098(99)00214-9.

[20] KANJANAWANISHKUL, K. Motion control of a wheeled mobile robot using model predictive control: A survey. *KKU Research Journal.* 2012, vol. 17, iss. 5. pp. 811–837. ISSN 0859-3957.

[21] KOTHARE, S. L. D. and M. MORARI. Contractive model predictive control for constrained nonlinear systems. *IEEE Transactions on Automatic Control.* 2000, vol. 45, iss. 6, pp. 1053–1071. ISSN 0018-9286. DOI: 10.1109/9.863592.

About Authors

Kiattisin KANJANAWANISHKUL was born in Trang, Thailand, in 1977. He received the B.Eng. in Electrical Engineering from Prince of Songkla University, Thailand in 2000. He received the M.Sc. in Mechatronics from University of Siegen, Germany in 2006. He received his Ph.D. in Computer Science from University of Tuebingen, Germany in 2010. Since 2010, he has been employed at the Faculty of Engineering, University of Mahasarakham, Thailand. His research interests include cooperative and distributed control, model predictive control, intelligent control, multi-robot systems, and robotic motion control.

10

Impact of Measurement Dating Inaccuracies in the Monitoring of Bulk Material Flows

Lionel MAGNIS, Nicolas PETIT

The Systems and Control Centre, MINES ParisTech, 60 Bd Saint-Michel, 75272 Paris Cedex, France

lionel.magnis@mines-paristech.fr, nicolas.petit@mines-paristech.fr

Abstract. This paper discusses the negative impact of errors in the dating of information gathered across a distributed network of sensors to be treated by a centralized monitoring algorithm. In this contribution, an example of flow monitoring serves as basis for the analysis. We consider an estimator setup for loss detection. Using a simple probability model, we determine the variance of this estimator and show how it is impacted by the dating uncertainty. A mitigating solution is proposed. Further extensions are discussed.

Keywords

Data synchronization, filtering, loss detection, network monitoring.

1. Introduction

In this article, we wish to stress the negative role on data analysis algorithms of dating inaccuracies of measurements. We base our discussion on a particular case of bulk material flow monitoring, but the reasoning could be generalized to other cases, and spur interest in this underestimated problem. To understand the issue under consideration here, let us now focus on flow monitoring applications. In various industrial fields, conveyor systems (or pipes) are used to transport bulk material from one inlet point to an outlet point, over possibly long distances, see [16]. These systems are subject to product losses (e.g due to hardware ageing) or thefts. Besides their obvious economical impact, these losses are caused for other numerous issues such as environmental disasters (e.g. oil leaks) or cascaded unreliability in downstream production processes (e.g. in a supply chain).

To prevent these malicious effects, numerous loss detection systems have been developed. A detection method commonly employed uses the law of conservation of mass to relate measurements from sensors distributed along the transport path. The resulting class of "balancing methods" evaluates the deviation in each part of the path between measured inlet and outlet mass or volume flow. In the oil industry, this method, also known as compensated mass balance (when the variation of density is accounted for, see [12]), is very popular. Its main advantages are the relative simplicity for actual implementation, the ease of tuning, the ability to uncover small leaks taking place over long time periods, along with its fast reaction to major leaks, see [6]. Considering the sensing technologies available for bulk material flow measurement (mostly laser or ultrasonic based, e.g. [9], [18]), this approach can be generalized to a wide class of bulk material conveyors.

Applying the conservation of mass principle based on information from spatially distributed sensors requires a good synchronization of data. The return of experience from several remarkable applications reported in the literature has served to formulate recommendations concerning the data acquisition technology, see [6]: *i)* the employed remote terminal units retrieving information from in-situ instruments should allow fast data transfer to the centralized master monitoring system, *ii)* the data should be carefully time-stamped with accurate and synchronized clocks (e.g. using GPS clock or Rugby clocks which, unfortunately, can be hardly available and subjected to jamming in many areas, or even SMS over cellular networks). Unfortunately, these recommendations are far from the technical status observed in currently installed systems. This is not surprising, as the problem of clock synchronization over a network is a complex one, even under the assumption of perfect two-way communications across the network, see [10], [11], and for this reason has been identified as a bottleneck in several control and monitoring architectures as described in [2], [13], [15], [19].

Let us now describe in more details the implementation of dating methodologies. As detailed in [1], the variation in reporting times from one data acquisition device (DAD) situated in one field location to the distant centralized supervisory control and data acquisition system (SCADA) can be quite large. The discussed time-stamping is usually not performed at the level of the DAD but the centralized SCADA level. The SCADA creates the time-stamp when the data is received. This procedure yields uncertainty on the age of data that are collected at the centralized level.

In this paper, we investigate the effect of time uncertainty on flow monitoring problem. Aiming at producing an analysis of the observations formulated by field practitioners, we conduct investigations on a "toy problem". As will appear, the model we propose shows that time uncertainty can produce false alarms in loss detection algorithms, which are usually considered as particularly annoying for production engineers. This underestimated problem can be as troublesome as the usual noises corrupting measurements. We believe that the conclusions that we draw here are sufficiently alarming to spur a general interest in this datation problem.

The paper is organized as follows. In Section 2. we formulate a simple transport model of a bulk material conveyor, the flow of which is monitored thanks to measurements produced by one inlet flow sensor and one outlet flow sensor. The measurements are produced almost periodically, due to the effect of a varying and unknown lag impacting each sample. Based on a probabilistic model of the lag value distribution, we establish the variance of the error introduced in the balance equation in Section 3. This balance relation serves as criterion for product-loss detection, and we relate the probability of detection and of false alarms to it. In Section 4. we determine flow pattern allowing one to minimize this variance. Simulations are presented in Section 5. stressing the role of data timing uncertainty, which, essentially increases the likelihood of false alarms.

2. Model and Problem Statement

Here, consider a bulk material conveyor with one input sensor DAD_I and one output sensor DAD_O as represented in Fig. 1. Each DAD gives a local measurement (q_I or q_O) of the material flow. Very generally, the sensors could be volume or mass flow meters (in the case of a pipeline as in the oil industry), or ultrasonic sensor (in the case of solid material, see [18], as in the mining or process industry [7]). The conveyor is assumed to generate a flow q having constant velocity v with respect to a fixed reference frame. Under this simplifying

consideration, q is solution of a simple delay equation. This hypothesis is typically satisfied by conveyor belts used for solid materials. The situation is more complex for liquid pipelines, for which the water-hammer equation is usually considered, see [3] and [17], or for multiphase flow, see [8]. However, the approach advocated in this article can be extended, at the expense of including physics-based transformations, and possibly, reaction terms.

For this "toy problem", we wish to detect the occurrence of product losses by monitoring the sensor values. Note l the length between the two measurement locations. In accordance with the assumptions above, our simple model states that, in the absence of any product loss, q_I and q_O are related by the delay equation:

$$q_O\left(t + \frac{l}{v}\right) = q_I(t). \qquad (1)$$

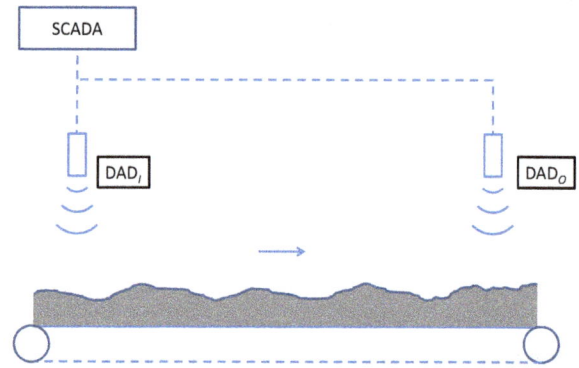

Fig. 1: Bulk material conveyor: two networked sensors communicate information to the SCADA.

For convenience, we assume that a loss has a linear effect on q_O, so that a fraction of the input flow is lost. The delay equation becomes:

$$q_O\left(t + \frac{l}{v}\right) = \lambda q_I(t), \qquad (2)$$

where λ is a parameter in $]0, 1]$ representing the product loss. Considering that a loss can randomly appear along the transport path, λ is the realization of a random variable Λ with values in $]0, 1]$, the occurrence of a loss being equivalent to the random event $\Lambda < 1$.

A simple way to detect the occurrence of losses is to monitor the mass imbalance over a time window of width T:

$$\int_0^T \left(q_I(t) - q_O\left(t + \frac{l}{v}\right)\right) dt, \qquad (3)$$

which equals 0 if $\lambda = 1$. Consistently with common practice and implementations, the input and output measurements are sampled at a frequency $\nu_s = \frac{T}{N}$, transmitted to and processed by the SCADA. Due to ill synchronisation of sampling dates, buffering and various other sources of network processes, the sampling time at which the measurements q_O and q_I are processed by the SCADA may differ. Taking $T = 1$ (without loss of generality) and the clock of the DAD$_I$ as a reference, the inlet sensor provides N input measurements of the form:

$$y_I[i] = q_I \left(\frac{i + \frac{1}{2}}{N} \right) + n_i, \ i = 0, \ldots, N - 1, \quad (4)$$

where n_i represents measurement noise. The synchronization discrepancies are modeled by a biased random time-shift (jitter) on the DAD$_O$ measurements.

Namely, the output measurements have the form:

$$y_O[i] = \lambda q_I \underbrace{\left(\frac{i + \frac{1}{2}}{N} + w_i \right)}_{\triangleq t_i} + n'_i, \ i = 0, \ldots, N - 1, \quad (5)$$

where n'_i represents measurement noise, and w_i is the realization of a zero-mean random variable W_i and we have assumed that the delay $\frac{l}{v}$ and the known average communication lag have been compensated by the appropriate time-shift.

A typical loss detection algorithm compares a discrete version of Eq. (3) such as:

$$b \triangleq \frac{1}{N} \sum_{i=0}^{N-1} (y_I[i] - y_O[i]), \quad (6)$$

against a threshold value b^*, rising a loss flag when $b \geq b^*$.

In the following, we study in details the effects of the law of the time uncertainty random variables W_i on this detection algorithm.

3. Imbalance Estimator

3.1. Preliminary Assumptions

Consider the two following assumptions.

Assumption 1. The W_i have support in $[-\delta, \delta]$ with $\delta < \frac{1}{2N}$. Thus, for any realization of the W_i, one has $0 < t_0 < \ldots t_{N-1} < 1$.

Assumption 2. q_I is continuous on $[0, 1]$ and affine on every $[\frac{i}{N}, \frac{i+1}{N}]$.

The parameter δ scales the time uncertainty. Note a_i the slope of $\frac{q_I}{N}$ on $[\frac{i}{N}, \frac{i+1}{N}]$. We have, for all i,

$$y_O[i] = \lambda q_I \left(\frac{i + \frac{1}{2}}{N} \right) + N\lambda a_i w_i + n'_i. \quad (7)$$

Without loss of generality, the total amount of bulk material entering the conveyor over the time window between 0 and 1 is unitary, i.e.:

$$1 = \int_0^1 q_I(t)dt. \quad (8)$$

As q_I is affine on every $[\frac{i}{N}, \frac{i+1}{N}]$, we have, exactly,

$$1 = \frac{1}{N} \sum_{i=0}^{N-1} q_I \left(\frac{i + \frac{1}{2}}{N} \right), \quad (9)$$

and the imbalance estimator is:

$$b = 1 - \lambda - \lambda \sum_{i \in \mathcal{I}} a_i w_i + \frac{1}{N} \sum_{i=0}^{N-1} (n_i - n'_i), \quad (10)$$

where $\mathcal{I} = \{i = 0, \ldots, N - 1 \mid a_i \neq 0\}$.

In the following, we assume that $\mathcal{I}$ is nonempty, so that the flow is not constant on the considered time-window.

3.2. Estimator Probability Law

To emphasize the effect of time uncertainty, we first consider a noise-free case where $n_i = n'_i = 0$. Then, b appears as the realization of the random variable:

$$B = 1 - \Lambda - \Lambda \sum_{i \in \mathcal{I}} a_i W_i. \quad (11)$$

To establish the probability law of B, we assume that:

- Λ and the W_i are jointly independent,

- the W_i are identically distributed (IID) and have a continuous probability density function (*pdf*) f^W (with support $[-\delta, \delta]$),

- to account for the likelihood of a no-loss scenario, Λ has a mixed-law comprising a Dirac at $\lambda = 1$ and a continuous density h on an interval $[\alpha, \beta] \subset]0, 1]$, so that the *pdf* of λ is of the form:

$$f^\Lambda(\lambda) = ph(\lambda) + (1 - p)\delta_1(\lambda), \quad (12)$$

where $p = P(\Lambda < 1)$ is the probability of occurrence of a loss.

By the formula of total probability, the *pdf* f^B of B can be recovered as:

$$f^B(b) = pf_L(b) + (1 - p)f_{\bar{L}}(b), \quad (13)$$

where f_L (respectively $f_{\bar{L}}$) is the *pdf* of B conditional to the loss event $\Lambda < 1$ (respectively the loss-free event $\Lambda = 1$). For $\lambda = 1$, we have $b = -\sum_{i \in \mathcal{I}} a_i w_i$. Hence,

$$f_{\bar{L}}(b) = \underset{i \in \mathcal{I}}{\circledast} \frac{1}{|a_i|} f^W \left(\frac{\cdot}{-a_i} \right)(b), \quad (14)$$

where $\circledast$ designates multiple convolution products. On the other hand, for any $\lambda \in [\alpha, \beta]$, $b = 1 - \lambda - \lambda \sum_{i \in \mathcal{I}} a_i w_i$. Hence,

$$f_L(b) = \int_\alpha^\beta \frac{1}{\lambda} f_{\bar{L}}\left(\frac{b - 1 + \lambda}{\lambda}\right) h(\lambda) d\lambda. \quad (15)$$

Example 1. *We take:*

$$N = 10, \ p = 0.5, \ a_i = \sin\left(\frac{i}{N}\right). \quad (16)$$

We assume that the W_i are independent Beta variables, see [14], of parameter $(2,2)$ with support in $[-\delta, \delta]$, namely:

$$f^W(w) = \frac{3}{4\delta^3}(\delta^2 - w^2)\mathbb{1}_{[-\delta, \delta]}(w) \quad (17)$$

and that, h is uniform on $[\alpha, \beta] = [0.6, 0.9]$. In Fig. 2, we represent $f_{\bar{L}}$, f_L and eventually f^B for $\delta = 0.01$ and $\delta = 0.04$. The larger the time uncertainty δ, the more the two modes of the pdf of B overlap.

3.3. Conditional Probability of a Loss Given the Measurement b

Consider the accuracy of measurement ε. A measured value b guarantees that $B \in I_\varepsilon \triangleq]b - \varepsilon, b + \varepsilon[$. We note:

$$p_L(b) \triangleq P_{B \in I_\varepsilon}(\Lambda < 1), \quad (18)$$

the conditional probability of a loss given $B \in I_\varepsilon$.

As illustrated in Fig. 2, the supports of $f_{\bar{L}}$ and f_L depend on the value of δ (and of α, β and the a_i). Indeed, note:

$$|a|_1 = \sum_{i=0}^{N-1} |a_i|. \quad (19)$$

According to Eq. (14) and Eq. (15), the respective supports $S_{\bar{L}}$ and S_L of $f_{\bar{L}}$ and f_L are, assuming $\delta|a|_1 \leq 1$:

$$S_{\bar{L}} = [-\delta|a|_1, \delta|a|_1], \quad (20)$$

$$S_L = [1 - \beta(1 + \delta|a|_1), 1 - \alpha(1 - \delta|a|_1)], \quad (21)$$

and we have:

$$p_L(b) = 1, \ \forall b \in S_L \setminus S_{\bar{L}}, \quad (22)$$

$$p_L(b) = 0, \ \forall b \in S_{\bar{L}} \setminus S_L. \quad (23)$$

Hence, if $S_L \cap S_{\bar{L}}$ is empty, the value of $p_L(b)$ is either 0 or 1 and the measure of b indicates a loss without any ambiguity. Both supports are disjoint if and only if:

$$\beta(1 + \delta|a|_1) < 1 - \delta|a|_1. \quad (24)$$

Condition Eq. (24) fails to be met when the time uncertainty becomes too large, namely when:

$$\delta \geq \frac{1 - \beta}{|a|_1(1 + \beta)}. \quad (25)$$

This is illustrated in Fig. 2. In such a case, a measure of $b \in S_L \cap S_{\bar{L}}$ is ambiguous. The Bayes formula [14] yields:

$$\begin{aligned} p_L(b) &= \frac{pP_{\Lambda<1}(B \in I_\varepsilon)}{pP_{\Lambda<1}(B \in I_\varepsilon)p + (1-p)P_{\Lambda=1}(B \in I_\varepsilon)} \\ &= \frac{p \int_{I_\varepsilon} f_L(x)dx}{p \int_{I_\varepsilon} f_L(x)dx + (1-p) \int_{I_\varepsilon} f_{\bar{L}}(x)dx} \\ &= \frac{pf_L(b)}{pf_L(b) + (1-p)f_{\bar{L}}(b)} + O(\varepsilon^2). \end{aligned} \quad (26)$$

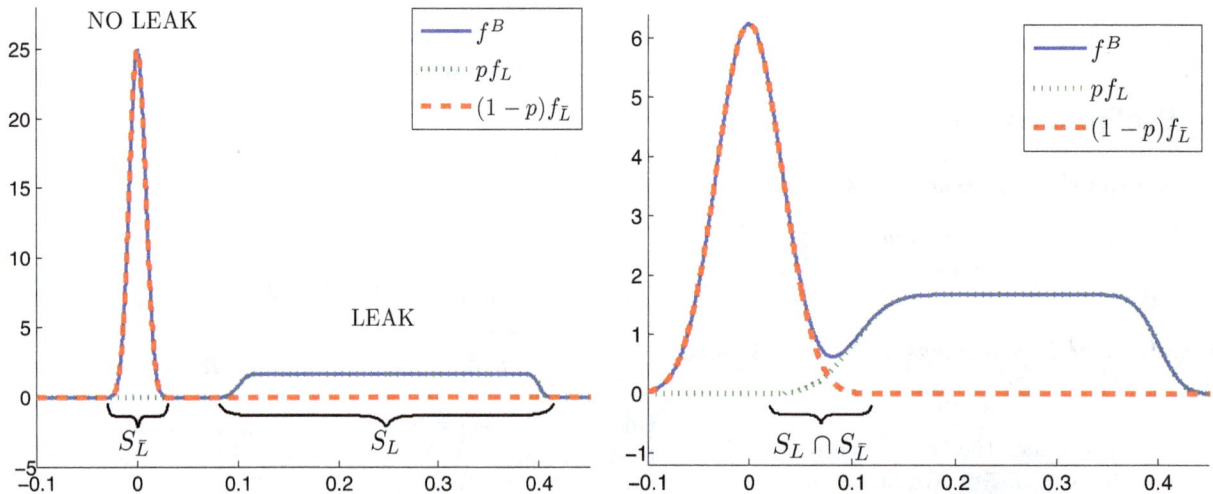

Fig. 2: *pdf* of B with parameters of Exm. 1 for $\delta = 0.01$ (left) and $\delta = 0.04$ (right). On the right, the overlap generated by the time uncertainty causes some ambiguity.

This probability is represented in Fig. 3 for the parameter values of Exm. 1 and various values of δ. The probability p_L varies from 0 to 1 with a slope, which gets steeper as δ decreases. In the extreme case $\delta = 0$, the probability is a step from 0 to 1, in which case the measurement b allows to identify a loss without ambiguity. The bigger δ, the further from this ideal case, the harder it is to identify a loss.

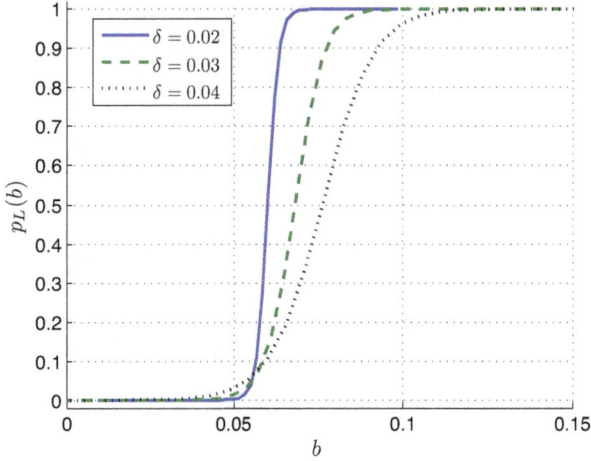

Fig. 3: Probability of a loss for values of b in the ambiguous zone.

3.4. Loss Detection and False Alarms

For a given threshold value b^*, we are interested in:

- the probability p_D of detecting a loss,

- the probability p_F of an alarm being false.

We have:

$$p_D = P_{\Lambda < 1}(b \geq b^*) = \int_{b^*}^{\infty} f_L(b)db, \quad (27)$$

$$p_F = P_{b \geq b^*}(\Lambda = 1) = \frac{(1-p)\int_{b^*}^{\infty} f_{\bar{L}}(b)db}{\int_{b^*}^{\infty} f^B(b)db}. \quad (28)$$

Ideally, one wants p_D as close to 1 as possible to detect most losses, and p_F as low as possible to avoid false alarms. These probabilities decrease as the threshold grows. We represent p_D and p_F in Tab. 1 for various values of these parameters.

3.5. Impact of Measurement Noise

For realism, we now add measurement noise to the model and assume that, for all i, n_i (respectively n_i') is the realization of a random variable N_i (respectively N_i') and that the N_i (respectively N_i') are IID zero-mean Gaussian variables with standard deviation σ_0 (respectively σ_1). We also assume that all the random variables of the problem are jointly independent. With these assumptions, Eq. (11) becomes:

$$B = 1 - \Lambda - \Lambda \sum_{i \in \mathcal{I}} a_i W_i + \frac{1}{N} \sum_{i=0}^{N-1} (N_i - N_i'), \quad (29)$$

where:

$$\frac{1}{N} \sum_{i=0}^{N-1} (N_i - N_i') \quad (30)$$

is a zero-mean Gaussian variable with standard deviation:

$$\sqrt{\frac{\sigma_0^2 + \sigma_1^2}{N}}. \quad (31)$$

Hence, the pdf f^B computed in Section 3.2. is simply convolved with a Gaussian pdf, and the same applies to $f_{\bar{L}}$ and f_L. As a result, p_D and p_F are altered as is reported in Tab. 1, with $\sigma_0 = 0.1$ and $\sigma_1 = 0.05$.

To sum-up the information gathered in Tab. 1, we can conclude that:

- the tuning of b^* results from a trade-off between the detection capabilities and the desired reliability of the loss-detection algorithm. This tuning is however difficult, as p_D and p_F also depend on the time uncertainty scaled by δ,

- for any fixed threshold value b^*, the performance of the loss-detection algorithm deteriorates with δ.

Tab. 1: Performances of loss-detection algorithm deteriorate with time uncertainty δ, p_D (probability of loss detection), p_F (probability of false alarm).

Threshold, case without noise	$\delta = 0.02$		$\delta = 0.03$		$\delta = 0.04$	
	p_D	p_F	p_D	p_F	p_D	p_F
$b^* = 0.03$	1.000	0.024	1.000	0.086	1.000	0.137
$b^* = 0.07$	0.999	0.000	0.996	0.001	0.991	0.011
$b^* = 0.11$	0.951	0.000	0.944	0.000	0.936	0.000
Threshold, case with noise						
$b^* = 0.03$	0.999	0.167	0.998	0.180	0.996	0.194
$b^* = 0.07$	0.984	0.031	0.981	0.043	0.976	0.059
$b^* = 0.11$	0.927	0.002	0.922	0.004	0.917	0.009

4.   Choice of an Optimal Input Flow Pattern

Time uncertainty has no effect under steady flow conditions, as the a_i are all zero and, irrespective of the time uncertainty, the measurements will be identical (up to noise). The detrimental effect of time uncertainty on the detection algorithm performance will appear for transient flow patterns. Then, a natural question is to determine a way to alleviate these effects. As will appear, an active control strategy brings a possible solution. In this section we assume that the conveyor has an actuator at the input point so that one can choose the input pattern q_I. The problem we consider is the scheduling of the flow for a sudden overload consisting of a unitary amount of bulk material spread over a unitary time. We investigate the choice of an "optimal" input pattern q_I with respect to loss detection.

Consider a fixed value for λ. The expectancy of B conditional to $\Lambda = \lambda$ is exactly the imbalance. Indeed, one has:

$$\mathbb{E}\left(1 - \lambda - \lambda\sum_{i\in\mathcal{I}} a_i W_i + \frac{1}{N}\sum_{i=0}^{N-1}(N_i - N_i')\right) \tag{32}$$

$$= 1 - \lambda.$$

The variance of B conditional to $\Lambda = \lambda$ is in Eq. (44), where σ_W is the standard deviation of the IID W_i. Decreasing σ seems promising for loss-detection, as the bias estimator will all the more accurately represent the true imbalance $1 - \lambda$. We have, for all i,

$$q_I(t) = q_I(0) + Na_i\left(t - \frac{i}{N}\right) +$$

$$+ \sum_{j=0}^{i-1} a_j, \forall t \in \left[\frac{i}{N}, \frac{i+1}{N}\right]. \tag{33}$$

Starting from and returning to a steady flow q_n, we consider equations:

$$q_I(0) = q_I(1) = q_n, \int_0^1 (q_I(t) - q_n)dt = 1, \tag{34}$$

which directly translate into two affine constraints bearing on the a_i:

$$\underbrace{\sum_{i=0}^{N-1} a_i = 0}_{\triangleq g_1(a)}, \quad \underbrace{N + \sum_{i=0}^{N-1} ia_i = 0}_{\triangleq g_2(a)}. \tag{35}$$

Thus, in view of Eq. (44), we consider the following problem.

Problem 1. *Find* $a = (a_0, \ldots, a_{N-1})$ *minimizing*

$$|a|_2 = \sqrt{\sum_i |a_i|^2} \tag{36}$$

under the affine constraints (35).

Solution 1. *Problem 1 has a unique solution* $a^\#$ *given by*

$$a_i^\# = \frac{6}{N+1} - \frac{12\,i}{N^2-1}, \quad \forall i = 0, \ldots, N-1. \tag{37}$$

Note $q^\#$ *the corresponding flow. The associated variance is:*

$$\sigma^2 = \frac{12N}{N^2-1}\lambda^2\sigma_W^2 + \frac{\sigma_0^2 + \sigma_1^2}{N}. \tag{38}$$

Proof 1. *As* $|.|_2$ *is convex and radially unbounded, it reaches a unique minimum under the affine constraints. Note* $f(a) \triangleq \frac{|a|_2^2}{2}$. *Then* $a^\#$ *satisfies the Lagrangian equation [4]:*

$$\nabla f(a) = \lambda_1 \nabla g_1(a) + \lambda_2 \nabla g_2(a)$$

$$\Leftrightarrow a_i = \lambda_1 + i\,\lambda_2, \;\forall i. \tag{39}$$

Injecting these relations into the constraint $g_1(a) = 0$ *yields:*

$$N\lambda_1 + \lambda_2 \sum_{i=0}^{N-1} i = 0 \;\Leftrightarrow\; \lambda_1 = \frac{1-N}{2}\lambda_2. \tag{40}$$

Injecting:

$$a_i = \left(\frac{1-N}{2} + i\right)\lambda_2, \quad \forall i \tag{41}$$

into the constraint $g_2(a) = 0$ *yields:*

$$\lambda_2\left(\frac{1-N}{2}\sum_{i=0}^{N-1} i + \sum_{i=0}^{N-1} i^2\right) = -N \tag{42}$$

$$\Leftrightarrow \lambda_2 = \frac{-12}{N^2-1}.$$

Hence, $a^\#$ *satisfies, for all* i:

$$a_i^\# = \frac{-12}{N^2-1} \times \frac{1-N}{2} - \frac{12}{N^2-1}\,i \tag{43}$$

$$= \frac{6}{N+1} - \frac{12\,i}{N^2-1}.$$

Then calculation of the variance is straightforward.

$$\sigma^2 \triangleq \text{var}\left(1 - \lambda - \lambda \sum_{i=0}^{N-1} a_i W_i + \frac{1}{N}\sum_{i=0}^{N-1}(N_i - N_i')\right) = \left(\lambda \sigma_W \sqrt{\sum_i |a_i|^2}\right)^2 + \frac{\sigma_0^2 + \sigma_1^2}{N}, \quad (44)$$

Interestingly, note that Problem 1 is equivalent to minimizing $\int_0^1 \dot{q}_I(t)^2 dt$ under Ass. 2. If this assumption is relaxed and smooth flow patterns are considered, one needs to solve the following straightforward calculus of variations problem.

Problem 2. *Minimizing* $\int_0^1 \dot{q}_I(t)^2 dt$ *under constraints:*

$$q_I(0) = q_I(1) = q_n, \quad \int_0^1 (q_I(t) - q_n) dt = 1. \quad (45)$$

Solution 2. *Problem 2 has a unique solution given by:*

$$q_I(t) = q^*(t) \triangleq q_n + 6t(1 - t). \quad (46)$$

The value of the minimum is:

$$\int_0^1 (\dot{q}^*(t))^2 dt = 12. \quad (47)$$

Proof 2. *Note* $Q(t) = \int_0^t (q_I(\tau) - q_n) d\tau$. *Problem 2 is equivalent to the auxiliary problem of finding Q minimizing*

$$J(Q) \triangleq \int_0^1 \ddot{Q}^2(t) dt, \quad (48)$$

under the constraints:

$$Q(0) = 0, \quad Q(1) = 1, \quad \dot{Q}(0) = 0, \quad \dot{Q}(1) = 0. \quad (49)$$

The corresponding Euler-Lagrange equation is [5]:

$$\frac{d^2}{dt^2}\frac{d}{d\ddot{Q}}\ddot{Q}^2 = 0 \Leftrightarrow \frac{d^4}{dt^4}Q = 0. \quad (50)$$

Hence, Q is of the form $Q(t) = c_3 t^3 + c_2 t^2 + c_1 t + c_0$. The constraints give a unique solution: $Q^*(t) = 2t^3 - 3t^2$. As J is convex, Q^* is the unique solution to the auxiliary problem. Hence, Problem 2 also has a unique solution given by:

$$q^*(t) = q_n + \dot{Q}^*(t)$$
$$= q_n + 6t(1-t). \quad (51)$$

The calculation of the corresponding value $\int_0^1 (\dot{q}^*(t))^2 dt$ is straightforward.

5. Simulation Results

We now study the performance of the loss detection algorithm on a theft scenario simulation for various input patterns. In Fig. 4, we represent a conveyor belt

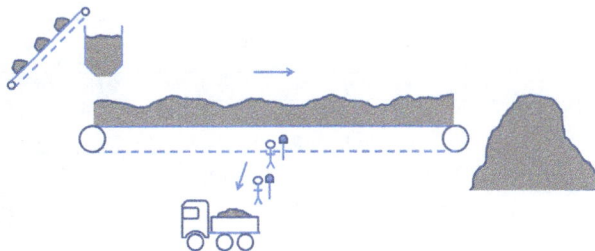

Fig. 4: Theft scenario.

connecting a production site to a storage facility. The flow of bulk material is steady with nominal value q_n except for punctual overloads (of unitary time, without loss of generality) randomly appearing with probability p_1. In such a case, the shape of the overloads is controlled by a flow input pattern q_I. Somewhere along the conveyor, a group of thieves may steal bulk material. One would like to detect this robbery.

We simulate N_{simu} unitary windows. On every window, we consider two theft scenarios:

- Basic theft: the thieves act randomly (for example whenever they are free of surveillance) with probability p. When they do so, they reroute a fraction $1 - \lambda$ of the flow (steady or overload), λ following the *pdf* f^Λ.

- Smart theft: the thieves act with the same *modus operandi* but only on the overloaded windows.

We use the following parameters: $N = 20$, $\delta = 0.08$, $\alpha = 0.6$, $\beta = 0.9$, $p = 0.3$, $\sigma_0 = \sigma_1 = 0.1$, $b^* = 0.09$, $N_{simu} = 100000$, $p_1 = 0.2$.

We compare the performance of the loss detection algorithm on those two cases for four different overload controlled input pattern represented in Fig. 5. The considered flows are the optimal patterns computed in Section 4. , as well as a reference piecewise affine pattern q^a and a reference smooth pattern q^s. As expected, we observe a strong similarity between both optimal patterns, $q^\# \simeq q^*$. It is easy to show that $q^\#$ converges uniformly to q^* as N goes to infinity.

The rate of loss detection and of false alarms for the four flow patterns and the two theft strategies are gathered in Tab. 2. The smart theft strategy is clearly more efficient from the thief's viewpoint. For all the input patterns the losses are less detected, and the rate of false alarms is much higher, which, in turn, deteriorates the reliability of the theft surveillance. Also, the

Tab. 2: Impact of the input flow pattern and the theft strategy on the algorithm performance.

time uncertainty: $\delta = 0.08$	random theft		smart theft	
	detections	false alarms	detections	false alarms
q^a (ref.)	96.8 %	3.1 %	95.3 %	11.0 %
$q^{\#}$ (opt.)	97.0 %	1.5 %	95.9 %	5.9 %
q^s (ref.)	96.8 %	2.2 %	95.7 %	8.4 %
q^* (opt.)	97.2 %	1.6 %	96.6 %	5.8 %
no time uncertainty: $\delta = 0$				
q^a (ref.)	97.3 %	0.5 %	97.3 %	2.2 %
$q^{\#}$ (opt.)	97.2 %	0.6 %	97.6 %	2.5 %
q^s (ref.)	97.1 %	0.5 %	97.4 %	2.4 %
q^* (opt.)	97.2 %	0.5 %	97.4 %	2.2 %

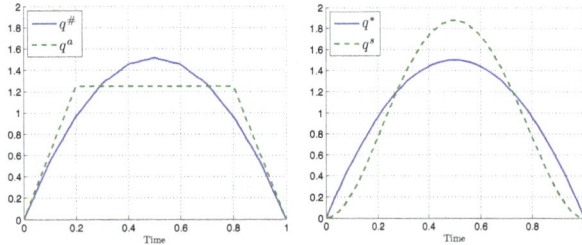

Fig. 5: Optimal and reference flow input patterns, piecewise affine (left) and smooth (right).

optimal patterns yield better overall performance than the reference ones, especially regarding false alarms. This difference is however smaller than the impact of theft strategy. The false alarms are partly due to measurement noise and partly to time uncertainty δ. To emphasize the contribution of time uncertainty on the algorithm performance, the same simulation has been run with $\delta = 0$.

6.   Conclusions and Perspectives

The study conducted in this article has shown that the problem of accurate data timing, which surprisingly has not generated significant theoretical studies before, is of true importance for online monitoring applications. The analysis has been performed on a very simple model, allowing one to derive explicit formulas for estimator variance and probabilities of detection and false alarms. Solutions to mitigate this malicious effect have been derived. For the "naive" average imbalance estimator, the variance is scaled by the L^2 norm of the input flow. More generally, investigations shall be focused on determining to which extent this conclusion holds for more advanced estimation methods such as (extended) Kalman filtering or state observers, in various contexts. Also, applying the same methodology to more complex flow dynamics such as water hammer equation for liquid pipelines is a topic for future research.

References

[1] API 1130 *Computational Pipeline Monitoring for Liquids Pipelines*. Washington: American Petroleum Institute, 2002.

[2] BARS, R., P. COLANERI, L. DUGARD, F. ALLGOWER, A. KLEIMENOV and C. SCHERER. Trends in theory of control system design. In: *Proc. 17th Internetional Federation of Automatic Control World Congress*. Seoul: IFAC, 2008, pp. 2144–2155. ISBN 978-3-902661-00-5. DOI: 10.3182/20080706-5-KR-1001.00363.

[3] BEGOVICH, O., A. NAVARRO, E. N. SANCHEZ and G. Besancon. Comparison of two detection algorithms for pipeline leaks. In: *CCA - 16th IEEE International Conference on Control Applications*. Singapore: IEEE, 2007, pp. 777–782. ISBN 978-1-4244-0442-1. DOI: 10.1109/CCA.2007.4389327.

[4] BOYD, S. and L. VANDENBERGHE. *Convex Optimization*. Cambridge: Cambridge University Press, 2004. ISBN 0-521-83378-7.

[5] BRYSON, A. E. and Y. HO. *Applied Optimal Control: optimization, estimation, and control*. Carlsbad: Hemisphere Publishing Company, 1975. ISBN 0-89116-228-3.

[6] DUDEK, M. Liquid Leak Detection Focused on Theft Protection. In: *OnePetro* [online]. 2005. Available at: https://www.onepetro.org/conference-paper/PSIG-05A4.

[7] ENSMINGER, D. and L. J. BOND. *Ultrasonics fundamentals, technologies, and applications*. Boca Raton: CRC Press, 2011. ISBN 978-0-8247-5889-9.

[8] FALCIMAIGNE, J. and S. DECARRE. *Multiphase production: Pipeline Transport, Pumping and Metering*. Paris: Editions OPHRYS, 2008. ISBN 978-2-7108-0913-5.

[9] FRADEN J. *Handbook of modern sensors: physics, designs and applications*. New York:

Springer, 2010. ISBN 978-1-4419-6465-6. DOI: 10.1007/978-1-4419-6466-3.

[10] FRERIS, N. M., V. S. BORKAR and P. R. KUMAR. A mode-based approach to clock synchronization. In: *48th IEEE Conference on Decsion and Control.* Shanghai: IEEE, 2009, pp. 5744–5749. ISBN 978-1-4244-3871-6. DOI: 10.1109/CDC.2009.5399516.

[11] FRERIS, N. M., S. R. GRAHAM and P. R. KUMAR. Fundamental Limits on Synchronizing Clocks Over Networks. *IEEE Transactions on Automatic Control.* 2011, vol. 56, iss. 6, pp. 1352–1364. ISSN 0018-9286. DOI: 10.1109/TAC.2010.2089210.

[12] GEIGER, G. State-of-the-Art in Leak Detection and Localisation. *Oil Gas European Magazine.* 2006, vol. 32, iss. 4, pp. 193–198. ISSN 0342-5622.

[13] HOKAYEM, P. F. and M. W. SPONG. Bilateral teleoperation: An historical survey. *Automatica.* 2006, vol. 119, iss. 12, pp. 2035–2057. ISSN 0005-1098. DOI: 10.1016/j.automatica.2006.06.027.

[14] MONTGOMERY, D. C. and G. C. RUNGER. *Applied Statistics and Probability for Engineers.* Jefferson City: Wiley, 2010. ISBN 978-0470053041.

[15] MURRAY, R., K. ASTROM, S. BOYD, R. BROCKETT and G. STEIN. Future directions in control in an information-rich world. *Control Systems.* 2003, vol. 23, iss. 2, pp. 20–33. ISSN 1066-033X. DOI: 10.1109/MCS.2003.1188769.

[16] PERRY, R. H. and D. W. GREEN. *Perry's Chemical engineers handbook.* New York: McGraw-Hill, 2007. ISBN 978-0702040429.

[17] RAHIMAN, W., B. LI, B. WU and Z. DING. Circle Criterion Based Nonlinear Observer Design For Leak Detection In Pipelines. In: *ICCA - International Conference on Control and Automation.* Guangzhou: IEEE, 2007, pp. 2993–2998. ISBN 978-1-4244-0818-4. DOI: 10.1109/ICCA.2007.4376911.

[18] SONBUL, O., P. POPEJOY and A. N. KALASHNIKOV. Ultrasonic Sensor array for remote sensing of profiles of bulk materials. In: *I2MTC - International Instrumentation and Measurement Technology Conference.* Graz: IEEE, 2012, pp. 1794–1797. ISSN 1091-5281. DOI: 10.1109/I2MTC.2012.6229189.

[19] ZEZULKA, F., Z. BRADAC, P. FIEDLER and M. SIR. Trends in Automation - investigation in Newtork Control Systems and Sensor Networks. In: *10th IFAC Workshop on programmable devices and embedded system.* Gliwice: International Federation of Automatic Control, 2010, pp. 109–113. ISBN 978-3-902661-95-1. DOI: 10.3182/20101006-2-PL-4019.00022.

About Authors

Lionel MAGNIS is Ph.D. candidate in Mathematics and Control at MINES ParisTech. He was born in 1985 in Saint-Mande, France. He obtained his Master's Degree from MINES ParisTech in 2008. His research interests includes the theory of automatic control, signal processing and observer design.

Nicolas PETIT is Professor at MINES ParisTech and head of Centre Automatique et Systemes. He was born in 1972 in Paris, France. He graduated from Ecole Polytechnique in 1995 (X92), and obtained his Ph.D. in Mathematics and Control at MINES ParisTech in 2000 and his Habilitation a Diriger des Recherches from Universite Paris 6 in 2007. His research interests includes the theory of automatic control, distributed parameter systems, constrained trajectory generation, delay systems and observer design. On the application side, he is active in industrial process control. He has developed the controllers of several industrial chemical reactors, and the patented softwares ANAMEL 4 and 5, currently in use for closed-loop control of blending devices in numerous refineries at the TOTAL company.

11

Practical Aspects of Primal-Dual Nonlinear Rescaling Method with Dynamic Scaling Parameter Update

Richard ANDRASIK

Department of Mathematical Analysis and Applications of Mathematics,
Faculty of Science, Palacky University Olomouc,
st. 17. listopadu 12, 771 46 Olomouc, Czech Republic

andrasik.richard@gmail.com

Abstract. *Primal-dual nonlinear rescaling method with dynamic scaling parameter update (PDNRD) is an optimization method from a class of nonlinear rescaling techniques. Previous work does not discuss practical aspects of PDNRD method such as the explanation and the setting of the parameters. To complete this framework, the parameters were described. Moreover, PDNRD method was applied on two quadratic programming problems with quadratic constraints and recommendations about the setting of the parameters were made.*

Keywords

Convex optimization, dynamic scaling parameter update, nonlinear rescaling method, parameters setting.

1. Introduction

The optimization theory is developing in parallel with the appearance of real-life problems and with the need to solve them. Thanks to that, some types of problems gained a special status among the others and many tools to solve these problems have been developed. An example is the linear programming (LP). In 1947, George Dantzig introduced the simplex method for solving LP [3]. However, the simplex method has a strong competitor – interior-point methods, especially primal-dual interior-point methods [15].

In the field of the nonlinear programming (NLP), interior-point methods have been also applied. However, the situation in NLP is more complicated in comparison with LP calculations and interior point methods are sometimes experiencing numerical difficulties. This fact motivated Roman Polyak and Igor Griva to design an alternative method based on the nonlinear rescaling (NR) theory.

Nowadays, NR methods can solve large-scale NLP problems with thousands of variables and constraints. They were successfully used to the radiotherapy treatment planning and are applied at some hospitals in USA and Europe [1].

The basic idea of NR methods is a nonlinear transformation of constraint functions to improve the properties of Lagrangian. Originally, the modified barrier methods [8] were introduced along with few modified barrier functions. Afterwards, the log-sigmoid function was also considered usable for NR [9], [10]. Consequently, the pieces of knowledge were refined, put together and a generalization of these techniques led to the concept of NR methods and NR functions. Similar to progress with interior-point methods, the primal-dual nonlinear rescaling (PDNR) method was developed [11].

PDNR method is locally convergent with the Q-linear convergence rate. To improve these properties, PDNR method can be combined with another optimization method (e.g. the primal-dual path-following method) to obtain the global convergence [11]. Another way how to improve the convergence of PDNR method is a dynamic scaling parameter update [5] together with some globalization strategy (e.g. a step length computation). Recently, generalizations and other improvements [12], [13] were developed to improve the asymptotic convergence rate and to reduce the computational effort. The main purposes of this paper are to describe in detail parameters of PDNRD method and to give recommendations about their setting.

The paper is organized as follows. First, the convex optimization problem is stated and basic assumptions are discussed. Then, NR functions are defined and the key idea of NR method is explained. Afterwards, basic primal-dual variant of NR method is presented. Next, PDNRD method is explained and its parameters are described. Finally, numerical experiments with different parameter settings were made and the results are presented in Section 7.

2. Statement of the Problem

We consider the convex optimization problem

$$
\begin{cases}
\text{minimize} & f(x), \quad x \in \Re^n, \\
\text{subject to} & c_i(x) \geq 0, \quad i = 1, \ldots, r.
\end{cases} \tag{1}
$$

Function f is convex and functions c_i are concave, $\forall i = 1, \ldots, r$. Let $S \subseteq \Re^n$ be the admissible set of problem Eq. (1). For simplicity we define mapping $c : \Re^n \to \Re^r$ as

$$
c(x) = (c_1(x), c_2(x), \ldots, c_r(x))^{\mathrm{T}}, \quad \forall x \in \Re^n.
$$

We suppose that:

- Functions f, c_i, $\forall i = 1, \ldots, r$, are at least twice continuously differentiable on the set $\Re^n$.

- The optimal set $X^* = \operatorname{Argmin}\{f(x); x \in S\}$ is bounded and not empty.

- The Slater condition holds.

For problem Eq. (1) we define the Lagrangian

$$
L(x; \lambda) = f(x) - \sum_{i=1}^{r} \lambda_i c_i(x). \tag{2}
$$

Due to assumption (C), Karush-Kuhn-Tucker's (KKT) conditions can be used to test the optimality. If $\hat{x} \in X^*$ then there is a vector $\hat{\lambda} \in \Re^r$ such that

$$
\begin{aligned}
\nabla_x L(\hat{x}; \hat{\lambda}) &= 0, \\
\hat{\lambda} &\geq 0, \\
\hat{\lambda}_i c_i(\hat{x}) &= 0, \quad \forall i = 1, \ldots, r.
\end{aligned} \tag{3}
$$

Conversely, if a pair $(\hat{x}, \hat{\lambda}) \in S \times \Re^r$ satisfies Eq. (3) then $\hat{x} \in X^*$.

3. Nonlinear Rescaling Functions

First, we define functions that will be used to transform constraints of problem Eq. (1).

Definition 1. Twice continuously differentiable function $\psi : (t_0; +\infty) \to \Re$, where $-\infty < t_0 < 0$, satisfying conditions

$$
\begin{aligned}
&(i) && \psi(0) = 0, \psi'(0) = 1, \\
&(ii) && \psi'(t) > 0, \forall t \in (t_0; +\infty), \\
&(iii) && \psi''(t) < 0, \forall t \in (t_0; +\infty), \\
&(iv) && \exists a > 0 : \psi(t) \leq -at^2, \forall t \in (t_0; 0), \\
&(v) && \exists b > 0 : \psi'(t) \leq bt^{-1}, \forall t > 0, \\
&(vi) && \exists c > 0 : \psi''(t) \geq -ct^{-2}, \forall t > 0
\end{aligned}
$$

is called NR function.

Remark 1. It follows from (ii) and (iii) that NR function ψ is increasing and concave on the whole domain. Because of (ii) and (v) it is true that $\lim_{t \to +\infty} \psi'(t) = 0$. Similarly from (iii) and (vi) it holds that $\lim_{t \to +\infty} \psi''(t) = 0$.

For example, the exponential transformation, the modified logarithmic function and the hyperbolic barrier function defined by the following formulas

$$
\begin{aligned}
\psi_1(t) &= 1 - e^{-t}, \\
\psi_2(t) &= \ln(t+1), \\
\psi_3(t) &= \frac{t}{t+1}.
\end{aligned}
$$

are nonlinear rescaling functions [11].

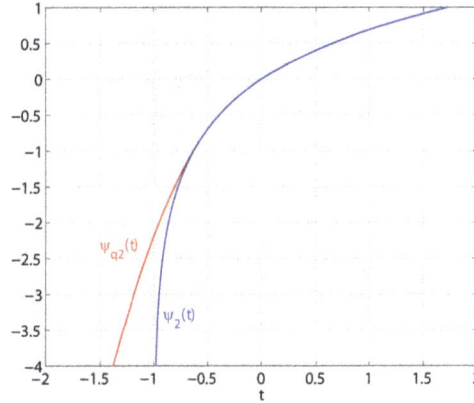

Fig. 1: Graphs of functions ψ_2 (blue) and ψ_{q2} (red), $\tau = -\frac{1}{2}$.

Functions ψ_i, $i = 2, 3$, can be modified so that $\psi_i \in \mathcal{C}^2(\Re)$. The function ψ_1 is already twice continuously differentiable and the following modification is not necessary, of course it can be done. For a given parameter $\tau \in (-1; 0)$ the quadratic extrapolation is defined by the relations

$$
\psi_{q_i}(t) = \begin{cases}
\psi_i(t), & \text{for } t \geq \tau, \\
q_i(t) = A_i t^2 + B_i t + C_i, & \text{for } t < \tau,
\end{cases}
$$

whereas coefficients of the function q_i can be determined from the formulas

$$
\psi_i(\tau) = q_i(\tau), \quad \psi_i'(\tau) = q_i'(\tau), \quad \psi_i''(\tau) = q_i''(\tau).
$$

From this equations we obtain

$$
\begin{aligned}
A_i &= \tfrac{1}{2}\psi_i''(\tau), \\
B_i &= \psi_i'(\tau) - \tau\psi_i''(\tau), \\
C_i &= \psi_i(\tau) - \tau\psi_i'(\tau) + \tfrac{1}{2}\tau^2\psi_i''(\tau).
\end{aligned}
$$

4.   Nonlinear Rescaling Approach

NR methods are based on the idea to transform problem Eq. (1) using a nonlinear rescaling function ψ defined on the whole real axis to the equivalent problem

$$
\begin{cases}
\text{minimize} & f(x), \quad x \in \Re^n, \\
\text{subject to} & k^{-1}\psi(kc_i(x)) \geq 0, \ i = 1, \ldots, r,
\end{cases}
\tag{4}
$$

From the definition of NR function, it is obvious that problems Eq. (1) and Eq. (4) have the same admissible sets, because

$$
\psi(kc_i(x)) \geq 0 \Leftrightarrow c_i(x) \geq 0, \ \forall i = 1, \ldots, r.
$$

Hence, the optimal sets are also the same. Positive real number k is the scaling parameter, which can be fixed or dynamically enlarged during the iteration process, see Algorithm 3.

The Lagrangian for equivalent problem Eq. (4) is given by the following formula:

$$
\mathcal{L}(x; \lambda, k) = f(x) - k^{-1} \sum_{i=1}^{r} \lambda_i \psi(kc_i(x)).
\tag{5}
$$

For any $r \in \mathbb{N}$ we denote

$$
\begin{aligned}
\Re^r_+ &= \{v \in \Re^r; \ v_i \geq 0\}, \\
\Re^r_{++} &= \{v \in \Re^r; \ v_i > 0\}.
\end{aligned}
$$

Suppose for a while that we know the solution of the dual problem $\lambda^* \in \Re^r_+$. Then it is sufficient to minimize the function $\mathcal{L}(x; \lambda^*, k)$ in the primal variable x. The constrained optimization problem would be transformed to an unconstrained optimization problem.

Since the Lagrange multipliers λ^* are not known, we estimate them and update them in every step of the method - just like the primal problem. In consequence, the constrained optimization problem is converted to a sequence of unconstrained optimization problems. Newton's method or its variant is applied in each step to minimize the Lagrangian $\mathcal{L}$ in the primal variable.

Algorithm 1. (The basic concept of NR methods)

Let $k > 0$ be a scaling parameter. Initial approximations $x^0 \in \Re^n$ and $\lambda^0 \in \Re^r_{++}$ are given. We suppose that an approximation $(x^s, \lambda^s) \in \Re^n \times \Re^r_{++}$, $s \in \mathbb{N}_0$,

is known already. We find the next primal-dual pair (x^{s+1}, λ^{s+1}) using the following formulas

$$
\begin{aligned}
x^{s+1} &: \quad \nabla_x \mathcal{L}(x^{s+1}; \lambda^s, k) = 0, \\
\lambda_i^{s+1} &= \psi'\left(kc_i(x^{s+1})\right)\lambda_i^s, \quad i = 1, \ldots, r.
\end{aligned}
\tag{6}
$$

The NR method converges for any fixed but large enough $k > 0$ under the standard second order optimality conditions [5].

Remark 2. If $\lambda^s \in \Re^r_{++}$, then also $\lambda^{s+1} \in \Re^r_{++}$. This property follows from condition (ii) of NR function ψ. In other words, NR methods are interior-point methods in the dual variable.

The Algorithm 1 is well defined due to the following theorem.

Theorem 1. *Suppose that X^* is bounded. For any given $(\lambda, k) \in \Re^r_{++} \times \Re_{++}$ there exists one and only one $\hat{x} \in \Re^n$ such that*

$$
\mathcal{L}(\hat{x}; \lambda, k) = \min_{x \in \Re^n} \mathcal{L}(x; \lambda, k).
$$

Proof: [10], page 206.

The main purpose of NR is to improve properties of the Lagrangian. The classical Lagrangian L – as a connection between the constrained and the unconstrained optimization – does not always work, because the existence of the unconstrained Lagrange minimizer is unknown in general. On the other hand, the unconstrained minimizer of the Lagrangian $\mathcal{L}$ always exists (according to Theorem 1). Moreover, NR dramatically sharpens the reaction of the Lagrangian to the constraint violation, which has an impact on the computations.

In comparison to interior-point methods, there is no "infinite wall". NR methods are exterior point methods. Also, the unbounded increase of the scaling parameter is not needed to guarantee the convergence of the method.

We illustrate the influence of NR on the Lagrangian on the following simple example.

Example 1. The Lagrangian of the following problem

$$
\begin{cases}
\text{minimize} & x, \quad x \in \Re^1, \\
\text{subject to} & x + 1 \geq 0,
\end{cases}
\tag{7}
$$

and the Lagrangian of the equivalent problem are considered.

$$
\begin{cases}
\text{minimize} & x, \quad x \in \Re^1, \\
\text{subject to} & k^{-1}\psi_{q_2}\left(k(x+1)\right) \geq 0,
\end{cases}
\tag{8}
$$

We take a look on the graphs of both Lagrangians, which are defined by the formulas

$$L(x;\lambda) = x - \lambda(x+1),$$
$$\mathcal{L}(x;\lambda,k) = x - k^{-1}\lambda\psi_{q_2}(k(x+1)). \quad (9)$$

NR fundamentally changes the shape of the Lagrangian. The sharp reaction of the Lagrangian $\mathcal{L}$ to the constraint violation is obvious in Fig. 2.

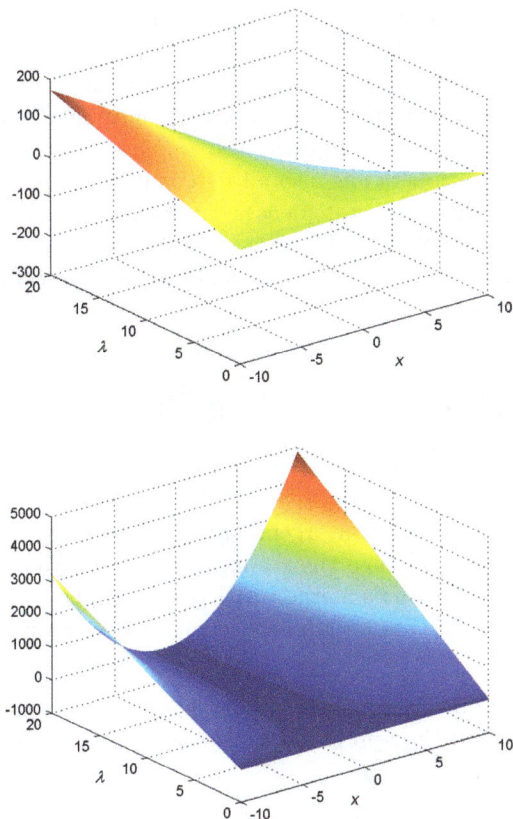

Fig. 2: The Lagrangian for the problem Eq. (7) - on the top - and for the equivalent problem Eq. (8) with $k = 1$ - on the bottom.

We project the graphs of the both Lagrangians to the phase plane in Fig. 3. Suppose that we want to find Lagrange minimizer for $\lambda = 2$. After substitution, formulas Eq. (9) have the form:

$$L(x;2) = -x - 2,$$
$$\mathcal{L}(x;1,k) = x - k^{-1}2\psi_{q_2}(k(x+1)).$$

The function $L(x;2)$ is a decreasing linear function. We cannot find the Lagrange minimizer of it, because it is unbounded below.

If we rearrange the first equation in formulas Eq. (9) in the following way:

$$L(x;\lambda) = (1-\lambda)x - \lambda,$$

it is apparent that for any $\lambda \neq 1$ we cannot find the Lagrange minimizer. On the other hand, we know from Theorem 1 that the Lagrangian $\mathcal{L}$ has a minimizer for any $\lambda > 0$. This is the main contribution of NR.

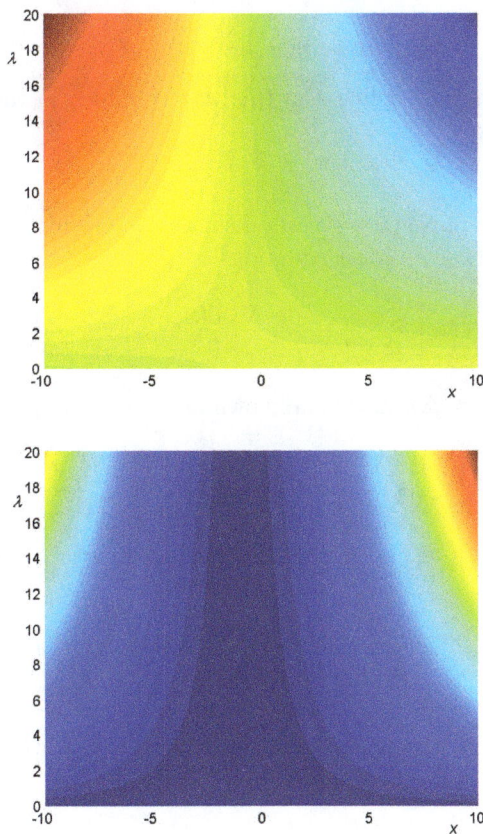

Fig. 3: The projection of the Lagrangian for the problem Eq. (7) - on the top - and for the equivalent problem Eq. (8) with $k = 1$ - on the bottom - to the phase plane.

5. Primal-dual Nonlinear Rescaling Method

The update of the Lagrange multipliers is clear from equations Eq. (6). When we want to calculate x^{s+1}, we must solve a nonlinear system of equations using Newton's method. Although this approach works well far from the solution, primal-dual variant of the NR method performs better in the neighborhood of the solution [5]. In Section 6, we will combine these two techniques to obtain a stable and globally convergent method.

Let approximations $x \in \Re^n$, $\lambda \in \Re^r_{++}$ be known. We suppose that

$$\hat{x} = x + \Delta x,$$
$$\hat{\lambda} = \bar{\lambda} + \Delta\lambda, \quad (10)$$

where $(\Delta x, \Delta \lambda)$ is a primal-dual Newton step and $\bar{\lambda}$ is a predictor of the Lagrange multipliers, which is given by the following formula

$$\bar{\lambda} = \Psi'(kc(x))\lambda,$$

where

$$\Psi'(kc(x)) = \operatorname{diag}(kc_i(x))_{i=1}^r.$$

The reason, why define $\bar{\lambda}$ in this way is hidden in the equality

$$\nabla_x \mathcal{L}(x; \lambda, k) = \nabla_x L(x; \bar{\lambda}).$$

One step of Newton's method consists of solving the system of $(n + r)$ linear equations

$$N(\cdot) \begin{bmatrix} \Delta x \\ \Delta \lambda \end{bmatrix} = \begin{bmatrix} -\nabla_x L(x; \bar{\lambda}) \\ 0 \end{bmatrix}, \qquad (11)$$

where a pair $(\Delta x, \Delta \lambda)$ is unknown and

$$N(\cdot) = \begin{bmatrix} \nabla_{xx}^2 L(x; \bar{\lambda}) & -\nabla c(x)^{\mathrm{T}} \\ -k\Psi''(kc(x))\Lambda\nabla c(x) & \mathbf{I}_r \end{bmatrix}.$$

Symbols $\Psi''(kc(x))$, Λ denotes diagonal matrices defined as

$$\begin{aligned} \Psi''(kc(x)) &= \operatorname{diag}(\psi''(kc_i(x)))_{i=1}^r, \\ \Lambda &= \operatorname{diag}(\lambda_i)_{i=1}^r. \end{aligned}$$

If $N(\cdot)$ is sparse, we can use numerical algebra techniques for sparse matrices [15] to solve the system Eq. (11).

In the opposite case, we express $\Delta \lambda$ from the second equation of the system Eq. (11) and we substitute it to the first equation. Now, we must solve only a system with n equations, instead of a system of $(n + r)$ equations. The system has the following form

$$M(\cdot)\Delta x = -\nabla_x L(x; \bar{\lambda}), \qquad (12)$$

where

$$M(\cdot) = \nabla_{xx}^2 L(x; \bar{\lambda}) - k\nabla c(x)^{\mathrm{T}}\Psi''(kc(x))\Lambda\nabla c(x).$$

Remark 3. It is obvious that $M(\cdot)$ is a symmetric matrix. It can be shown [11], page 120 that if the second order optimality conditions hold true at the point $(x^*; \lambda^*)$, then $M(\cdot)$ is a positive definite matrix for all $(x; \lambda)$ sufficiently close to the optimal primal-dual pair, when the scaling parameter is sufficiently large.

Algorithm 2. (One step of PDNR method)

Let $k \in \Re_{++}$ be fixed. The approximations $x \in \Re^n$, $\lambda \in \Re_{++}^r$ are given.

- We compute a dual predictor

$$\bar{\lambda} = \Psi'(kc(x))\lambda.$$

- We calculate the Newton step $(\Delta x, \Delta \lambda)$ from system Eq. (11), or

- we find out the primal Newton step Δx from formula Eq. (12) and then we compute the dual Newton step using the following relation

$$\Delta \lambda = k\Psi''(kc(x))\Lambda\nabla c(x)\Delta x.$$

- We calculate the new primal-dual approximation

$$x := x + \Delta x, \quad \lambda := \bar{\lambda} + \Delta \lambda.$$

The Algorithm 2 describes only one step of PDNR method. Although, the input vector of multipliers belongs to $\Re_{++}^r$, it is not guaranteed that the updated vector of multipliers has also all its components positive. In contrast, this statement is true for Algorithm 1, according to Remark 2. Therefore, PDNR method cannot stand alone in a general case and the improvement is needed (see Algorithm 3).

6. Dynamic Scaling Parameter Update

To obtain a higher convergence rate of the method, we dynamically change the scaling parameter. Moreover, we use Newton's method with a step length (e.g. the backtracking line search algorithm) to solve formulas Eq. (6).

We introduce a function which measures the distance between the approximation (x, λ) and the solution (x^*, λ^*).

Definition 2. The function $\nu : \Re^n \times \Re^r \to \Re_+$, defined as follows

$$\begin{aligned} \nu(x, \lambda) = \max\{\ &\|\nabla_x L(x; \lambda)\|, \\ &-\min_{1 \le i \le r} c_i(x), \\ &-\min_{1 \le i \le r} \lambda_i, \\ &\textstyle\sum_{i=1}^r |\lambda_i c_i(x)|\ \}, \end{aligned} \qquad (13)$$

is called the merit function.

Formula (13) is motivated by KKT conditions. From the first order optimality conditions it follows that

$$\nu(\hat{x}, \hat{\lambda}) = 0 \Leftrightarrow (\hat{x}, \hat{\lambda}) \in X^*.$$

For known primal-dual pair (x, λ) we set the parameter k according to the relation

$$k = \nu(x, \lambda)^{-1/2}. \qquad (14)$$

It is obvious that if a primal-dual sequence $\{(x^s, \lambda^s)\}_0^{+\infty}$ tends to (x^*, λ^*), then $\nu(x^s, \lambda^s) \to 0^+$ and also $k_s \to +\infty$ when $s \to +\infty$.

To prevent from a singularity in system Eq. (11), we solve the following regularized system

$$N_k(\cdot)\begin{bmatrix} \Delta x \\ \Delta\lambda \end{bmatrix} = \begin{bmatrix} -\nabla_x L(x;\bar\lambda) \\ 0 \end{bmatrix}, \qquad (15)$$

where the primal-dual pair (x,λ) is an approximation of solution, k is the scaling parameter, $\bar\lambda = \Psi'(kc(x))\lambda$ is the dual predictor of the Lagrange multipliers and

$$N_k(\cdot) = \begin{bmatrix} \nabla_{xx}^2 L(x;\bar\lambda) + \frac{1}{k^2}\mathbf{I}_n & -\nabla c(x)^{\mathrm{T}} \\ -k\Psi''(kc(x))\Lambda\nabla c(x) & \mathbf{I}_r \end{bmatrix}.$$

It is important to remark that the regularization has no effect to the 1.5-superlinear convergence rate of PDNRD method [5].

In a comparison to interior-point methods, the inverse matrix to $N_k(\cdot)$ exists at the optimal point for every $k \in \Re_{++}$.

Assume that the approximation (x,λ) of the point (x^*,λ^*) is known. First, we use PDNRD method. If there is a superlinear decrease of the merit function, we have found the next approximation. In the opposite case, we use the primal Newton step Δx (calculated during the use of PDNRD method) to minimize the function $\mathcal{L}(x;\lambda,k)$, where λ and k are fixed. We apply the backtracking line search method to guarantee the global convergence in the minimization process. In this way, we obtained the globally convergent PDNRD method with the 1.5-superlinear convergence rate.

Algorithm 3. (The globally convergent PDNRD method)

An initial approximation $x^0 \in \Re^n$ is given. An accuracy parameter $\varepsilon > 0$ and an initial scaling parameter $k \in \Re_{++}$ are given. Parameters $q \in (0;1)$, $\eta \in (0;0.5)$, $\omega > 1$, $\sigma > 0$ and $\theta > 0$ are also given. Set $x := x^0$, $\lambda := (1,1,\ldots,1) \in \Re^r$, $\lambda_g := \lambda$ and $H := \nu(x,\lambda)$.

- If $H \leq \varepsilon$, then stop, output (x,λ).

- Find $\bar\lambda$ and $(\Delta x, \Delta\lambda)$ from primal-dual system Eq. (15) with known (x,λ) and set

$$\hat x := x + \Delta x, \ \hat\lambda := \bar\lambda + \Delta\lambda, \ \hat H := \nu(\hat x, \hat\lambda).$$

- If $\hat H \leq \min\left\{H^{3/2-\theta}, 1-\theta\right\}$, then set

$$x := \hat x, \ \lambda := \hat\lambda, \ H := \hat H, \ k := \max\left\{\frac{1}{\sqrt{H}}, k\right\}$$

and go to step 1.

- Find $\alpha \in (0;1\rangle$ so that it holds

$$\mathcal{L}(x+\alpha\Delta x;\lambda_g,k) - \mathcal{L}(x;\lambda_g,k)$$
$$\leq \eta\alpha\Delta x^{\mathrm{T}}\nabla_x\mathcal{L}(x;\lambda_g,k),$$

using the backtracking line search algorithm.

- Set

$$x := x + \alpha\Delta x, \ \hat\lambda := \Psi'(kc(x))\lambda_g.$$

- If

$$\|\nabla_x\mathcal{L}(x;\lambda_g,k)\| \leq \frac{\sigma}{k}\|\hat\lambda - \lambda_g\|,$$

then go to step 8.

- Find $(\Delta x, \Delta\lambda)$ from primal-dual system Eq. (15) with known (x,λ_g) and go to step 4.

- If $\nu(x,\hat\lambda) \leq qH$, then set

$$\lambda := \hat\lambda, \ \lambda_g := \hat\lambda, \ H := \nu(x,\hat\lambda), \ k := \max\left\{\frac{1}{\sqrt{H}}, k\right\}$$

and go to step 1.

- Set $k := \omega k$ and go to step 7.

7. Numerical Experiments

From the finite element approximation of contact problems of the linear elasticity with the friction in three space dimensions arise a minimization problem

$$\begin{cases} \text{minimize} & \frac{1}{2}x^{\mathrm{T}}Ax - x^{\mathrm{T}}b, \quad x \in \Re^n \\ \text{subject to} & g_i^2 - x_{i+m}^2 - x_{i+2m}^2 \geq 0, \ i \in \mathcal{I}, \\ & x_i - l_i \geq 0, \ i \in \mathcal{I}, \end{cases} \tag{16}$$

where $n = 3m$ is the number of variables, $A \in \Re^{n\times n}$ is a symmetric and positive definite matrix, $b \in \Re^n$, $g \in \Re_+^m$, $l \in \Re^m$ and $\mathcal{I} = \{1,2,\ldots,m\}$. This is a convex programming problem so we can use NR approach to solve it. We use the function ψ_{q_2} with $\tau = \frac{1}{2}$ to rescale the conditions and subsequently we obtain the equivalent problem. The reason for this choice is that for a sufficiently large class of functions $c(x)$ is the function $-\ln c(x)$ self-concordant. If we apply Newton's method to a self-concordant function, we can say something more about the convergence of Newton's method [2]).

PDNRD method was tested on two model problems - the chord problem, which in contrary to Eq. (16) contains also unconstrained variables, and the steel brick problem, whose finite element approximation leads exactly to Eq. (16). All computations were performed in MATLAB on the PC Intel Core i7 (2.4 GHz) with 8 GB RAM. In tables below we reported the number of iterations (it), the number of solutions of the primal-dual system (n_S) and the solution time in seconds ($time$). If some data is missing, it means that the solution time was too long in the comparison to the other cases in the table.

7.1.　Chord Problem

We consider a problem

$$\min_{u \in \mathcal{K}} \mathcal{J}(u), \qquad (17)$$

where

$$\mathcal{J}(u) = \frac{1}{2}\int_0^1 \|u'(t)\|^2 \mathrm{d}t - \int_0^1 u(t)^{\mathrm{T}} f(t)\mathrm{d}t,$$

$$\mathcal{K} = \{\ \ u \in \left(H_0^1(0;1)\right)^2 :\ u_2(t) \geq 0, \forall t \in (0;0.5),$$
$$\|u(t)\| \leq 1.4, \forall t \in (0.5;1)\ \ \},$$

$$f(t) = \left(36\pi^2 \sin 6\pi t,\ -4\pi^2 \sin 2\pi t\right)^{\mathrm{T}}.$$

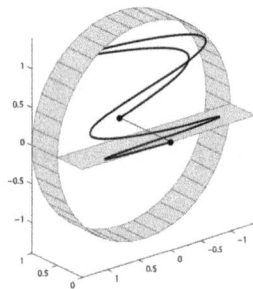

Fig. 5: The structure of the matrix $N_k(\cdot) \in \Re^{192 \times 192}$ (the chord problem). Black colour indicates 636 nonzero elements.

Tab. 1: The chord problem. PDNRD method with parameters $\omega = 10$, $\sigma = \frac{1}{2}k_{init}$, $\theta = 0.4$, $q = 0.5$, $\eta = 0.01$, $\varepsilon = 10^{-6}$.

n	r	k_{init} 10	10^2	10^3
64	32	10/31/0.173	4/23/0.118	7/7/0.021
128	64	7/50/0.238	6/33/0.126	3/79/0.351
256	128	8/60/0.515	8/37/0.454	3/143/1.099
512	256	11/65/1.989	5/49/2.037	7/150/5.036
1024	512	12/70/10.330	6/72/13.088	8/100/22.229
2048	1024	12/58/37.640	6/70/53.256	7/66/50.650
4096	2048	13/74/221.834	–	–
		$it/n_S/time$	$it/n_S/time$	$it/n_S/time$

Fig. 4: The chord deformation.

Minimization problem Eq. (17) describes a loaded chord fixed at the endpoints. The chord is partially above a plain and partially inside a cylindrical tube Fig. 4. The function $u(t)$ is the chord deflection. The chord problem was presented as a model problem in [6].

The objective function $\frac{1}{2}x^{\mathrm{T}}Ax - x^{\mathrm{T}}b$, $x \in \Re^n$ matches the convex quadratic functional $\mathcal{J}(u)$, linear constraints in problem Eq. (16) matches the constraint $u_2(t) \geq 0$, $\forall t \in (0;0.5)$ from the definition of the set $\mathcal{K}$ and quadratic constraints matches $\|u(t)\| \leq 1,4$, $\forall t \in (0.5;1)$.

First, we decided if the primal-dual system has a sparse matrix. We say that the matrix is sparse if it has at most 10 % nonzero elements. In Fig. 5, the structure of the matrix $N_k(\cdot) \in \Re^{192 \times 192}$ for $n = 128$ is depicted. This matrix has only 636 nonzero elements (1.7 %). Hence, $N_k(\cdot)$ is the sparse matrix. In the same way we could argue for other choices of n.

We solved the chord problem for different settings of parameters of PDNRD method. The main result from Tab. 1 is a non-increasing number of iterations and also a non-increasing number of solutions of the primal-dual system while increasing the number of variables. The best choice of the scaling parameter was $k_{init} = 10$ in

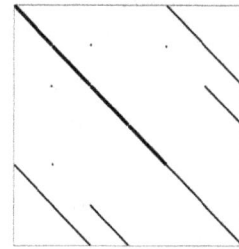

Tab. 2: The chord problem ($n = 1024$). PDNRD with parameters $\omega = 10$, $\sigma = \frac{1}{2}k_{init}$, $k_{init} = 10$, $\varepsilon = 10^{-6}$.

	$\eta = 0.01$	
q	$\theta = 0.1$	$\theta = 0.4$
0.1	6/49/8.015	6/49/8.054
0.5	10/65/11.554	12/70/10.518
0.9	30/99/16.347	31/77/13.548
	$\eta = 0.3$	
q	$\theta = 0.1$	$\theta = 0.4$
0.1	6/49/8.133	6/49/8.036
0.5	10/68/11.503	12/71/10.671
0.9	30/92/15.094	31/78/12.885
	$it/n_S/time$	$it/n_S/time$

this case, because of the best performance for larger problems.

From Tab. 2 it is obvious that the choice $q = 0.1$ was the best one. Parameters η and θ had no significant impact on the computation in this case.　■

7.2.　Steel Brick Problem

Let us consider a steel brick lying on the rigid obstacle. The brick occupies the domain $\mathcal{S} = (0;3) \times (0;1) \times$

$(0; 1)$. The boundary ∂S is divided into three parts

$$
\begin{aligned}
\Gamma_u &= \{0\} \times (0; 1) \times (0; 1), \\
\Gamma_c &= (0; 3) \times (0; 1) \times \{0\}, \\
\Gamma_p &= \partial S \backslash \left(\bar{\Gamma}_u \cup \bar{\Gamma}_c\right),
\end{aligned}
$$

on which different boundary conditions are given (see Fig. 6). The problem is described in the detail in [6].

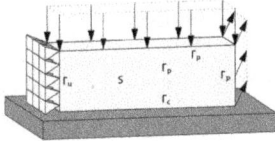

Fig. 6: The steel brick.

The progress was similar to the previous one considering the chord problem. First, the matrix $N_k(\cdot)$ was tested whether it is sparse or not. In Fig. 7, the structure of the matrix $N_k(\cdot) \in \Re^{150 \times 150}$ for $n = 90$ is depicted. This matrix contains 8340 nonzero elements (37 %). Hence, the matrix $N_k(\cdot)$ is not sparse and we solved the primal-dual system in a different way (see Section 5).

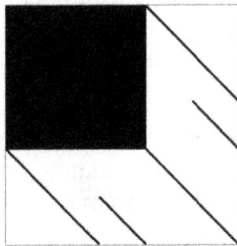

Fig. 7: The structure of matrix $N_k(\cdot) \in \Re^{150 \times 150}$ (steel brick problem). Black colour indicates 8340 nonzero elements.

The number of variables was denoted as $n = 3\,m$ and the number of conditions as $r = 2\,m$. The steel brick problem was solved for different choices of n. According to the data in Tab. 3, the best setting for the initial values of the scaling parameter was $k_{init} = 10$ in this problem. Number n_S rose with increasing k_{init}.

The results from Tab. 4 supports that the choice $\eta = 0.01$ is better than $\eta = 0.3$. The parameter θ has no significant impact on the computation.

Tab. 3: The steel brick problem. PDNRD method with parameters $\omega = 10$, $\sigma = \frac{1}{2} k_{init}$, $\theta = 0.4$, $q = 0.5$, $\eta = 0.01$, $\varepsilon = 10^{-6}$.

		k_{init}		
n	r	10	10^2	10^3
54	36	6/39/0.153	5/54/0.268	4/119/0.570
90	60	6/38/0.218	6/45/0.239	4/102/0.743
180	120	8/40/0.257	4/90/0.700	4/136/1.653
324	216	7/37/0.663	10/35/0.619	4/130/3.146
648	432	12/67/7.258	10/77/10.003	6/123/15.696
		$it/n_S/time$	$it/n_S/time$	$it/n_S/time$

Tab. 4: The steel brick problem ($n = 648$). PDNRD method with parameters $\omega = 10$, $\sigma = \frac{1}{2} k_{init}$, $k_{init} = 10$, $\varepsilon = 10^{-6}$.

	$\eta = 0.01$	
q	$\theta = 0.1$	$\theta = 0.4$
0.1	5/46/5.562	5/46/5.767
0.5	12/45/4.460	12/67/7.135
0.9	12/67/7.545	12/67/7.105
	$\eta = 0.3$	
q	$\theta = 0.1$	$\theta = 0.4$
0.1	5/44/5.593	5/44/5.482
0.5	11/41/4.398	12/62/8.820
0.9	12/62/8.761	12/62/8.718
	$it/n_S/time$	$it/n_S/time$

The characteristic property of PDNRD method is the "hot" start of the method (as it was observed and proved in [11]). "Hot" start means that from some point only one Newton step is needed to sufficiently shrink the distance between a current approximation and the solution. We can expect that a computational effort to solve the problem with the accuracy 10^{-12} will not be dramatically higher than with an accuracy 10^{-6}. Our results support this statement (compare Tab. 3 and Tab. 5).

Tab. 5: The steel brick problem. PDNRD method with parameters $\omega = 10$, $\sigma = \frac{1}{2} k_{init}$, $k_{init} = 10$, $\theta = 0.4$, $q = 0.5$, $\eta = 0.01$, $\varepsilon = 10^{-12}$.

n	r	$k_{init} = 10$
54	36	8/45/0.165
90	60	10/51/0.252
180	120	12/44/0.269
324	216	10/47/0.806
648	432	14/71/7.371
		$it/n_S/time$

7.3. Parameters

When using a numerical method the setting of the parameters is very important. Experience with numerical experiments helps us to find the optimal setting. Based on the tests of PDNRD method on examples from Sections 7.1 and 7.2, we made considerations about a suitable setting of the parameters.

Factor ω

The factor ω affects the rate of the increase of the scaling parameter. For $\omega \in \langle 5; 20 \rangle$ we obtain almost the same results. So we set $\omega = 10$. However, even for the other choices $\omega > 1$ there are not any significant changes. At most, it may happen that it takes a few extra steps of the method.

Parameter σ

The choice of the parameter σ is less important than the choice of the ratio between σ and the initial choice of the scaling parameter k_{init}. It is the fraction $\frac{\sigma}{k}$ which decides about the number of inner iterations (and thus about the number of Newton steps in the damped phase of Newton's method). It is clear that for $\sigma >> k$ there are too little inner steps. On the other hand, for $\sigma << k$ there is too many of them. According to the results of the numerical experiments, the choice $\sigma = \frac{1}{2} k_{init}$ is suitable.

Parameter θ

The parameter θ affects whether the inner solver runs or not. If we set $\theta = 0$, the inner solver runs whenever the 1.5-superlinear decrease of the merit function was not achieved. Choosing $\theta = 0.5$, we are saying that we are satisfied with only linear decrease of the merit function. Therefore, it is wise to set $\theta \in \langle 0; 0.5 \rangle$. It appears (see Tab. 2 and Tab. 4) that the method does not depend on this parameter.

In the condition, which is related to the parameter θ, the term $\min \left\{ H^{3/2-\theta}, 1 - \theta \right\}$ is calculated. We can ask why do not simply use the term $H^{3/2-\theta}$ instead of the previous one. As we know, the classical Newton method is effective only in the neighborhood of the solution, so if we are "far" from the solution the damped Newton method is better suited to use. The expression "far" means that the merit function is greater than $1 - \theta$.

Factor q

The number of inner steps is influenced by the fraction $\frac{\sigma}{k}$ together with the factor q. This factor also affects how often is the scaling parameter increased. Due to the way in which is this parameter used in PDNRD method, it is needed to set $q \in (0; 1)$. Moreover, we must choose the factor q so that the method does not use too many inner steps. Based on the data in Tab. 2 and Tab. 4, any factor $q \in (0; 1)$ was suitable. However, the best results were obtained by setting $q = 0.1$.

Parameter η

The backtracking line search parameter η is usually chosen in the range from 0.01 to 0.3 (see [2] page 466). According to data in Tab. 2 and Tab. 4 the choice of the parameter $\eta \in \langle 0.01; 0.3 \rangle$ is arbitrary.

Initial value of the scaling parameter

Based on the results shown in Tab. 1 and Tab. 3, increasing the initial value of the scaling led to instabilities in computations. The best results in the both model examples were obtained with $k_{init} = 10$. Therefore, we recommend setting k_{init} around ten or less. Further reason to set k_{init} rather lower is that the scaling parameter can be increased during the computation, but it cannot be decreased.

Initial approximation

The computation also depends on the initial approximation. The number of steps can be decreased, if "lucky" initial approximation is chosen. In problems like a beam deflection or contact problems we choose an initial state of the system as the initial approximation, because the shape changes of a body are usually very small. Thus we set $x^0 = (0, 0, \ldots, 0)$.

8.   Conclusion

The practical aspects of PDNRD method were described. Especially, the meaning of the parameters of PDNRD method were explained in detail. PDNRD method was tested on two quadratic programming problems with quadratic constraints (the chord problem and the steel brick problem). Based on the tests, the recommendations about setting the parameters were made.

The solution of the primal-dual system is the most expensive operation, thus the number of solutions of the primal-dual system determines the total complexity of computations. It was found out that the increasing number of variables in both presented problems has not a consequence in the increasing number of solutions of the primal-dual system (see Tab. 1 and Tab. 3). This fact supports the applicability of PDNRD method on problems of an arbitrary size.

Acknowledgment

This research was supported by the grant PrF 2013, 013 Mathematical models of the Internal Grant Agency of Palacky University Olomouc.

References

[1] ALBER, M. and R. REEMTSEN. Intensity modulated radiotherapy treatment planning by use of a barrier-penalty multiplier method. *Optimization Methods and Software*. 2007, vol. 22, iss. 3, pp. 391–411. ISSN 1055-6788. DOI: 10.1080/10556780600604940.

[2] BOYD, S. P. and L. VANDENBERGHE. *Convex optimization*. Cambridge: Cambridge University Press, 2004. ISBN 978-0-521-83378-3.

[3] DANTZIG, G. B. and L. VANDENBERGHE. *Linear programming and extensions*. Princeton, N.J.: Princeton University Press, 1998. ISBN 978-069-1059-136.

[4] GRIVA, I., S. G. NASH and A. SOFER. *Linear and nonlinear optimization*. 2nd ed. Philadephia: Society for Industrial and Applied Mathematics, 2009. ISBN 978-0-89871-661-0.

[5] GRIVA, I. and R. A. POLYAK. Primal-dual nonlinear rescaling method with dynamic scaling parameter update. *Mathematical Programming*. 2006, vol. 106, iss 2, pp. 237–259. ISSN 1436-4646. DOI: 10.1007/s10107-005-0603-6.

[6] KUCERA, R., J. MACHALOVA, H. NETUKA and P. ZENCAK. An interior-point algorithm for the minimization arising from 3D contact problems with friction. *Optimization Methods and Software*. 2013, vol. 28, iss. 6, pp. 1195–1217. ISSN 1029-4937. DOI: 10.1080/10556788.2012.684352.

[7] NOCEDAL, J. and S. J. WRIGHT. *Numerical optimization*. 2nd ed. New York: Springer, 2006, Springer series in operations research and financial engineering. ISBN 03-873-0303-0.

[8] POLYAK, R. A. Modified barrier functions (theory and methods). *Mathematical Programming*. 1992, vol. 54, iss. 1–3, pp. 177–222. ISSN 1436-4646. DOI: 10.1007/BF01586050.

[9] POLYAK R. A. Log-Sigmoid Multipliers Method in Constrained Optimization. *Annals of Opereration Research*. 2001, vol. 101, iss. 1–4, pp. 427–460. ISSN 1572-9338. DOI: 0.1023/A:1010938423538.

[10] POLYAK, R. A. Nonlinear rescaling vs. smoothing technique in convex optimization. *Mathematical Programming*. 2002, vol. 92, iss. 2, pp. 197–235. ISSN 1436-4646. DOI: 10.1007/s101070100293.

[11] POLYAK, R. A. Primal–dual exterior point method for convex optimization. *Optimization Methods and Software*. 2008, vol. 23, iss. 1, pp. 141–160. ISSN 1573-2878. DOI: 10.1080/10556780701363065.

[12] GRIVA, I. and R. A. POLYAK. 1.5-Q-superlinear convergence of an exterior-point method for constrained optimization. *Journal of Global Optimization*. 2008, vol. 40, iss. 4, pp. 679–695. ISSN 1573-2916. DOI: 10.1007/s10898-006-9117-x.

[13] POLYAK, R. A. and I. GRIVA. Proximal point nonlinear rescaling method for convex optimization. *Numerical Algebra, Control and Optimization*. 2011, vol. 1, iss. 2, pp. 283–299. ISSN 2155-3297. DOI: 10.3934/naco.2011.1.283.

[14] POLYAK, R. A. and M. TEBOULLE. Nonlinear rescaling and proximal-like methods in convex optimization. *Mathematical Programming*. 1997, vol. 76, iss. 2, pp. 265–284. ISSN 1436-4646. DOI: 10.1007/BF02614440.

[15] WRIGHT, S. J. *Primal-dual interior-point methods*. Philadelphia: Society for Industrial and Applied Mathematics, 1997. ISBN 08-987-1382-X.

About Authors

Richard ANDRASIK was born in Karvina, Czech republic, 1988. He received his M.Sc. from Palacky University Olomouc in 2012. Currently he is a Ph.D. student. His research interests include mathematical and physical models of fluid dynamics, optimization theory and numerical methods for solving partial differential equations.

12

Optimization of Modulation Waveforms for Improved EMI Attenuation in Switching Frequency Modulated Power Converters

Deniss STEPINS[1], Jin HUANG[2]

[1]Institute of Radio Electronics, Faculty of Electronics and Telecommunications, Riga Technical University,
16 Azenes Street, LV-1048 Riga, Latvia
[2]School of Electrical and Electronic Engineering, Huazhong University of Science and Technology,
1037 Luoyu Road, Wuhan 430074, China

deniss.stepins@rtu.lv, huangjin.mail@163.com

Abstract. *Electromagnetic interference (EMI) is one of the major problems of switching power converters. This paper is devoted to switching frequency modulation used for conducted EMI suppression in switching power converters. Comprehensive theoretical analysis of switching power converter conducted EMI spectrum and EMI attenuation due the use of traditional ramp and multislope ramp modulation waveforms is presented. Expressions to calculate EMI spectrum and attenuation are derived. Optimization procedure of the multislope ramp modulation waveform is proposed to get maximum benefits from switching frequency modulation for EMI reduction. Experimental verification is also performed to prove that the optimized multislope ramp modulation waveform is very useful solution for effective EMI reduction in switching power converters.*

Keywords

Electromagnetic interference, frequency modulation, optimization, switching power converter.

1. Introduction

Switching power converters (SPC) used in many electronic devices for electric power conversion are noticeable sources of electromagnetic interference (EMI) which can deteriorate normal operation of other electronic equipment. EMI both conducted and radiated must be reduced in order to meet international electromagnetic compatibility standards (e.g. CISPR 22) requirements. Traditional ways for conducted EMI reduction usually are input EMI filters, proper design of printed circuit boards, soft-switching techniques, etc. Recently two more novel conducted EMI reduction techniques have been proposed and used [1], [2], [3], [4], [5]. The first one is active EMI filtering using digital signal processing [4] and the second one is spread spectrum technique [3]. The spread spectrum technique which is usually based on switching frequency modulation (SFM) is very attractive technique mainly because it is simple to implement. By modulating the switching frequency peak conducted EMI levels can be easily suppressed using simple periodic modulation waveforms (such as sine, triangle, sawtooth, etc) as shown in Fig. 3. The main parameters of periodic SFM are switching frequency deviation Δf_{sw}, modulation frequency f_m and modulating waveform $m(t)$ with unitary amplitude. Of course, for modulating the switching frequency random and chaotic signals can be used, but the use of simple periodic modulation waveforms is more popular in practical SPC, mainly because negative effect of periodic SFM on peak-to-peak output voltage ripples is less visible and periodic SFM is simpler to implement than random or chaotic SFM [6], [7].

Benefits of periodic SFM for EMI reduction can be significantly reduced due to amplitude modulation of SPC currents caused by SFM [8]. The amplitude modulation leads to EMI spectrum sidebands distortion and asymmetry with respect to central switching frequency f_{sw0} and consequently EMI attenuation is lower than expected for the given value of modulation index [8], [9]. In order to increase effectiveness of SFM for EMI reduction in SPC (especially for higher values of Δf_{sw}), multislope ramp modulating waveform (MRMW) was proposed by S. Johnson and R. Zane [8].

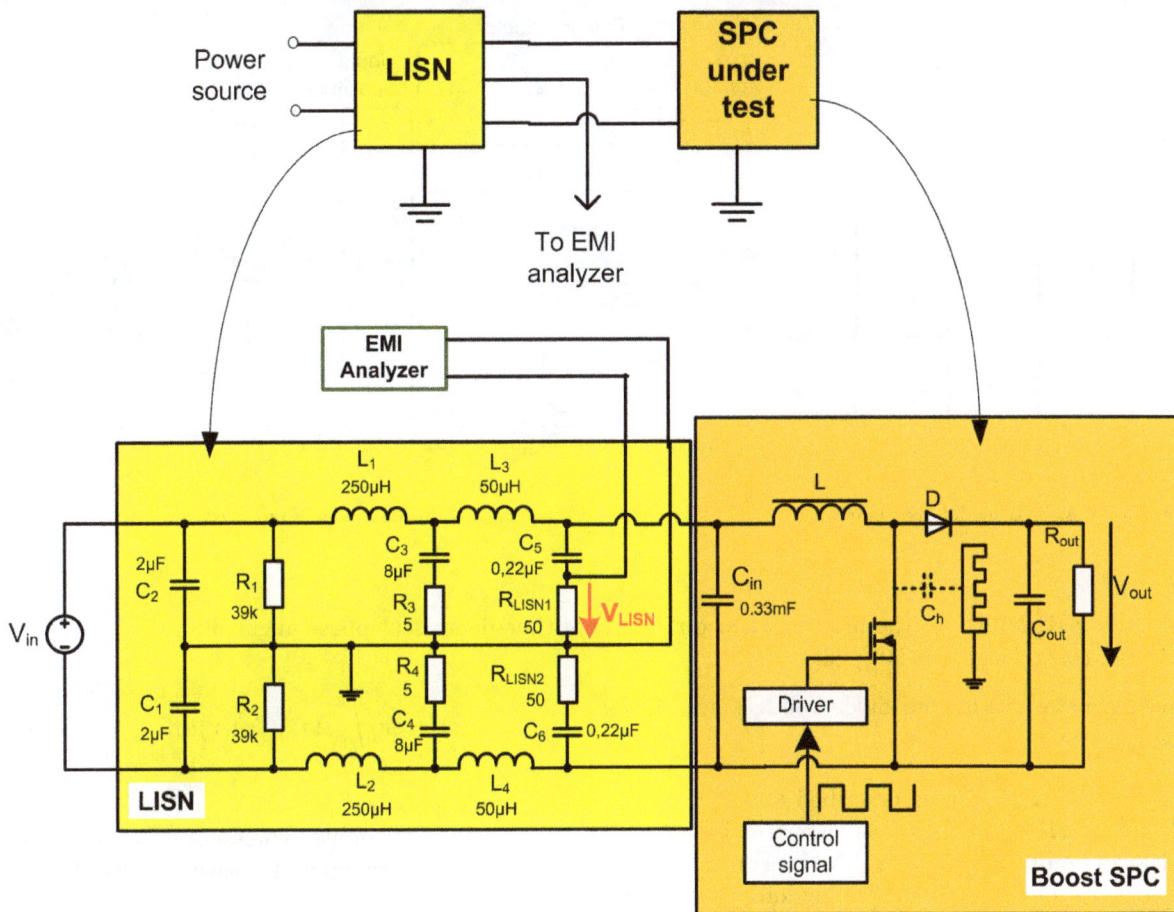

Fig. 1: Typical conducted EMI measurement setup (top) and boost SPC schematic diagram with LISN (bottom) used in the analysis.

Slopes of the MRMW are set by parameter t_0 (Fig. 5). As it is stated in [8], electronic ballast output current spectral distortion and output EMI can be effectively reduced when $t_0 = 0.35T_m$ (where T_m is modulation period). However as it will be proved in our paper, t_0 is the function of both Δf_{sw} and f_m. Thus $t_0 = 0.35T_m$ can be optimal only for one set of Δf_{sw} and f_m values (e.g. for $\Delta f_{sw} = 5$ kHz and $f_m = 500$ Hz), but for the other set of values of the parameters, it can even worsen EMI attenuation and be less effective than traditional ramp modulating waveform. Therefore optimization of MRMW is necessary. Lack of the theoretical analysis in [8] does not give a possibility to calculate optimum values of t_0 to get maximum EMI attenuation. Thus the main aim of the paper is to make the theoretical analysis of EMI spectrum and its attenuation due to the use of MRMW and propose a procedure for optimum t_0 values calculation in order to get full benefits from the use of MRMW for EMI reduction in SFM SPC.

The paper is organized as follows. In the section 2 effect of SFM on conducted EMI of boost converter is theoretically analyzed in details. In the section 3 theoretical analysis of conducted EMI and attenuation

due to the use of MRMW is presented, expressions to calculate EMI spectrum are derived and procedure to calculate optimum t_0 values is proposed. In the section 4 experimental verification of the theoretical results is performed. And finally, in the section 5 conclusions are given.

2. Effect of SFM on Conducted EMI

In the analysis boost DC-DC converter operating in continuous conduction mode will be used (Fig. 1). Since for conducted EMI measurements line impedance stabilization network (LISN) is used, it will be taken into account. To simplify the analysis several assumptions will be considered:

- conducted EMI dominates at lower frequencies, because fundamental switching frequency harmonic amplitude is more pronounced than others [5], [10],

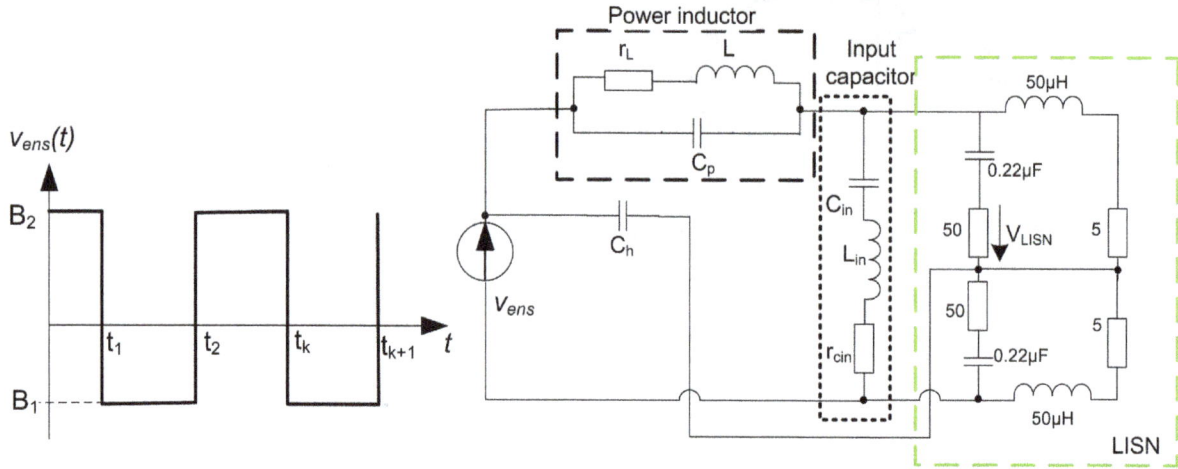

Fig. 2: Simple boost SPC model. For boost SPC: $B_1 = -V_{in}$; $B_2 = V_{out} - V_{in}$. Note the model is modified version taken from [11].

- conducted EMI levels are the highest when duty ratio D of control signal is 50 % [3],

- inductor, power switch and diode voltages are of rectangular shape.

Considering the assumptions described above, simple EMI model [11] can be used as shown in Fig. 2. The model can be used up to several MHz. For higher frequencies more complex models (taking into account other non-idealities) should be used. Using the model the transfer function between equivalent noise source V_{ens} and V_{LISN} can be derived [9]:

$$K_{EMI}(f) =$$

$$= \frac{50}{50 + Z_{C5}} \cdot \left(\frac{Z_e}{Z_{Ch} + Z_e} - \frac{Z_{Cin}}{2\left(Z_{Cin} + Z_L\right)} \right), \quad (1)$$

where $Z_e = (25 + Z_{C5}/2)(5 + Z_{L3})/(55 + Z_{C5} + Z_{L3})$; $Z_{C5} = 1/(j2\pi f C_5)$; $Z_{Ch} = 1/(j2\pi f C_h)$; $Z_{L3} = j2\pi f L_3$; C_h is parasitic capacitance between MOSFET drain and grounded heatsink; Z_{Cin} is real input capacitor complex impedance and Z_L is real power inductor complex impedance.

If f_{sw0}/f_m is an integer number then period of SFM voltage or current equals modulation waveform period T_m. In this case complex Fourier series coefficients for SFM V_{ens} (Fig. 2) can be derived as follows:

$$d_{sn} = -\frac{B_1}{j2\pi n} \sum_{i=1}^{N/2} \left[e^{-j2\pi n f_m t_{2i}} - e^{-j2\pi n f_m t_{2i-1}} \right]$$

$$- \frac{B_2}{j2\pi n} \sum_{i=1}^{N/2} \left[e^{-j2\pi n f_m t_{2i+1}} - e^{-j2\pi n f_m t_{2i}} \right], \quad (2)$$

where an integer number $N = 2f_{sw0}/f_m$; time instants t_k at which V_{ens} crosses zero (Fig. 2) can be calculated by solving the equation [12]:

$$\cos\left(2\pi f_{sw0}t + \theta(t)\right) = 0, \quad (3)$$

where time-dependent phase angle [3]:

$$\theta(t) = 2\pi \int_0^t \Delta f_{sw} \cdot m(\tau)d\tau. \quad (4)$$

By solving Eq. (3) simple expression to calculate t_k can be derived. For example, for sawtooth SFM it is [9]:

$$t_k = \frac{-f_{sw\,min} + \sqrt{f_{sw\,min}^2 + \Delta f_{sw}(2k - 1)/T_m}}{2\Delta f_{sw}/T_m}, \quad (5)$$

where minimum switching frequency $f_{sw\,min} = f_{sw0} - \Delta f_{sw}$; modulation index $\beta = \Delta f_{sw}/f_m$; T_m is $m(t)$ period.

In general f_{sw0}/f_m is not an integer number. In this case period of SFM voltage or current does not equal T_m but it equals KT_m, where positive integer K can be calculated as follows:

$$K = \frac{F}{f_{sw0} \cdot T_m}, \quad (6)$$

where F and K are least positive integers. Let us consider two examples:

- if $f_{sw0} = 80$ kHz and $f_m = 1$ kHz ($T_m = 1$ ms), then $K = 1$, $F = 80$ and SFM signal period is $T_m = 1$ ms,

- if $f_{sw0} = 80$ kHz and $f_m = 3$ kHz ($T_m = 1/3$ ms), then $K = 3$, $F = 80$ and SFM signal period is $3T_m = 1$ ms.

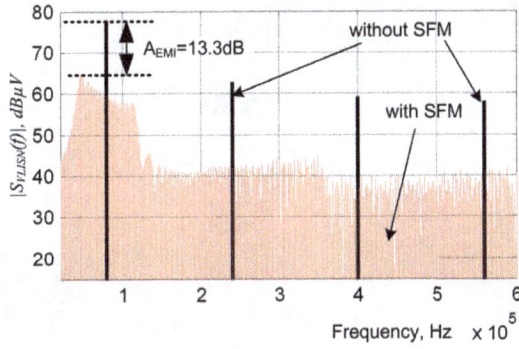

Fig. 3: V_{LISN} spectra before and after the use of SFM with sawtooth $m(t)$. Parameters: $f_m = 1$ kHz; $\Delta f_{sw} = 40$ kHz; $f_{sw0} = 80$ kHz; $C_h = 10$ pF; $V_{in} = 4$ V.

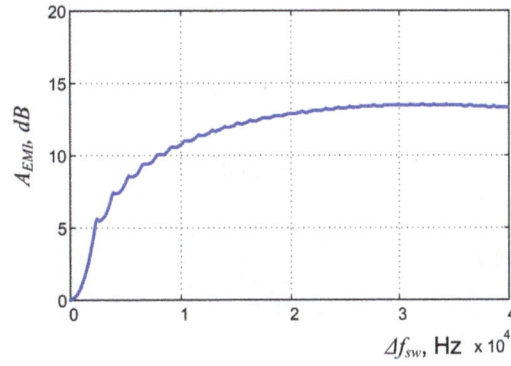

Fig. 4: A_{EMI} versus Δf_{sw}. Modulation waveform: sawtooth.

In general case (when f_{sw0}/f_m is not an integer) complex Fourier series coefficients for SFM V_{ens} can be calculated using the following expression:

$$
\begin{aligned}
d_{sn} = & \\
= & -\frac{B_1}{j2\pi n} \sum_{i=1}^{F} \left[e^{-j2\pi n \frac{f_m}{K} t_{2i}} - e^{-j2\pi n \frac{f_m}{K} t_{2i-1}} \right] \\
& -\frac{B_2}{j2\pi n} \sum_{i=1}^{F} \left[e^{-j2\pi n \frac{f_m}{K} t_{2i+1}} - e^{-j2\pi n \frac{f_m}{K} t_{2i}} \right].
\end{aligned} \tag{7}
$$

In order to calculate V_{LISN} (conducted EMI) spectrum the following Eq. (8) can be used

$$
|S_{VLISN}(f)| = 2 \left| d_{sn} K_{EMI} \left(\frac{f_m n}{K} \right) \right|, \tag{8}
$$

where $n = 1, 2, 3 \ldots$

As an example calculated V_{LISN} spectra of boost SPC with and without SFM are shown in Fig. 3. SFM leads to EMI spectrum spreading and in its turn to EMI attenuation. The attenuation (A_{EMI}) which is the difference in dB between maximum amplitude of unmodulated and SFM conducted EMI spectra in the frequency range of interest can be calculated as follows [9], [13]:

$$
A_{EMI} = 20 \log_{10} \left(\frac{\max |S_{VLISN}(f)|}{\max |S_{VLISN1}(f)|} \right), \tag{9}
$$

where S_{VLISN} and S_{VLISN1} are unmodulated and SFM V_{LISN} spectra. As an example, calculated A_{EMI} versus Δf_{sw} is shown in Fig. 4.

3.   Improving EMI Attenuation Using Optimization of MRMW

Fundamental switching frequency sideband can become highly asymmetrical with respect to f_{sw0} (Fig. 3)

due to parasitic amplitude modulation of SPC currents [8], [9]. As it is proved in [8], [9] this leads to degradation of benefits of SFM because EMI attenuation is lower than it is expected for the given set of values of Δf_{sw} and f_m. In order to increase effectiveness of the use of SFM for EMI reduction in SPC (especially for higher values of Δf_{sw}), multislope ramp modulating waveform (MRMW) can be used [8].

3.1.   Theoretical Analysis of MRMW

In this subsection comprehensive theoretical analysis of MRMW for conducted EMI reduction in boost SPC will be done.

As it can be seen from Fig. 5 slopes of the MRMW are set by parameter t_0. V_{LISN} spectrum can be calculated using Eq. (7) and Eq. (8). However it should be noted that in the case of MRMW (when $t_0 \neq 0.5 T_m$) central switching frequency is:

$$
f_{sw01} = f_{sw0} + \left(0.5 - \frac{t_0}{T_m} \right) \Delta f_{sw}. \tag{10}
$$

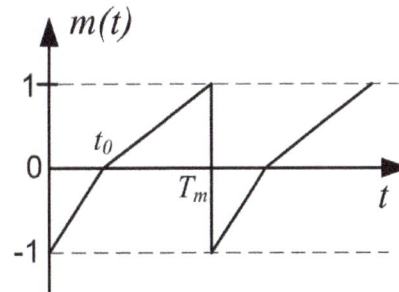

Fig. 5: MRMW.

Time instants t_K at which V_{ens} crosses zero (Fig. 2) can be calculated by solving the following Eq. (11):

$$
\cos \left(2\pi f_{sw01} t + \theta(t) \right) = 0, \tag{11}
$$

where time dependent phase angle can be derived using Eq. (4) as follows:

$$\theta(t) = 2\pi\Delta f_{sw} \begin{cases} \int_b^t \left(\dfrac{\tau - b}{t_0} - 1 - a\right) d\tau & \text{if } b \le t \le t_0 + b \\ -t_0\left(\dfrac{1}{2} + a\right) + \int_{t_0+b}^t \left(\dfrac{\tau - b - t_0}{T_m - t_0} - a\right) d\tau & \text{if } t_0 + b < t \le T_m p \end{cases}$$

$$= 2\pi\Delta f_{sw} \begin{cases} \dfrac{t^2}{2t_0} - t(1 + a + \dfrac{b}{t_0}) + \dfrac{b^2}{2t_0} + b(a+1) & \text{if } b \le t \le t_0 + b \\ -\dfrac{1}{2}t_0 + ab + \dfrac{t^2 + (t_0 + b)^2}{2(T_m - t_0)} - t\left(\dfrac{t_0 + b}{T_m - t_0} + a\right) & \text{if } t_0 + b < t \le T_m p \end{cases} \tag{12}$$

where $a = 0.5 - t_0/T_m$; $b = T_m(p-1)$; p is modulation period number($p = 1, 2, 3 \ldots K$).

Following the derivations given in App. A, time instants t_k can be derived as follows

$$t_k = \begin{cases} \dfrac{-b_1 + \sqrt{b_1^2 - 4a_1 c_1}}{2a_1} & T_m(p-1) \le t \le t_0 + T_m(p-1) \\ \dfrac{-b_2 + \sqrt{b_2^2 - 4a_2 c_2}}{2a_2} & t_0 + T_m(p-1) < t \le T_m p \end{cases}, \tag{13}$$

where $b_1 = f_{sw01} - \Delta f_{sw}(1 + a + b/t_0)$; $a_1 = \Delta f_{sw}/(2t_0)$; $c_1 = \frac{1}{4} - \frac{k}{2} + \Delta f_{sw}\left(\frac{T_m^2(p-1)^2}{2t_0} + T_m(p-1)(a+1)\right)$; $b_2 = f_{sw01} - \Delta f_{sw}\left(\frac{t_0 + T_m(p-1)}{T_m - t_0} + a\right)$; $a_2 = \frac{\Delta f_{sw}}{2(T_m - t_0)}$; $c_2 = \frac{1}{4} - \frac{k}{2} + \Delta f_{sw}\left(\frac{(t_0 + T_m(p-1))^2}{2(T_m - t_0)} + aT_m(p-1) - \frac{1}{2}t_0\right)$.

V_{LISN} spectra and EMI attenuation due to the use of MRMW can be calculated using Eq. (7), Eq. (8), Eq. (9), Eq. (10), Eq. (11), Eq. (12), Eq. (13). For all the calculations Matlab® software is used. As an example A_{EMI} as a function of both t_0/T_m and Δf_{sw} for $f_m = 1$ kHz is shown in Fig. 6. But in Fig. 7 A_{EMI} as a function of t_0/T_m for different values of f_m is also shown. Presented results clearly show that optimum t_0 for which EMI attenuation is maximum is function of both Δf_{sw} and f_m. That is why for a given set of Δf_{sw} and f_m values optimum t_0 value should be found. For example, if $f_m = 1$ kHz and $\Delta f_{sw} = 40$ kHz, then (according to Fig. 7) optimum value of t_0 is 0.23 and maximum A_{EMI} is 16.8 dB; if $f_m = 10$ kHz and $\Delta f_{sw} = 40$ kHz, then (according to Fig. 7) optimum value of t_0 is 0.34 and maximum A_{EMI} is 7.5 dB.

3.2.   MRMW Optimization Procedure

In order to get full benefits from the use of MRMW for EMI reduction procedure for optimum t_0 value calculations is to be proposed. The optimization procedure can be used not only for boost converter with different specifications but also for other power converter topologies. Since f_m and Δf_{sw} can affect not only EMI attenuation, but also other SPC parameters, such as peak-to-peak output voltage ripples, then recommendations for the choice of f_m and Δf_{sw} values

proposed in [8], [14] will be taken into account in the procedure.

The MRMW optimization procedure is as follows:

- choose f_m slightly higher than RBW (resolution bandwidth) of a spectrum or EMI analyzer [8],

- choose Δf_{sw} lower than Δf_{swcr} at which peak-to-peak output voltage ripples are equal to maximally allowable values [14],

- derive conducted EMI equivalent circuit model (with LISN),

- derive analytic expression for complex transfer function $\underline{K}_{EMI}(f)$ between equivalent noise source V_{ens} and V_{LISN},

- calculate values of B_1 and B_2 for V_{ens},

- change t_0 with a small step (e.g. 0.01), and using Eq. (7), Eq. (8), Eq. (9), Eq. (10), Eq. (11), calculate V_{LISN} spectrum and A_{EMI} for each value of t_0 in the range $0.1 - 0.9$,

- find global maximum for all A_{EMI} values calculated in the previous step and find optimum t_0 for which A_{EMI} is maximum,

- end.

For determining optimized t_0 value Matlab software algorithm has been implemented. Matlab code for the optimization procedure is shown in App. B.

In order to show usefulness of the optimization procedure proposed, let us consider one example. Let us assume that we have boost converter with the following specifications: $V_{in} = 4 \ldots 8$ V, $D_{max} = 50$ %,

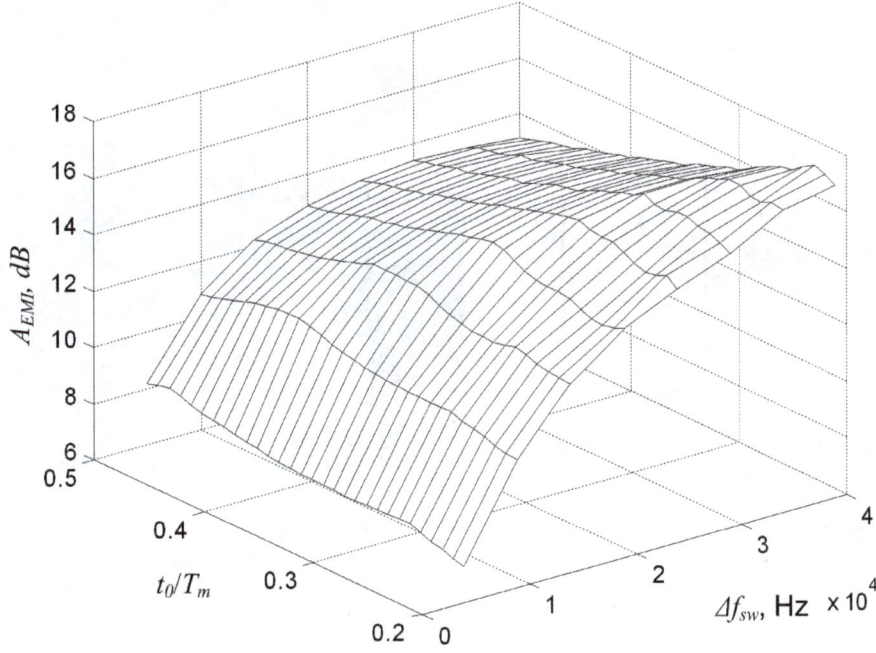

Fig. 6: Theoretically calculated A_{EMI} as a function of Δf_{sw} and t_0 Parameters: $f_m = 1$ kHz; $f_{sw0} = 80$ kHz.

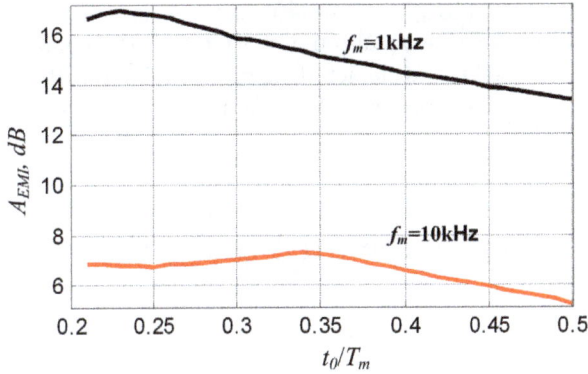

Fig. 7: Theoretically calculated A_{EMI} as a function of t_0/T_m for different f_m Parameters: $\Delta f_{sw} = 40$ kHz and $f_{sw0} = 80$ kHz.

$P_{out\,max} = 20$ W, $f_{sw0} = 80$ kHz, $L = 40$ μH, $C_{in} = 330$ μF, $C_{out} = 1000$ μF, output capacitor equivalent series resistance (ESR) $r_{Cout} = 0.033$ Ω, output load minimum resistance $R_{out} = 12$ Ω, maximally allowable peak-to-peak output voltage ripple value $V_{pp\,max} = 150$ mV. Let us assume that our task is to improve conducted EMI attenuation using optimized MRMW according to CISPR16 in Band A.

3.3. Solution

1) 1st Step: Choice of f_m

Since CISPR16 in Band A requires conducted EMI measurements to be done with RBW = 200 Hz and f_m should be higher than RBW, let us choose $f_m = 1$ kHz.

2) 2nd Step: Choice of Δf_{sw}

According to the procedure Δf_{sw} should be lower than Δf_{swcr} at which $V_{pp\,max}$ are equal to maximally allowable values. In order to calculate Δf_{swcr} expression for the $V_{pp\,max}$ calculation should be derived. The expression for the boost converter is as follows [15]

$$V_{pp\,max} = \left(\frac{V_{in\,max}}{(1-D_{max})^2 R_{out}} + \frac{V_{in\,max} D_{max}}{2 f_{sw} L} \right) r_{Cout} + \frac{V_{in\,max} D_{max}}{(1-D_{max}) R_{out} C_{out} f_{sw}}. \tag{14}$$

If f_{sw} is modulated then $f_{sw\,min} = f_{sw0} - \Delta f_{swcr}$ should be substituted in Eq. (14) instead of f_{sw}. It can be derived that

$$\Delta f_{swcr} = f_{sw0} - \frac{\frac{V_{in\,max} D_{max} r_{Cout}}{2L} + \frac{V_{in\,max} D_{max}}{(1-D_{max}) R_{out} C_{out}}}{V_{pp\,max} - \frac{V_{in\,max} r_{Cout}}{(1-D_{max})^2 R_{out}}}. \tag{15}$$

From Eq. (15) it can be obtained that $\Delta f_{swcr} = 40.4$ kHz. So Δf_{sw} is chosen to be 40 kHz.

3) 3rd Step

Boost converter conducted EMI equivalent circuit model (with LISN) is derived as shown in Fig. 2.

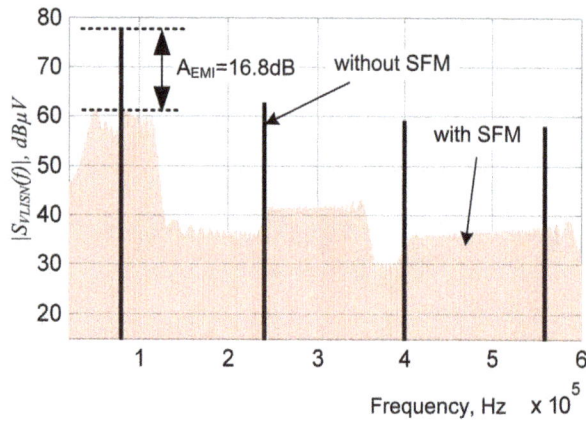

Fig. 8: V_{LISN} spectra before and after the use of SFM with optimized MRMW ($t_0 = 0.23T_m$). Other modulation and circuit parameters: the same as in Fig. 3. Note: V_{LISN} spectrum when non-optimized ramp modulation waveform is used is shown in Fig. 3.

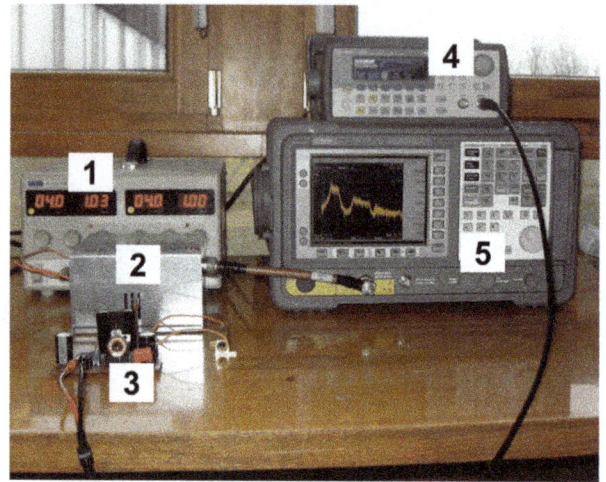

Fig. 10: Experimental setup picture. **1** DC power source TTI EL302D; **2** DC LISN (homemade); **3** boost converter under test; **4** arbitrary waveform generator Agilent 33220A; **5** spectrum analyzer Agilent E4402B.

4) 4th Step

Analytic expression for complex transfer function $\underline{K}_{EMI}(f)$ between equivalent noise source V_{ens} and V_{LISN} is derived according to Eq. (1).

5) 5th Step

$$B_1 = -V_{in}; \quad B_2 = V_{out} - V_{in}.$$

6) Final step

Using Matlab code shown in App. B optimum t_0 is calculated to be 0.23.

V_{LISN} spectrum after the optimization of MRMW is shown in Fig. 8. As it can be seen from Fig. 3, Fig. 8 and Fig. 13d the use of optimized MRMW gives

3.5 dB better A_{EMI} than the use of conventional ramp modulation waveform (when $t_0 = 0.5T_m$) and 1.7 dB better A_{EMI} than the use of MRMW proposed in [8] (when $t_0 = 0.35T_m$). Comparison of A_{EMI} versus Δf_{sw} for conventional ramp modulation waveform and optimized MRMW is depicted in Fig. 9. The results clearly show that optimized MRMW is more useful for higher Δf_{sw}.

4. Experimental Verification

4.1. Experimental Setup

For the experimental verifications a low power SFM boost SPC is used (Fig. 11). The converter operates in open-loop continuous conduction mode. Duty ratio D and input voltage of the converter can be varied. However in the experiments duty ratio of 50 % is chosen because from conducted EMI point of view it is the worst situation [3]. Input DC voltage of the

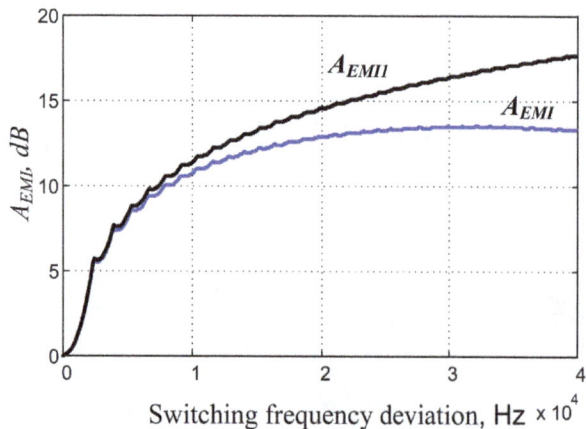

Fig. 9: Calculated A_{EMI} (when $t_0 = 0.5T_m$) and A_{EMI1} (when optimum t_0 is used) as a function of Δf_{sw} for $f_m = 1$ kHz and $f_{sw0} = 80$ kHz.

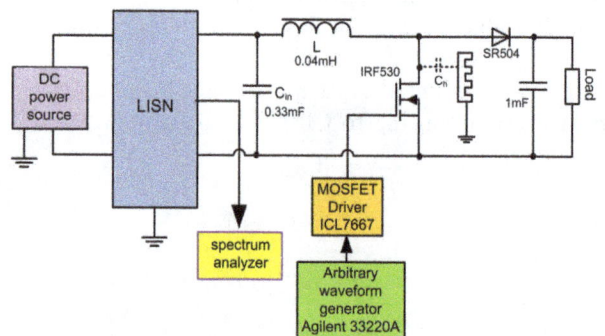

Fig. 11: Simplified schematic diagram of the experimental setup.

converter can be changed from 2 V to 8 V. Nominal switching frequency is 80 kHz. Nominal output power is 20 W. In the experiments power MOSFET is used with grounded external heatsink. Measured parasitic capacitance C_h between MOSFET drain and ground is of about 10 pF. Arbitrary waveform generator (AWG) Agilent 33220A controls the power MOSFET via MOSFET driver ICL7667. The control signal of the power MOSFET can be both unmodulated and switching frequency modulated. All the necessary SFM parameters (including t_0) can be set using the AWG which has built-in frequency modulator and waveform editor. MRMW with custom t_0 can be obtained using the waveform editor. Figure 12 depicts the photo of the AWG Agilent 33220A front panel and display.

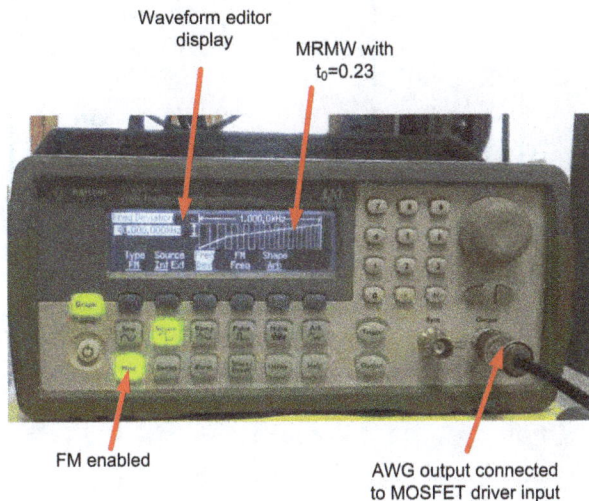

Fig. 12: AWG front panel and waveform editor display photo.

4.2. Experimental Results

Conducted EMI (V_{LISN} spectrum) is measured using Agilent E4402B spectrum analyzer with RBW=200 Hz and peak detector. Experiments have been done for different V_{in} and output power values. Comparison of the experimental and the theoretical V_{LISN} spectra when $V_{in} = 4$ V and $D = 50$ % is presented in Fig. 13. Experiments revealed that changing input voltage and output load does not have any noticeable influence on EMI attenuation and optimized t_0 value. Experimental results confirm theoretical predictions that the use of optimized MRMW can noticeably increase EMI attenuation: the use of optimized MRMW gives 3.8 dB better A_{EMI} than the use of conventional ramp modulation waveform (when $t_0 = 0.5T_m$) and 2 dB better A_{EMI} than the use of MRMW proposed in [8] (when $t_0 = 0.35T_m$). Experimental and theoretical results are in a good agreement: when Δf_{sw} was increased to 40 kHz, experimental EMI attenuation increased to 16.9 dB, but theoretical one to 16.8 dB.

5. Conclusion

Comprehensive theoretical analysis of SPC conducted EMI spectrum and attenuation presented in the paper shows that the use of multislope ramp modulation waveform can noticeably improve EMI reduction in switching-frequency-modulated switching power converters. However in order to get full benefits from the use of MRMW, optimum value of the parameter t_0 which controls the slopes of the waveform should be found. For this purpose the optimization procedure has been proposed and usefulness of the procedure has been verified experimentally using boost SPC. The procedure is based on the analytic expressions derived for conducted EMI spectrum and attenuation calculation. It has been shown that optimum value of parameter t_0 is the function of switching frequency deviation and modulation frequency.

Acknowledgment

Support for this work was provided by the Riga Technical University through the Scientific Research Project Competition for Young Researchers No. ZP-2014/9.

References

[1] MAINALI, K. and R. ORUGANTI. Conducted EMI Mitigation Techniques for Switch-Mode Power Converters: A Survey. *IEEE Transactions on Power Electronics*. 2010, vol. 25, iss. 9, pp. 2344–2356. ISSN 0885-8993. DOI: 10.1109/TPEL.2010.2047734.

[2] YAZDANI, M. R., H. FARZANEHFARD and J. FAIZ. Classification and Comparison of EMI Mitigation Techniques in Switching Power Converters - A review. *Journal of Power Electronics*. 2011, vol. 11, iss. 5, pp. 767–777. ISSN 1598-2092.

[3] GONZALEZ, D., J. BALCELLS, A. SANTOLARIA, J.-C LE BUNETEL, J. GAGO, D. MAGNON and S. BREHAUT. Conducted EMI Reduction in Power Converters by Means of Periodic Switching Frequency Modulation. *IEEE Transactions on Power Electronics*. 2007, vol. 22, iss. 6, pp. 2271–2281. ISSN 0885-8993. DOI: 10.1109/TPEL.2007.909257.

[4] HAMZA, D. and M. QIU. Digital Active EMI Control Technique for Switch Mode Power Converters. *IEEE Transactions on Electromagnetic Compatibility*. 2013, vol. 55, iss. 1, pp. 81–88. ISSN 0018-9375. DOI: 10.1109/TEMC.2012.2213590.

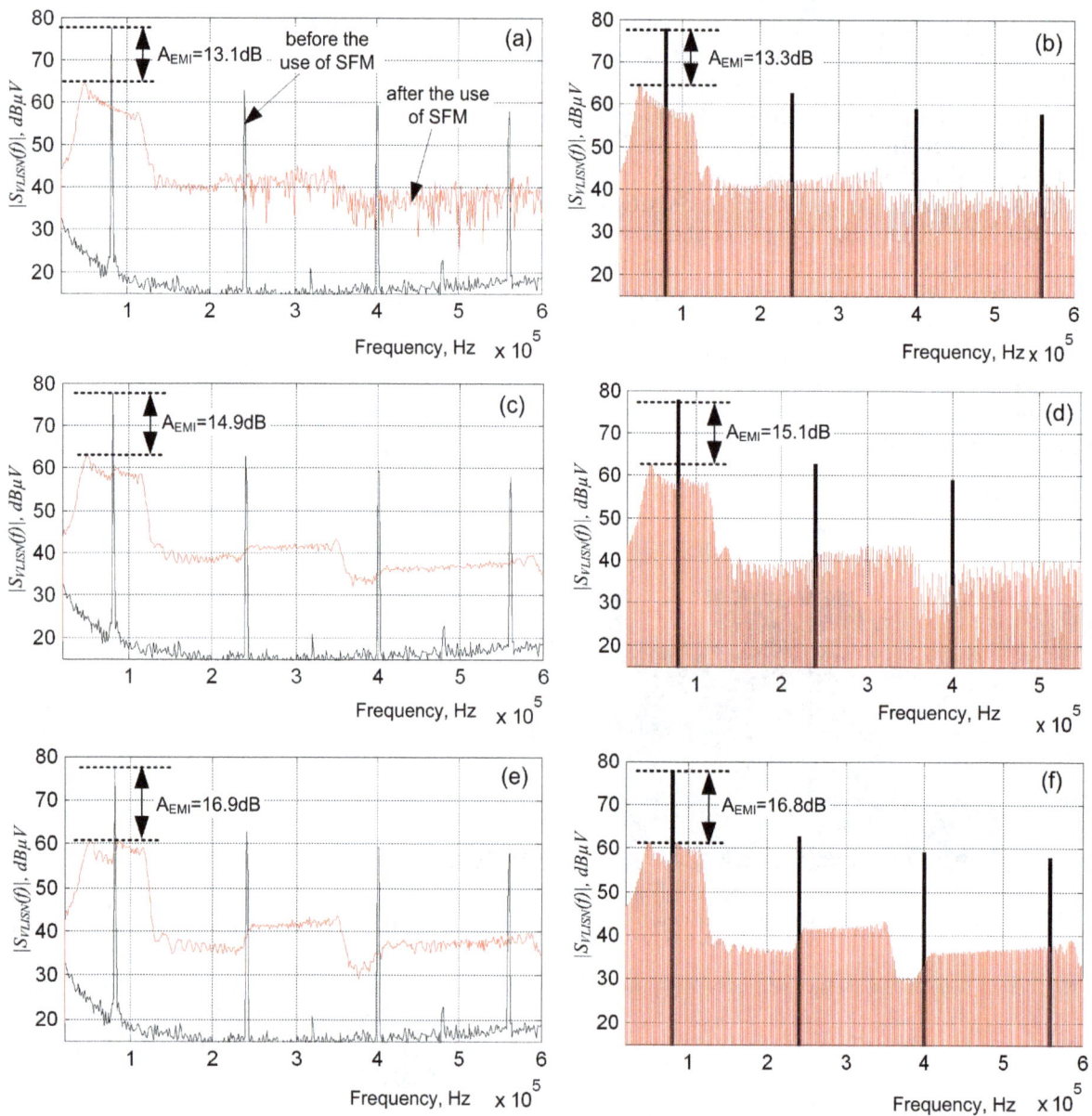

Fig. 13: Experimental (a), (c), (e) and theoretical (b), (d), (f) conducted EMI spectra in the frequency range $20 - 600$ kHz: (a), (b) conventional ramp ($t_0 = 0.5T_m$); (c), (d) modified ramp proposed in [8] ($t_0 = 0.35T_m$); (e), (f) optimized MRMW propsed in this paper (optimum $t_0 = 0.23T_m$). Spectrum analyzer parameters: RBW=200 Hz; peak detector. SPC and SFM parameters: $V_{in} = 4$ V; $D = 0.5$; $f_{sw0} = 80$ kHz; $f_m = 1$ kHz; $\Delta f_{sw} = 40$ kHz; $C_h = 10$ pF.

[5] CABUK, G. and S. KILINC. Reducing electromagnetic interferences in flyback AC-DC converters based on the frequency modulation technique. *Turkish Journal of Electrical Engineering & Computer Sciences*. 2012, vol. 20, iss. 1, pp. 71–86. ISSN 1300-0632.

[6] TSE, K., H. CHUNG, R. NG and R. HUI. An Evaluation of the Spectral Characteristics of Switching Converters with Chaotic-Frequency Modulation. *IEEE Transactions on Industrial Electronics*. 2003, vol. 50, iss. 1, pp. 171–181. ISSN 0278-0046. DOI: 10.1109/TIE.2002.807659.

[7] JANKOVSKIS, J., D. STEPINS, S. TJUKOVS

and D. PIKULINS. Examination of Different Spread Spectrum Techniques for EMI Suppression in dc/dc Converters. *Electronics and Electrical Engineering*. 2008, vol. 86, iss. 6, pp. 60–64. ISSN 1392-1215.

[8] JOHNSON, S. and R. ZANE. Custom spectral shaping for EMI reduction in high-frequency inverters and ballasts. *IEEE Transactions on Power Electronics*. 2005, vol. 20, iss. 6, pp. 1499–1505. ISSN 0885-8993. DOI: 10.1109/T-PEL.2005.857565.

[9] STEPINS, D. Conducted EMI of Switching Frequency Modulated Boost Converter. *Electrical,*

Control and Communication Engineering. 2013, vol. 3, iss. 1, pp. 12–18. ISSN 2255-9159. DOI: 10.2478/ecce-2013-0009.

[10] BALCELLS, J., D. GONZALES, J. GAGO, A. SANTOLARIA, J.-C. LE BUNETEL, D. MAGNON and S. BREHAUT. Frequency modulation techniques for EMI reduction in SMPS. In: *Proceedings of the 9th European Conference on Power Electronics and Applications.* Dresden: IEEE, 2005, pp. 1–8. ISBN 90-75815-09-3. DOI: 10.1109/EPE.2005.219198.

[11] MUSZNICKI, P., J.-L. SCHANEN, P. GRANJON and P. CHRZAN. Better understanding EMI generation of power converters. In: *Proceedings of IEEE Power Electronics Specialists Conference (PESC2005).* Recife: IEEE, 2005, pp. 1052–1056. ISBN 0-7803-9033-4. DOI: 10.1109/PESC.2005.1581758.

[12] HARDIN, K., J. FESSLER and D. BUSH. Spread Spectrum Clock Generation for the Reduction of Radiated Emissions. In: *Proceedings of IEEE International Symposium on Electromagnetic Compatibility.* Chicago: IEEE, 1994, pp. 227-231. ISBN 0-7803-1398-4. DOI: 10.1109/ISEMC.1994.385656.

[13] BARRAGAN, L., D. NAVARRO, J. ACERO, I. URRIZA and J. BURDIO. FPGA Implementation of a Switching Frequency Modulation Circuit for EMI Reduction in Resonant Inverters for Induction Heating Appliances. *IEEE Transactions on Industrial Electronics.* 2008, vol. 55, iss. 1, pp. 11–20. ISSN 0278-0046. DOI: 10.1109/TIE.2007.896129.

[14] STEPINS, D. Effect of frequency modulation on input current of switch-mode power converter. In: *Proceedings of 39th Annual Conference of the IEEE Industrial Electronics Society, IECON 2013.* Vienna: IEEE, 2013, pp. 683–688. ISBN 978-1-4799-0224-8. DOI: 10.1109/IECON.2013.6699217.

[15] HAUKE, B. Basic calculation of a boost converter's power stage. In: *Texas Instruments: Application Report* [online]. 2009. Available at: http://www.ti.com/lit/an/slva372c/slva372c.pdf.

About Authors

Deniss STEPINS received the B.Sc., M.Sc. (with honors) and Dr.Sc.ing degrees in electronics from Riga Technical University, Riga, Latvia, in 2004, 2006 and 2011 respectively. He is currently a senior research fellow and lecturer in the Institute of Radio Electronics, Riga Technical University. He has been involved in several research projects on examination of spread spectrum technique for switching power converters, improvement of power magnetic components and EMI filter optimization. His research interests include EMI reduction techniques applied to switching power converters, control of switch-mode power converters, planar magnetic components and three-phase EMI filters. He is currently an IEEE and an IEEE Industrial Electronics Society member.

Jin HUANG received the B.Sc. degree from Shanghai Jiaotong University, Shanghai, China, in 1992, and the M.Sc. and Ph.D. degrees from the Huazhong University of Science and Technology, Wuhan, China, in 2003 and 2009, respectively. In 2000, he joined Huazhong University of Science and Technology as a Lecturer. He is currently an Associate Professor in the School of Electrical and Electronic Engineering. His research interests include the control technique, EMC and reliability of power electronics devices.

Appendix A   Derivation of Eq. (13)

By substituting Eq. (10) and Eq. (12) into Eq. (11) the following trigonometric equations can be obtained

$$
\begin{cases}
\cos\left[2\pi f_{sw01}t + 2\pi\Delta f_{sw}\left(\dfrac{t^2}{2t_0} - t\left(1 + a + \dfrac{b}{t_0}\right) + \dfrac{b^2}{2t_0} + b(a+1)\right)\right] = 0 \\
\text{if } b \le t \le t_0 + b, \\[2ex]
\cos\left[2\pi f_{sw01}t + 2\pi\Delta f_{sw}\left(-\dfrac{1}{2}t_0 + ab + \dfrac{t^2 + (t_0+b)^2}{2(T_m - t_0)} - t\left(\dfrac{t_0+b}{T_m - t_0} + a\right)\right)\right] = 0 \\
\text{if } t_0 + b < t \le T_m p.
\end{cases}
\tag{A1}
$$

Assuming that $\arccos(0) = -\pi/2 + \pi k$, where $k = 1,2,3\ldots$, the following equation can be obtained from Eq. (A1).

$$
\begin{cases}
2\pi f_{sw01}t + 2\pi\Delta f_{sw}\left(\dfrac{t^2}{2t_0} - t\left(1 + a + \dfrac{b}{t_0}\right) + \dfrac{b^2}{2t_0} + b(a+1)\right) = -\dfrac{\pi}{2} + \pi k \\
\text{if } b \le t \le t_0 + b, \\[2ex]
2\pi f_{sw01}t + 2\pi\Delta f_{sw}\left(-\dfrac{1}{2}t_0 + ab + \dfrac{t^2 + (t_0+b)^2}{2(T_m - t_0)} - t\left(\dfrac{t_0+b}{T_m - t_0} + a\right)\right) = -\dfrac{\pi}{2} + \pi k \\
\text{if } t_0 + b < t \le T_m p.
\end{cases}
\tag{A2}
$$

From Eq. (A2) one can get the following quadratic equations

$$
\begin{cases}
\dfrac{\Delta f_{sw}}{2t_0}t^2 + \left(f_{sw01} - \Delta f_{sw}\left(1 + a + \dfrac{b}{t_0}\right)\right)t + \left(\dfrac{1}{4} - \dfrac{k}{2} + \Delta f_{sw}\left(\dfrac{b^2}{2t_0} + b(a+1)\right)\right) = 0 \\
\text{if } b \le t \le t_0 + b, \\[2ex]
\dfrac{\Delta f_{sw}}{2(T_m - t_0)}t^2 + \left(f_{sw01} - \Delta f_{sw}\left(\dfrac{t_0+b}{T_m - t_0} + a\right)\right)t + \\
+ \left(\dfrac{1}{4} - \dfrac{k}{2} + \Delta f_{sw}\left(\dfrac{(t_0+b)^2}{2(T_m - t_0)} + ab - \dfrac{1}{2}t_0\right)\right) = 0 \\
\text{if } t_0 + b < t \le T_m p.
\end{cases}
\tag{A3}
$$

Finally Eq. (13) can be derived by solving Eq. (A3).

Appendix B   Matlab code for optimum t_0 calculation

```
delta=40e3; % enter switching frequency deviation
fsw=80e3; % enter central switching frequency
fm=1e3; % enter modulation frequency
Tm=1/fm; K=5; Uin=4; D=0.5; % enter input voltage and duty ratio
L=40e-6; % enter power inductor inductance
Ch=10e-12; % enter parasitic capacitance Ch between MOSFET drain and ground
Cin=330e-6; Lp=10e-9; esr=0.06; % enter parameters of input capacitor
L3=50e-6; R3=5; C5=0.22e-6; % enter LISN parameters
Uout=Uin/(1-D); % calculation of output voltage


%-----------------------------------------------------------------------------
for p=1:40 % calculation of time instants tk according to Eq. (13)
t0=0.1*Tm+0.01*Tm*p;
to(p)=t0;
f=1; a1=delta/2/t0; a2=delta*0.5/(Tm-t0); a=0.5-t0/Tm; t1=0; fsw1=fsw+a*delta;
M=K*2*fsw1/fm;
for m=1:M
```

```
if t1<t0+Tm*(f-1)
  b1=fsw1-delta*(1+a+Tm*(f-1)/t0);
  c1=0.25-m/2+((Tm.^2*(f-1).^2)/2/t0+Tm*(f-1)*(a+1))*delta;
  tk(m)=(-b1+sqrt(b1.^2-4*a1*c1))./(2*a1);
  t1=(-b1+sqrt(b1.^2-4*a1*c1))./(2*a1);
if t1>t0+Tm*(f-1)
  b2=fsw1-delta*((t0+Tm*(f-1))/(Tm-t0)+a);
  c2=0.25-m/2+delta*(-0.5*t0+a*Tm*(f-1)+0.5./(Tm-t0).*(t0+Tm*(f-1)).^2);
  tk(m)=(-b2+sqrt(b2.^2-4*a2*c2))./(2*a2);
  t1=(-b2+sqrt(b2.^2-4*a2*c2))./(2*a2);
end
else
  b2=fsw1-delta*((t0+Tm*(f-1))/(Tm-t0)+a);
  c2=0.25-m/2+delta*(-0.5*t0+a*Tm*(f-1)+0.5./(Tm-t0).*(t0+Tm*(f-1)).^2);
  tk(m)=(-b2+sqrt(b2.^2-4*a2*c2))./(2*a2);
  t1=(-b2+sqrt(b2.^2-4*a2*c2))./(2*a2);
if t1>Tm*f
  f=f+1;
  b1=fsw1-delta*(1+a+Tm*(f-1)/t0);
  c1=0.25-m/2+((Tm.^2*(f-1).^2)/2/t0+Tm*(f-1)*(a+1))*delta;
  tk(m)=(-b1+sqrt(b1.^2-4*a1*c1))./(2*a1);
  t1=(-b1+sqrt(b1.^2-4*a1*c1))./(2*a1);
end
end
end
%------------------------------------------------------------------------
%------------------------------------------------------------------------
for  n=200:1000 %calculation of complex Fourier series coefficients dsn according
to Eq. (2)
  for i=1:M/2
    ann(i)=Uout*(exp(-j*n*2*pi*fm/K*tk(2*i))-exp(-j*n*2*pi*fm/K*(tk(2*i-1))))./
    (j*2*pi*n);
  end
an=sum(ann);

%------------------------------------------------------------------------
w=2*pi*fm*n/K;
Z=(j*w*L3+R3).*(50+1./(j*w*C5))./2./((j*w*L3+R3+50+1./(j*w*C5)));    %enter LISN
impedance Ze
KEMI=-(1./(j*Cin*w)+esr+j*Lp*w)./(j*L*w)./2+Z./(Z+1./(j*Ch*w));  %enter KEMI from
EMI model
VLISN(n)=abs(2*an.*KEMI); %calculation of VLISN spectrum
end


Z1=(j*2*pi*fsw*L3+R3).*(50+1./(j*2*pi*fsw*C5))./2./((j*2*pi*fsw*L3+R3+50+1./
(j*2*pi*fsw*C5)));
KEMI1=-(1./(j*Cin*2*pi*fsw)+esr+j*Lp*2*pi*fsw)./(j*L*2*pi*fsw)./2+Z1./(Z1+1./
(j*Ch*2*pi*fsw));
cc1=4*Uin/pi;
VLISN1=abs(cc1.*KEMI1); %calculation of VLISN 1st harmonic amplitude without SFM


AEMI(p)=max(abs(20*log10(VLISN1./(max(VLISN))))); %calculation of AEMI vs t0
end


[x,y]=max(AEMI); %determination of max AEMI (finding global maximum)
optimumt0=to(y)./Tm %optimum t0 calculation
```

13

Novel 3D Modelling System Capturing Objects with Sub-Millimetre Resolution

Adam CHROMY[1,2,3]*, Ludek ZALUD*[1,2,3]

[1]CEITEC - Central European Institute of Technology, Brno University of Technology,
Technicka 3058/10, 616 00 Brno, Czech Republic
[2]International Clinical Research Center, St. Anne's University Hospital Brno,
Pekarska 53, 656 91 Brno, Czech Republic
[3]Department of Control and Instrumentation, Faculty of Electrical Engineering and Communication,
Brno University of Technology, Technicka 3082/12, 616 00 Brno, Czech Republic

adam.chromy@ceitec.vutbr.cz, zalud@feec.vutbr.cz

Abstract. *This paper presents novel 3D modelling system providing three-dimensional shaded models of objects, with representing brightness and smoothness of it's surface. Proposed device is based on laser scanner and robotic manipulator, which empowers high flexibility since it is programmable in 6 degrees of freedom. Exchangeable laser scanners empower capability of building models of both tiny detailed structures and large object. Device is primarily intended to use at health care domain, where it brings lot of expenses savings compared to presently used scanning systems, but it could be used in many other applications. Scanning system constitution, operation and parameters are discussed.*

Keywords

3D modelling, 3D scanning, laser scanner, robotic manipulator.

1. Introduction

Three-dimensional scanning systems are very frequently requested devices today, its market is rapidly growing and its development also move forward very fast. Hand in hand with these new technologies, also their new applications appear. 3D computer models are more and more used in situations, where state of any object must be preserved in permanent, time-invariant state. In this case, colour-covered 3D models seems to be the best modality [20]. Also the domain of object cloning is rapidly growing presently. There are lots of 3D printers available on the market now and each of them requires tool for building 3D model to be printed [14]. Finally, computer 3D models are,

due to its plasticity, becoming more and more used for visualization of objects, which are unreachable (contaminated, dangerous areas) [18], [24] or environments, which are invisible without invasive surgery (human inner structures) [12]. Another objects are visible, but its important details are too tiny (human outer structures) and it is necessary to enlarge them plastically [9]. All these applications requires the same device: 3D scanner.

In present state, 3D scanners are based on several technologies. Contact 3D scanners uses probe to follow surface of scanned object [10]. The most of available 3D scanners are contact-less and uses laser beam, which serves as a base for 3D model computation. Some scanners are based on distance measurements (time-of-flight, triangulation, modulated light), some detects deformation of projected pattern (structured light) [10]. All these scanners has common feature – they have to know its position. The most approaches uses linear gantry systems (precise 3D scanners) or ICP computation (hand-held scanners) as a source of scanners position [16].

This paper presents 3D modelling system based on novel constitution, which uses combination of 6 DOF industrial robotic manipulator and laser scanner. This solution provide high flexibility as ICP based scanning systems [16] together with high precision and reliability of gantry systems [13].

Result of this work is high accurate and very flexible 3D modelling system, useful for many different applications in medical domain such as monitoring of tissue recovery process, body measurements for rehabilitation purposes or ergonomic splints design; but also at other domains from archiving of historical materials in museums, through design, models for computer games and industrial inspection to 3D object cloning.

As mentioned above, proposed device is based on combination of robotic manipulator and laser scanner. High accuracy is reached by using precise manipulator with accurate laser scanner and high flexibility is caused by programmable scanning trajectory in six degrees of freedom and by replaceability of laser scanner, which provides possibility of scanning both tiny and large structures.

The main purpose, what this device has been developed for, is capturing models of parts of human body. Presently, building such models for these purposes is done by very uneconomical way: there is a detailed model created by Magnetic Resonance Imaging, which contains a lot of information about inner structures. But in many cases, these information are useless for our purposes – just information about outer surface is required. Magnetic Resonance Imaging is very expensive device as well as its operation, creating these models takes a long time and by building our models, patients who need MRI images for more important reasons are blocked.

Scanning system described in this paper is much cheaper as well as its operation is cheaper and creating three-dimensional models is faster.

2. 3D Scanner Architecture

Proposed 3D modelling system can be divided into two subsystems: the first is Robotic 3D Scanner, what is a system, that receives desired scanning trajectory as its input and outputs three-dimensional point cloud. It is composed from commercially available manipulator and scanner, equipped with software fully developed by us in .NET framework. Second is 3D modelling software, which specifies scanning trajectory and processes data from 3D scanner into the form of final 3D model. This software is also made in .NET and fully developed by ourselves.

This chapter describes architecture of the first mentioned subsystem, second subsystem is decribed in chapter 3.

Robotic 3D scanner consists of industrial robotic manipulator and planar laser scanner, which is mounted directly on the manipulator's end-point (Fig. 1) and can be easily unmounted by unscrewing fixing screws and unplugging Ethernet and power cable. It allows easy change of used laser scanner in dependence on object size and complexity. Both devices are connected by Ethernet interface to computer, which is equipped with software for controlling both devices.

Controlling software moves with laser scanner along predefined trajectory, and measures distance profiles from each point at scanning trajectory. Since position

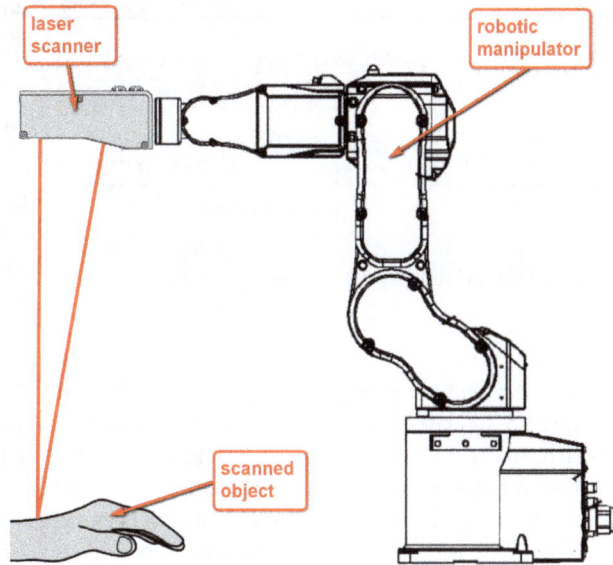

Fig. 1: Overview of Robotic 3D Scanner physical constitution.

and looking direction of scanner is known at each point of trajectory, measured data can be transformed to default coordinate system and produced as point cloud (Fig. 2). This approach is more explained in following text.

If scanned object is a large structure, the measuring range must be as wide as possible, so laser scanner based on TOF principle is used. On the other hand, if scanned object is a tiny structure, the accuracy is decisive factor, so we use scanner based on triangulation principle [21].

2.1. Laser Scanner Measurements

As mentioned above, this system uses different laser scanners according to the measured object. Each of them have different controlling mechanism, but we want to have unified interface, which doesn't matter on the scanner currently connected. As a result of this, C# driver with unified interface has been created for each used scanner. This approach provides independence on particular device.

At each point from scanning trajectory, distance profile is captured. Every point from this profile contains information about distance from laser scanner, rotation of measuring mechanism, reflected intensity of laser beam, and in case of some triangulation scanners, also width of reflected beam.

Distance and rotation is used for computing position of measured point in default coordinate system, what is described in section 2.3. Reflected intensity and reflection width determines object's material parameters and are used for surface colouring at fi-

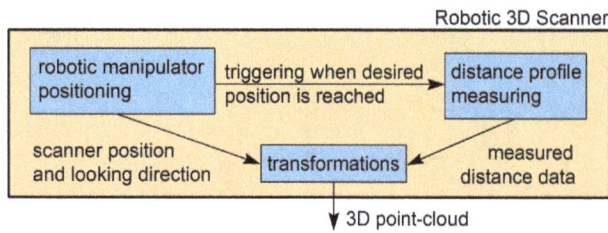

Fig. 2: Basic idea of Robotic 3D Scanner operation.

nal model. As shown on Fig. 3, intensity of reflected light refers to lightness of surface colour, and reflection width refers to smoothness or sleekness of surface. Smooth surfaces reflects narrow beam, so they are darker coloured than background paper, which is more coarse. The dependence of beam width on smoothness clearly illustrates engraved letter on metallic surface - they causes changes in reflected width, but the intensity is the same.

It has to pointed out, that reflected intensity refers to amplitude of particular frequency of laser emitter, not to lightness as is visible by eye. It is clear on two plastic film on left: first one is black, what means that this surface absorbs all frequencies including red, what is the colour of laser beam, so, the reflected intensity is low.

Second one is blue, what means, that blue frequency is dominant in reflected light, but also other frequencies are present since the colour is not exactly monochro-

matic. Presence of red frequency causes higher value of reflected intensity than at black material.

2.2. Robotic Manipulator Control

Robot ordinarily executes program, which has been written in manufacturer-designed scripting language SPEL+ and uploaded into the Control Unit before starting robot's operation. It doesn't allow the flexible reaction on external events received from surrounding systems, what is important for our purposes. However, SPEL+ includes instructions for receiving and transmitting data over Ethernet, so this could be solved by solution shown on Fig. 4.

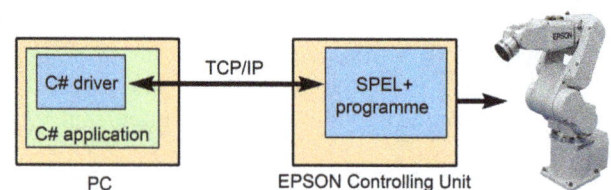

Fig. 4: Scheme of manipulator's driver solution.

There is a SPEL+ programme which receives commands over TCP/IP and performs desired actions. On the other side of Ethernet link, there is a C# driver which receives requests over its interface and transmits commands to the SPEL+ program. The real-time manipulator control (by C# application) is then empowered [8].

Each command sent from C# driver is acknowledged by robot after it is finished. There are also commands, that return actual position of robotic arm, so there is full control of robot's operation. Driver implement's most of commands available in SPEL+ language, including commands for continuous path navigation through points, which are defined in 6 coordinates - three Cartesian axis defining position of robotic end-point and three rotation coordinates describing its RPY rotation [3]. There is not necessary to compute inverse kinematics, since it is computed inside the Control Unit.

2.3. 3D Point-Cloud Computing

Kernel part of Robotic 3D Scanner represents software, which block scheme is illustrated on Fig. 5 as a block called 3D Scanner.

It receives desired scanning trajectory in form of list of points, where each of them is described by 6 coordinates. Output of this block is a 3D point cloud, what is a set of points described by three Cartesian coordinates and values representing additional informations like a

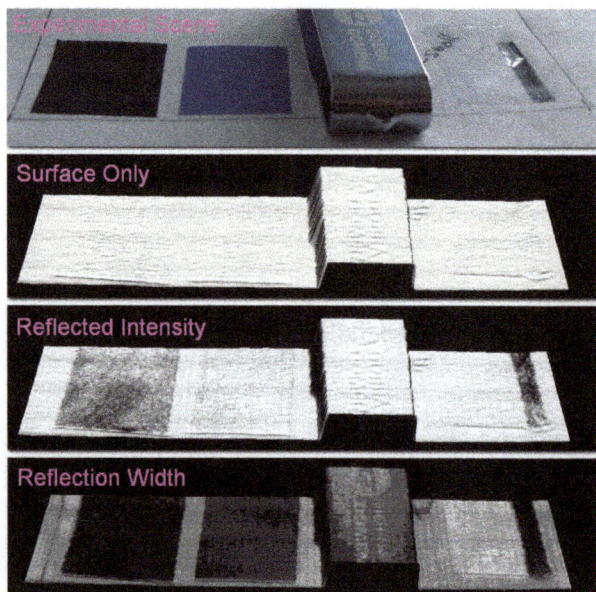

Fig. 3: Visualization of surface data, reflected intensity and reflection width of different materials - from left: two smooth plastic films, metallic polished surface with engraved letters and transparent tape on paper.

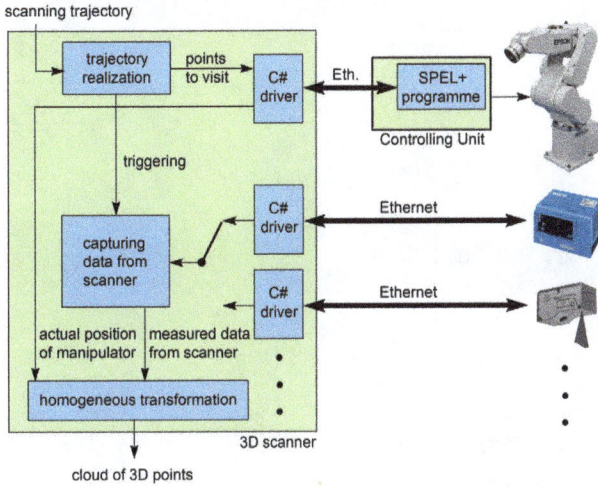

Fig. 5: Block scheme of scanning system with accent on kernel software.

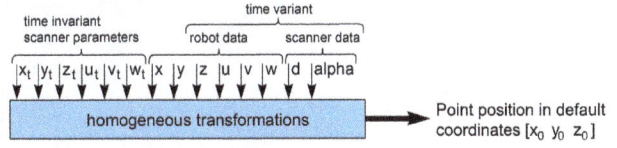

Fig. 6: Inference mechanism of homogeneous transformations block.

reflection intensity and width. This block encapsulates several others blocks:

Trajectory Realization block receives list of points, which should be visited by scanner and from which measuring should be performed. Each point is defined by x, y, z coordinates and by three angles of rotation according to Roll-Pitch-Yaw convention [3]. This block consequently instructs manipulator's driver with point to be moved to, and waits for its response. When desired point is reached, driver passes confirmation to Trajectory Realization block, which triggers capturing data from scanner. In the same time, request for real precise position of robotic manipulator, and send these data to Homogeneous Transformation block.

Capturing Data from Scanner block acquires data from laser scanner via scanner's driver. Scanner measures distances along the linear profile and data are acquired in form of distance and angle of actual rotation of sweeping mechanism. It is possible to switch between several laser scanners and its drivers, depending on character of actually scanned object. Measured values are also passed to Homogeneous Transformation block.

Homogeneous Transformation block provides mathematical transformations of acquired data into the form of point cloud. It computes position of measured point in default uniform coordinate system (x_0, y_0, z_0) using data from manipulator encoders and data from laser scanner (Fig. 6).

As an input of this block, there are constants values x_t, y_t, z_t, u_t, v_t and w_t, which describes position of laser scanner relative to robotic manipulator's end-point. These values are time invariant. Other Input values x, y, z, u, v and w represents position of manipulator's end-point at the time of profile capturing using laser scanner, and they are different for each profile.

Values d and α represents measured data from laser scanner.

Let's divide entire system into particular systems, where each of them has its own coordinate system (Fig. 7). System 0 is default coordinate system in which measurement results should be presented. System M is own coordinate system of manipulator. Its origin is placed in the centroid of bottom mounting plate. System E is system with origin at the manipulator's end-point. System S is coordinate system of laser scanner and finally, system L is system of measuring laser beam emitter. Entire homogeneous transformation (H_{0L}) from measured values to point in default coordinate system could be defined as sequence of essential transformations between neighbour coordinate systems:

$$H_{0L} = H_{0M}H_{ME}H_{ES}H_{SL}. \tag{1}$$

Then homogeneous coordinates of measured point in coordinate system 0 (P_0) are defined as:

$$P_0 = H_{0L}P_L, \tag{2}$$

where P_L represents measured point in coordinate system L and is defined in Eq. (3).

In Eq. (3), d is distance of point from laser emitter. This value is returned by laser scanner.

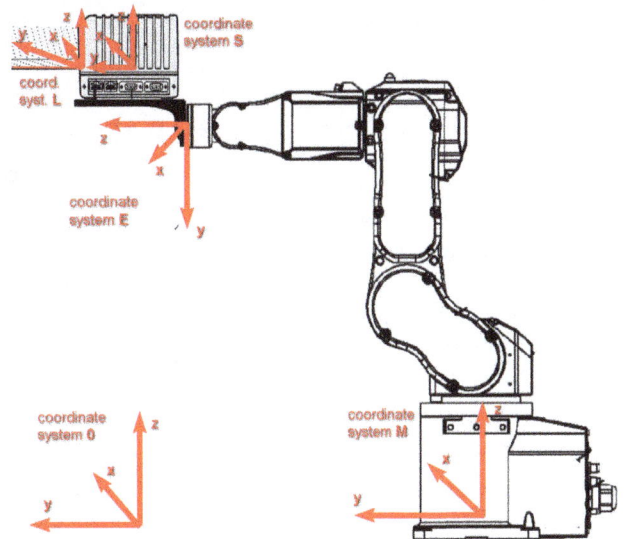

Fig. 7: Overview of coordinate systems used in homogeneous transformation block.

$$P_L = \begin{bmatrix} d \\ 0 \\ 0 \\ 1 \end{bmatrix}. \tag{3}$$

$$H_{SL} = \begin{bmatrix} \cos(180 - \alpha) & -\sin(180 - \alpha) & 0 & 0 \\ \sin(180 - \alpha) & \cos(180 - \alpha) & 0 & 0 \\ 0 & 0 & 1 & 0 \\ 0 & 0 & 0 & 1 \end{bmatrix}. \tag{4}$$

$$H_{ES} = \begin{bmatrix} c_u c_v & c_u s_v s_w - c_w s_u & s_u s_w + c_u c_w s_v & x_t \\ c_v s_u & c_u c_w + s_u s_v s_w & c_w s_u s_v - c_u s_w & y_t \\ -s_v & c_v s_w & c_v c_w & z_t \\ 0 & 0 & 0 & 1 \end{bmatrix}. \tag{5}$$

$$H_{ME} = \begin{bmatrix} c_u c_v & c_u s_v s_w - c_w s_u & s_u s_w + c_u c_w s_v & x \\ c_v s_u & c_u c_w + s_u s_v s_w & c_w s_u s_v - c_u s_w & y \\ -s_v & c_v s_w & c_v c_w & z \\ 0 & 0 & 0 & 1 \end{bmatrix}. \tag{6}$$

Let's derive particular homogeneous transformations which entire transformation H_{0L} (Eq. (4)).

There is a sweeping mechanism, which rotates with this laser source in xy plane of scanner's coordinate system S. Coordinate systems S of each used scanner are shown on Fig. 8.

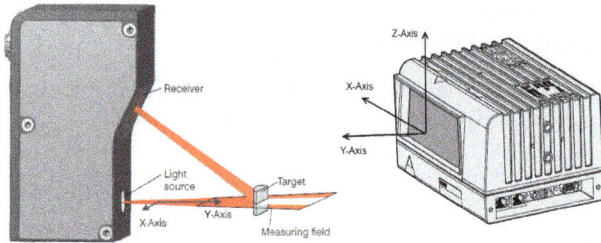

Fig. 8: Example of coordinate systems S of both scanners used at our particular application of proposed 3D scanning system [2], [1].

Because system L is system S rotated along z axis by angle α, homogeneous transformation between coordinate systems L and S is defined by matrix H_{SL} (Eq. (4)), where α is instant deviation from negative x axis of coordinate system S and is also returned by laser scanner.

This laser scanner is mounted on the manipulator's end-point. Centroid of manipulator's end-point defines origin of coordinate system E. Homogeneous transformation from system E to S is combination of translation in all axes and rotation described by RPY model [22] (Eq. (5)), where $c_u = \cos(u_t)$, $s_u = \sin(u_t)$, $c_v = \cos(v_t)$, ... and x_t, y_t and z_t defines translation

of system S in system E along appropriate axis, u_t is roll, v_t is pitch and w_t is yaw of system S in coordinate system E. Parameters x_t, y_t, z_t, u_t, v_t and w_t describe mounting of laser scanner and could be acquired from documentation of each laser scanner and its mounting holder.

This homogeneous matrix defines dimensions and constitution of scanner. Be aware, that in our case, only these parameters are different for different scanners. Other parameters (and consequently other matrices H_{ab}) are same for every scanner.

Actual position of manipulator's endpoint (eg. origin of system E) is placed inside own manipulator's coordinate system M. Homogeneous transformation from system M to E is once again combination of translation in all axes (x, y and z) and rotation described by RPY model (u is roll, v is pitch and w is yaw) [22] (Eq. (6)), where $c_u = \cos(u)$, $s_u = \sin(u)$, $c_v = \cos(v)$, ... x, y and z is actual position of manipulator's endpoint in own manipulator's coordinate system M, u is its roll, v is pitch and w is yaw. These parameters are time-variant and could be acquired from manipulators driver.

The coordinate system of manipulator M is placed inside a default coordinate system 0, which we want to transform measured points into. The homogeneous transformation from system 0 to M is H_{0M}. In our case, system 0 is identical with system M. If it is not, another matrix similar to H_{ME} could be used.

2.4. Scanning System Accuracy

Term accuracy can be defined as maximal distance between computed (measured) position of point relative to true position of point at 99.7 % of measurements ($\pm 3\sigma$)[4]. Overall accuracy of proposed 3D scanning system Δ_P it then defined as composition of robotic manipulator positioning uncertainty Δ_M, laser scanner measuring uncertainty Δ_S and uncertainty Δ_H caused by precision of holder connecting scanner and manipulator's end-point:

$$\Delta_P = \Delta_M + \Delta_S + \Delta_H. \qquad (7)$$

Enumeration of Δ_H is not possible, since its value depends on present distance of laser scanner from object. But be aware of fact, that precision of currently mounted holder is constant, time-invariant value, so it can be significantly minimizable by calibration, which is described in the following section. Considering calibrated device, overall accuracy describes equation:

$$\Delta_P = \Delta_M + \Delta_S + \Delta_C, \qquad (8)$$

where both Δ_M and Δ_S are values specified in documentation of each device. Last parameter Δ_C is a residual uncertainty in case, when calibration has been performed. By calibration process (which is discussed in following section) the value of Δ_H is significantly decreased, so $\Delta_C \ll \Delta_H$. In the worst case, value of Δ_C could be defined as in Eq. (10), so the overall accuracy of device according to [4] is then given by equation:

$$\Delta_{Pmax} = 3(\Delta_M + \Delta_S). \qquad (9)$$

Note, that proposed uncertainty estimation defines the worst case (3σ) and it is experimentally verified, that in real operation, overall accuracy is usually much more better.

In our research, we use robotic manipulator Epson C3 with accuracy of end-point placing 0.02 mm [3], laser scanner MicroEpsilon ScanCONTROL 2750-100 with accuracy 0.04 mm [2] and laser scanner SICK LMS with accuracy 4 mm [1].

For most of capturing sessions, the high resolution is the important factor, so we use MicroEpsilon scanner and according to Eq. 9, overall accuracy in this case is 0.18 mm. But this laser scanner disposes with measuring range only 10 cm long [2], so we can capture just in proximity of manipulator's operating area. In case of capturing larger structures (e.g. as buildings), we use laser scanner SICK LMS400. In this configuration, overall accuracy is lower (12.06 mm), but measuring range is up to 3 m from robotic end-point and these structures are usually so big, that this resolution is sufficient.

2.5. Calibration

The aim of calibration procedure is to compensate deviations of real dimensions of mounting holder compare to dimensions stated in its documentation. If these deviations are compensated, then equation for overall accuracy Eq. (8) is valid. In fact, it is looking for 6 calibration constants determining translation and rotation of scanner in 6-DOF caused by imprecision of mounting holder.

Proposed method is based on principle, that if some scene is firstly captured from one direction, and then the same scene is captured from another direction, deviations in mounting holder dimensions causes shifting and deforming of scene, but in both cases differently. From differences among these two images of the same scene, we can compute desired calibration constants. This approach would work in case of any scene, but for the simplification of image processing, we use scene of cuboid laying on the flat ground [7].

Significant advantage of this method is, that it is based on general object, without any requirements on its precision. If we use common calibration method based on reference object with known dimensions, it will be very difficult to manufacture reference object with enough precision.

Detailed description of computing of each calibration parameter can be found at [7], here it is not presented in order to be concise.

Residual Error after Calibration

Calibration procedure search for identical points in two scans of the same scene and evaluates such transformation, which merges these points together. Lets imagine the worst case of this – there is only one point measured from 2 places. Each representation of point is measured with maximal deviation from true value $\Delta = \Delta_M + \Delta_S$, so if we merge there points together, we can cause residual error Δ_C, which maximal value is:

$$\Delta_C = 2(\Delta_M + \Delta_S). \qquad (10)$$

Each uncertainty of Δ_C is composed from uncertainty type A and type B [19]. When number of measurements increases, the uncertainty type A is being minimized and measured value converges in direction to true value. Because calibration procedure uses several thousands of points, also residual error Δ_C after calibration will be smaller than at Eq. (10).

As a result of this, the overall accuracy will be usually better than maximal guaranteed error defined in Eq. (9), what is a limit value of error.

Evaluating Benefits of Calibration

To be able to evaluate benefits of calibration, the criterion function has been introduced. For each point from one scan, distance to every point from second scan is computed and square of distance to the nearest point is added to the sum:

$$f_{err}(I_1, I_2) = \sum_{i=1}^{Count(I_1)} \min(\| I_1(i)I_2(j) \|)^2 \quad (11)$$

$$j \in (1; Count(I_2)),$$

where I_1 and I_2 are scans which correlation we are looking for and $\| xy \|$ is a distance between points x and y. The task of calibration procedure is to compute such combination of calibration constants, which provides the smallest value of f_{err}.

Beside the criterion function values, also average-absolute error Δ_{abs} and root-mean-square error Δ_{abs} were computed in order to provide illustrative demonstration of calibration influence using equations [17]:

$$\Delta_{abs} = \frac{1}{N} \sum_{i=1}^{Count(I_1)} \min(\| I_1(i)I_2(j) \|) \quad (12)$$

$$j \in (1; Count(I_2)),$$

$$\Delta_{RMS} = \sqrt{\frac{1}{N-1} \sum_{i=1}^{Count(I_1)} \min(\| I_1(i)I_2(j) \|)^2} \quad (13)$$

$$j \in (1; Count(I_2)).$$

Influence of calibration is demonstrated in Fig. 9 and in Tab. 1.

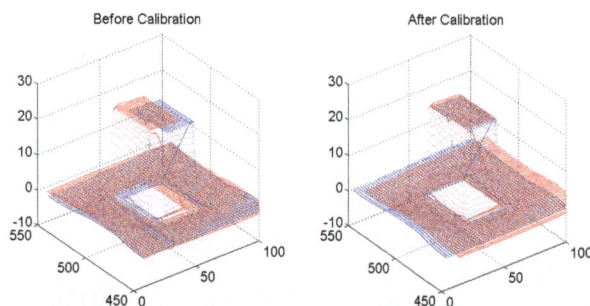

Fig. 9: Influence of calibration. On the left, point-cloud measured without previous calibration is presented. On the right, there is a point-cloud measured by calibrated device. On both figured, two scans of one scene captured from different positions are presented. One scan is displayed in blue, second in red.

Tab. 1: Illustrating influence of calibration. Values of criterion function demonstrates benefits of proposed calibration procedure.

	Before Calibration	After Calibration
f_{err} [−]	60 215.61	76.41
Δ_{abs} [mm]	2.907	0.085
Δ_{RMS} [mm]	4.044	0.144

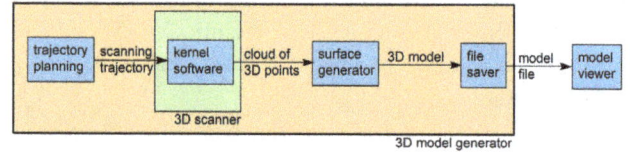

Fig. 10: Block scheme of entire scanning system focused on higher software layer enclosing the kernel 3D scanner.

3.   3D Modelling By Robotic Scanner

Previous chapter deals with kernel scanning system, which produces cloud of points in three dimensional space. But it is not the entire system, since the visualization of objects in form of point cloud is not illustrative enough; there are also other modules as shown on Fig. 10, which processes generated point cloud to form of three-dimensional shaded model, covered with texture representing reflected intensity or reflection width data (examples in section 3.5.). Beside this, user friendly environment for scanning trajectory defining has been provided, introducing simple scripting language for trajectory definition.

3.1.   Trajectory Planning

Trajectory planning module generates scanning trajectory in form of list of points to be visited. Each point

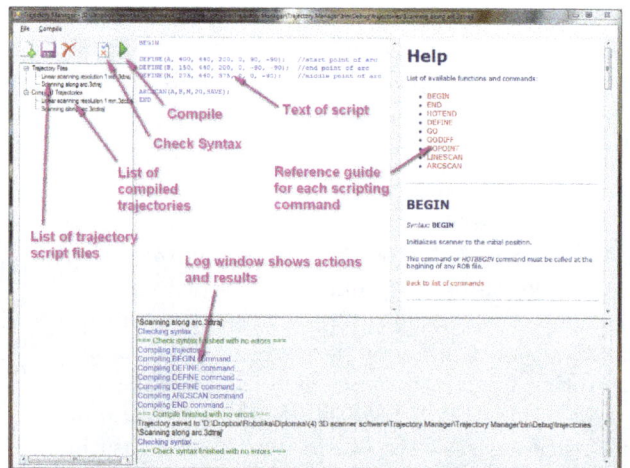

Fig. 11: Screenshot of Trajectory Manager, the tool for scanning trajectory definition and management.

is defined by three Cartesian coordinates and three rotation angles, which uniquely determines position of manipulator's end-point with mounted scanner and direction, where is scanner looking to. At each point robotic manipulator stops, triggers capturing of single profile by laser scanner and after that, continues to the next point.

This approach brings high flexibility, because scanning trajectory could be adjusted according to the type of scanned objects, and because of 6 DOF movements of manipulator, is practically limited only by its dimensions. Comparing to common three-dimensional scanners, which usually uses sliding carriages similar to carriage in common optical 2D scanner, it is much more flexible.

Possibility of defining directly each point of trajectory provides flexibility, but low comfort for users defining trajectory. To make work with device more comfortable, there is also environment, which allows defining trajectory using scripting approach (Fig. 11). There are several commands for essential geometrical shapes, which trajectory could be assembled from, such as moving along line between two points and capturing defined number of profiles, moving along the arc defined by three points, etc. These commands are translated to the list of points by Trajectory Planning processor and then sent to the Trajectory Realization block.

To see programmed trajectory before it is sent to manipulator, in order to avoid possible malfunction in case of trajectory wrong defined, there is a visualization tool, showing position of manipulator's end-point and direction, where scanner looks at (Fig. 12).

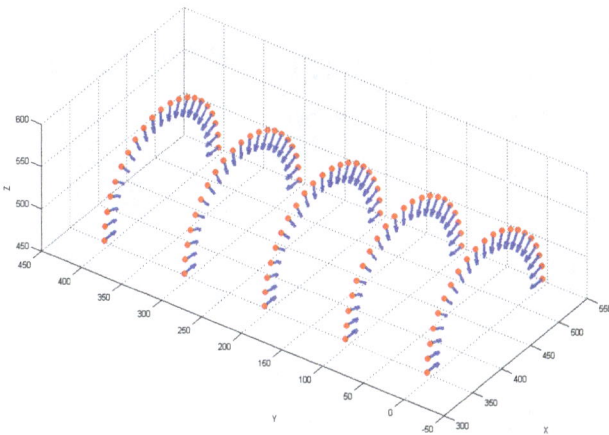

Fig. 12: Screenshot of trajectory visualization tool. Dots represents scanner positions and arrows directions, where scanner looks at.

3.2. Surface Generation

Surface Generator module processes measured point cloud to form of three-dimensional shaded and texture covered model. This process consists in determination of three points on the surface in neighbourhood, which can be connected into the triangle, what is an essential primitive of graphic devices [11].

In this 3D modelling system, we use two different methods: one of them is a Delaunay Triangulation [6]. This is a common mesh generation method, which is usable for totally unorganized sets of points [5] and is significantly slow. Because of that, we use it just in cases, where measured sample is a complex non-convex object or consists of many unconnected parts.

But in most cases, the sample is solid one-piece object without any cavities. For these purposes, we developed significantly faster method, which is based on polyhedral terrain model [5]. It uses not only pure point cloud data, but also additional information about order in which they were scanned and which point of scanning trajectory they were scanned from.

Since object meets defined preconditions, each of its particular small areas of surface can be considered as terrain map projected into the plane perpendicular to direction where scanner looks at. And in this view, the nearest neighbours on the surface are points, which were scanned just in sequence, one after another.

If we also define scanning trajectory in a way, that its tangent is always perpendicular to measuring plane of laser scanner, the nearest neighbours on the surface are also points, which were scanned with same ordering number in two subsequently captured profiles. This condition is satisfied in most cases, because in most cases it is the trajectory with best view to object.

From these two findings about neighbours, we can formulate links between points as:

$$C(P_{i,j}) = \{P_{i,j+1}, P_{i,j-1}, P_{i+1,j}, P_{i-1,j}, \\ P_{i+1,j+1}, P_{i-1,j+1}, P_{i+1,j-1}, P_{i-1,j-1}\}, \quad (14)$$

where $C(P_{i,j})$ is a set of connections from point $P_{i,j}$ to points inside the set, i is the number of profile and j is the number of sample in profile. All these sets of connections creates desired triangle-consisted surface, displayable by graphic device.

3.3. Saving Models

Created 3D models are saved to file in format, which is based on XML architecture. Root element encapsulates several child elements:

- Metadata: XML formatted metadata information about creator, time of creation, configuration, with which the model was captured, etc. Each information is saved in separate XML element.

- Vertices: Base64 encoded binary data containing 3D coordinates of points belonging to model. Each coordinate is saved in 4 bytes of float format.

- Indices: Base64 encoded binary data containing indices of points defining the triangles. Each index is saved in 4 bytes of integer format. Vertices and indices are separated the same way as in graphic device, what empowers very easy file reading — vertex and index buffers are simply copied into the graphic device's memory.

- Trajectory: Text encoded scanning trajectory for purposes of post-processing, eg. recomputing of data acquired with uncalibrated system, etc.

- Reflected Intensity: Base64 encoded binary data containing reflected intensity of points belonging to model. Each value is saved in 2 bytes of short integer format.

- Reflection Width: Base64 encoded binary data containing reflected intensity of points belonging to model. Each value is saved in 2 bytes of short integer format.

The Base64 encoding of large data blocks is used from compress reasons of resulting file. Due to modular structure of file, this format is easily extensible, some block can be missing and also supports backward compatibility. The advantage of this structure is also easy loading of data in different viewing modes (more at following section).

3.4. Model Viewer

An integral part of proposed 3D modelling system is the viewer of created models (Fig. 13). As same as all other software applications related to this 3D system, it has been created on our own to keep full control of its operation.

Among basic features for viewing of 3D models (like rotating, moving or scaling), allows switching between different viewing modes:

- **Surface Only**: Surface of object is covered with uniform colour to express pure spatial representation of object (Fig. 15).

- **Reflectivity**: Surface of object is covered with colour expressing intensity of reflected laser beam (Fig. 16).

- **Refl. Width**: Surface of object is covered with colour expressing width of reflected laser beam (Fig. 17).

Both Reflectivity and Reflection Width data can be displayed in both Grayscale and Rainbow colour representation.

3.5. Examples of Generated 3D Models

Examples of resulting shaded models generated by this 3D modelling system are in Fig. 14, Fig. 15, Fig. 16, Fig. 17 and Fig. 18. Figure 14, Fig. 15, Fig. 16 and Fig. 17 were captured using precise triangulation scanner MicroEpsilon scanCONTROL 2750-100 with measuring accuracy 0.04 mm [2]. Visibility of pleats of skin

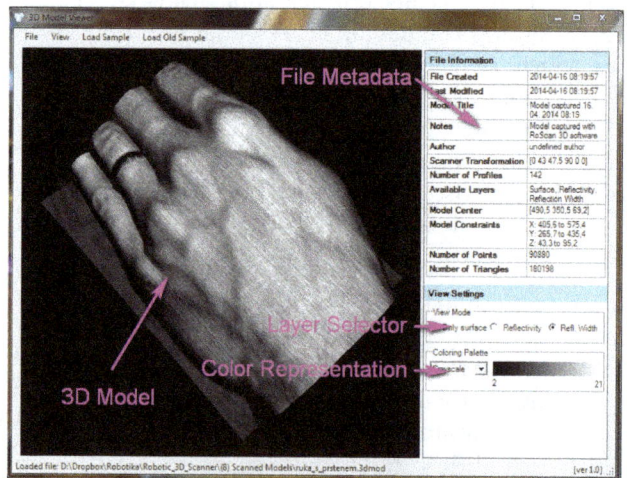

Fig. 13: Screenshot of 3D Model Viewer, tool for displaying 3D models.

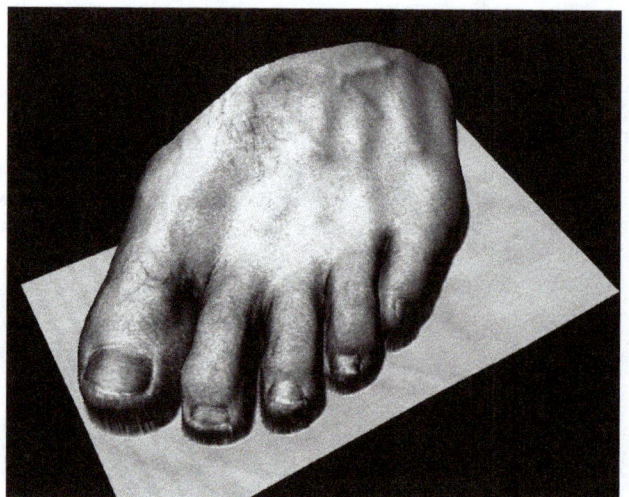

Fig. 14: 3D model of foot with reflectivity based colouring.

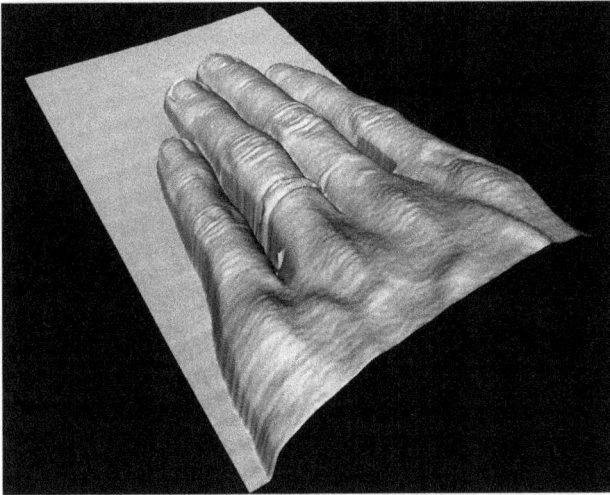

Fig. 15: 3D model of hand with ring on finger – pure shaded surface model.

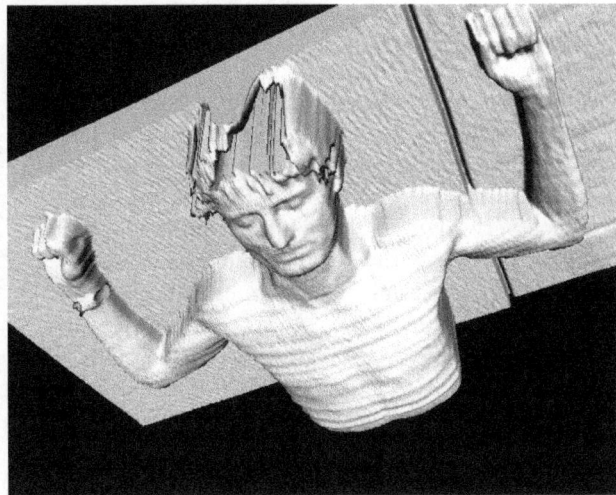

Fig. 18: Surface covered model of human body created with SICK LMS400 scanner.

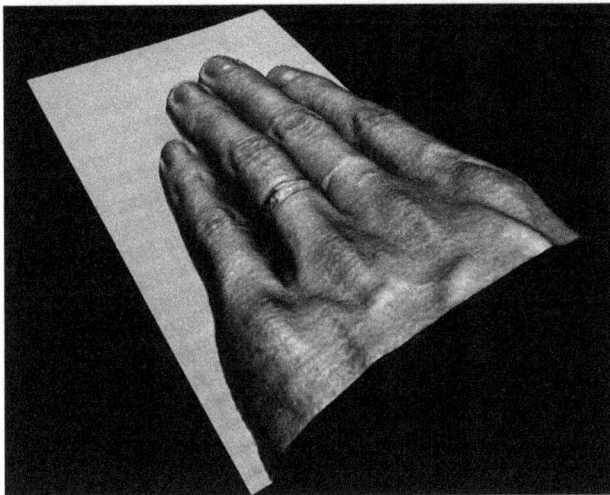

Fig. 16: 3D model of hand with ring on finger – reflectivity expressed by Grayscale colouring.

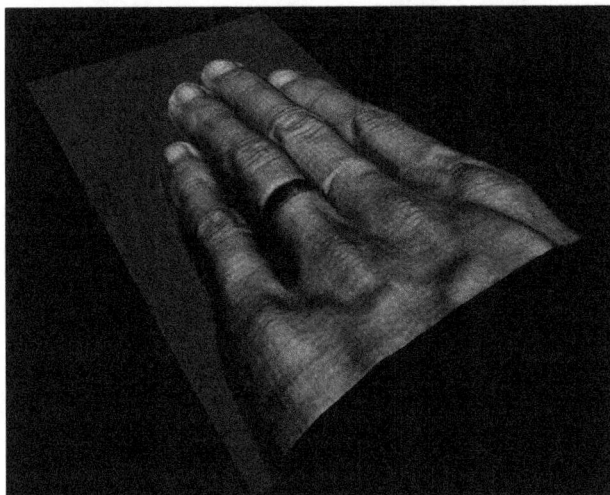

Fig. 17: 3D model of hand with ring on finger – reflection width expressed by Grayscale colouring.

on knuckles or distinguishability of particular vessels proves very high resolution of proposed 3D scanner.

Figure 15 shows model of human hand figured in pure surface model mode. Figure 16 displays 3D model, covered by colour related to surface reflectivity. Its clear, that reflectivity-driven colouring of surface turns model into more friendly visualization than in case of pure surface. The golden ring on the finger is not significantly visible, because its brightness is similar to brightness of skin around. Bigger differences between ring and skin shows Fig. 17 with colouring related to reflection width (lightest colour means highest divergence of laser beam). The narrowest beam is reflected from shiny ring, bigger beam divergence causes skin, and the widest beam is detected at nails, what is caused by its diffuse structure.

Figure 14 shows, that hairs does not affect surface of model, but they are detectable at reflectivity image since they absorb/diffuse laser beam.

There is also a model of entire body captured by TOF laser scanner SICK LMS400 with measuring accuracy 4 mm [1] presented (Fig. 18). The lower accuracy causes surface corrugation. Waves on patient's belly are caused by non zero time between measurement of particular profiles, what is a common problem when scanning living creatures. Possible solution could be speeding up of whole scanning process.

4. Conclusions

This paper presents novel 3D modelling system based on robotic manipulator and laser scanner. Due to use of ones of the most accurate devices, scanning system is capturing models with submillimetre resolution, however, proposed solution is universal and can be repeated with any robotic manipulator and with any scanner.

Essential part of entire 3D modelling system is 3D scanner producing cloud of spatial measurements. Due to robotic manipulator's motion in 6 degrees of freedom, its scanning trajectory can be almost arbitrary, what brings very high flexibility of 3D modelling system usage. Together with exchangeable laser scanners, proposed system is an universal 3D modelling tool capable of capturing almost any object – in case of tiny structures, we can use triangulation-based scanners, if high accuracy is not so important for us (in case of large structures), we can use TOF laser scanner.

If the used laser scanner provides besides distance data also intensity of reflected beam, the 3D model can be coloured according to this additional information, what makes resulting 3D models more authentic to the original object.

In future work, this system could be extended by data fusion of 3D surface model and texture from CCD camera [23], [15]. We are also planning to implement functions for computing object volume and surface area from 3D model, what could be useful for medical domain.

In case of use of this system in health care domain, it could bring significant savings in acquisition costs and operation costs with keeping quality of service at the same level, since in many cases, this 3D scanning system can supply Magnetic Resonance Imaging, what is one of the most expensive diagnostic tool available.

This device is developed in cooperation with St. Anne's University Hospital, where final Robotic 3D Scanner should be installed and used for several diagnostic examinations.

Acknowledgment

This work was supported by the project CEITEC - Central European Institute of Technology (CZ.1.05/1.1.00/02.0068) financed from European Regional Development Fund, by the project FNUSA-ICRC (CZ.1.05/1.1.00/02.0123) financed from European Regional Development Fund and by project FEKT-S-14-2429 "The research of new control methods, measurement procedures and intelligent instruments in automation" financed from internal grant system BUT.

References

[1] LMS 400 laser measurement sensor operating instructions. *SICK AG Waldkirch, Auto Ident,*

Reute Plant. Nimburg Strasse 11, 792 76 Reute, Germany, 2006. Available at: `https://www.mysick.com/saqqara/pdf.aspx?id=im0010698`.

[2] Instruction manual scanCONTROL. MICRO-EPSILON MESSTECHNIK GmbH & Co. KG, Königbacher Strasse 15, 94496 Ortenburg, Germany, 2008. Available at: `http://www.micro-epsilon.com/download/manuals/man--scanCONTROL-2700--en.pdf`.

[3] EPSON ProSix Manipulator manual. *SEIKO EPSON CORPORATION.* 2009–2013. Available at: `http://robots.epson.com/admin/uploads/product_catalog/files/EPSON_C3_Robot_Manual(R7).pdf`.

[4] International vocabulary of metrology — Basic and general concepts and associated terms (VIM). *Working Group 2 of the Joint Committee for Guides in Metrology (JCGM/WG 2).* 2008. Available at: `file:///C:/Users/Steewo/Downloads/JCGM_200_2008.pdf`.

[5] BERG, M. and M. KREVELD. *Computational Geometry Algorithms and Applications. Second, Revised Edition.* Berlin: Springer, 2000. ISBN 978-366-2042-458.

[6] CHEN, M.-B. A Parallel 3D Delaunay Triangulation Method. In: *2011 IEEE Ninth International Symposium on Parallel and Distributed Processing with Applications.* Busan: IEEE, 2011, pp. 52–56. ISBN 978-0-7695-4428-1. DOI: 10.1109/ISPA.2011.52.

[7] CHROMY, A. *3D Scanning with Proximity Planar Scanner.* Brno, 2013. Master thesis. Brno University of Technology.

[8] CHROMY, A., P. KOCMANOVA and L. ZALUD. Creating Three-Dimensional computer models using robotic manipulator and laser scanners. In: *12th IFAC Conference on Programmable Devices and Embedded Systems, Programmable devices and systems.* Velke Karlovice: Elsevier, 2013, pp. 268–273. ISBN 978-3-902823-53-3. DOI: 10.3182/20130925-3-CZ-3023.00037.

[9] CHROMY, A. and L. ZALUD. Robotic 3D scanner as an alternative to standard modalities of medical imaging. *SpringerPlus.* 2014, vol. 3, iss. 1, pp. 13–17. ISSN 2193-1801. DOI: 10.1186/2193-1801-3-13.

[10] CURLESS, B. From range scans to 3D models. *ACM SIGGRAPH Computer Graphics.* 1999, vol. 33, iss. 4, pp. 38–41. ISSN 0097-8930. DOI: 10.1145/345370.345399.

[11] GROOTJANS, R. *Xna 3.0 game programming recipes: a problem-solution approach*. Berkeley: Apress, 2009. ISBN 14-302-1855-X.

[12] HU, J., Y. CAO, T. WU, D. LI and H. LU. High-resolution three-dimensional visualization of the rat spinal cord microvasculature by synchrotron radiation micro-CT. *Medical Physics*. 2014, vol. 41, iss. 10, pp. 150–158. ISSN 0094-2405. DOI: 10.1118/1.4894704.

[13] KIM, K.-W., K.-G. KIM, J.-H. KIM, J.-H. MIN, H.-S. LEE and J. LEE. Development of the 3D volumetric micro-CT scanner for preclinical animals. *3D Research*. 2011, vol. 2, iss. 2, pp. 38–44. ISSN 2092-6731. DOI: 10.1007/3DRes.02(2011)5.

[14] KLEIN, S., M. AVERY, G. ADAMS, S. POLLARD and S. SIMSKE. From scan to print: 3D printing as a means for replication. *HP Laboratories Technical Report*. 2014. Available at: http://www.hpl.hp.com/techreports/2014/HPL-2014-30.pdf.

[15] KOCMANOVA, P. and L. ZALUD. Spatial calibration of TOF camera, thermal imager and CCD camera. In: *Mendel 2013 19th International Conference on Soft Computing*. Brno: Brno University of Technology, 2013, pp. 343–348. ISBN 978-80-214-4755-4.

[16] KOFMAN, J. and K. BORRIBANBUNPOTKAT. Hand-held 3D scanner without sensor pose tracking or surface markers. *High Value Manufacturing: Advanced Research in Virtual and Rapid Prototyping*. Leiria: CRC Press, 2013, pp. 429–434. ISBN 978-1-315-81741-5. DOI: 10.1201/b15961-79.

[17] LEHMANN, E. and G. CASELLA. *Theory of point estimation*. 2nd ed. New York: Springer, 1998. ISBN 03-879-8502-6.

[18] NEJDL, L., J. KUDR, K. CIHALOVA, D. CHUDOBOVA, M. ZUREK, L. ZALUD, L. KOPECNY, F. BURIAN, B. RUTTKAY-NEDECKY, S. KRIZKOVA, M. KONECNA, D. HYNEK, P. KOPEL, J. PRASEK, V. ADAM and R. KIZEK. Remote-controlled robotic platform ORPHEUS as a new tool for detection of bacteria in the environment. *ELECTROPHORESIS*. 2014, vol. 35, iss. 16, pp. 2333–2345. ISSN 1522-2683. DOI: 10.1002/elps.201300576.

[19] PALENCAR, R., F. VDOLECEK and M. HALAJ. Nejistoty v mereni II: nejistoty primych mereni. *Automa*. 2001, vol. 2001, no. 10, pp. 52–56. ISSN 1210-9592.

[20] PARK, J. H., J. KUDR, K. CIHALOVA, D. CHUDOBOVA, M. ZUREK, L. ZALUD, L. KOPECNY, F. BURIAN, B. RUTTKAY-NEDECKY, S. KRIZKOVA, M. KONECNA, D. HYNEK, P. KOPEL, J. PRASEK, V. ADAM and R. KIZEK. Digital Restoration of Seokguram Grotto: The Digital Archiving and the Exhibition of South Korea's Representative UNESCO World Heritage. In: *2012 International Symposium on Ubiquitous Virtual Reality*. Adaejeon: IEEE, 2012, pp. 26–29. ISBN 978-1-4673-2258-4. DOI: 10.1109/ISUVR.2012.23.

[21] SHAN, J. and C. K. TOTH. *Topographic laser ranging and scanning: principles and processing*. Boca Raton: CRC Press/Taylor, 2009. ISBN 14-200-5142-3.

[22] SOLC, F. *Robotics, modelling and control of robots*. Brno: VUTIUM, 2004. ISBN 80-214-2618-7.

[23] ZALUD, L. and P. KOCMANOVA. Fusion of thermal imaging and CCD camera-based data for stereovision visual telepresence. In: *2013 IEEE International Symposium on Safety, Security, and Rescue Robotics (SSRR)*. Linkoping: IEEE, 2013, pp. 1–6. ISBN 978-1-4799-0879-0. DOI: 10.1109/SSRR.2013.6719344.

[24] ZALUD, L., P. KOCMANOVA, F. BURIAN and T. JILEK. Color and Thermal Image Fusion for Augmented Reality in Rescue Robotics. In: *8th International Conference on Robotic, Vision, Signal Processing and Power Applications*. Penang: Springer, 2013, pp. 47–56. ISBN 978-981458541-5. DOI: 10.1007/978-981-4585-42-2_6.

About Authors

Adam CHROMY is student of Ph.D. program at Brno University of Technology, researcher at Central European Institute of Technology (CEITEC) and researcher at International Clinic Research Centre (ICRC). His research is focused on laser scanners and biomedical applications of robotics and kybernetics.

Ludek ZALUD is associate professor at Brno University of Technology, senior researcher at Central European Institute of Technology (CEITEC) and senior researcher at International Clinic Research Centre (ICRC). His research is focused on field robotics and spatial data fusion. His team constructed well-known military teleoperated robot Orpheus.

14

A MAN AS THE REGULATOR IN MAN-MACHINE SYSTEMS

Marie HAVLIKOVA, Sona SEDIVA, Zdenek BRADAC, Miroslav JIRGL

Department of Control and Instrumentation, Faculty of Electrical Engineering and Communication,
Brno University of Technology, Technicka 3082/12, 616 00 Brno, Czech Republic

havlika@feec.vutbr.cz, sediva@feec.vutbr.cz, bradac@feec.vutbr.cz, xjirgl00@stud.feec.vutbr.cz

Abstract. *The aim of the paper is to present the role of the human element in regularly used man-machine systems (MMS). From the technical point of view, it is possible to denote the human being and the machine as two components of one system. In the second part of the article, the authors introduce the results obtained through simulations of the human driver model; these simulations are focused on the quality of the control process. The structure of the model facilitates the detection and analysis of human error identifiers.*

Keywords

Control, high-risk system, human driver, man-machine system, regulator, simulation models.

1. Introduction

The human operator is a powerful, universal and effective regulatory element capable of solving unexpected situations quickly and adapting itself flexibly to the various operating conditions that characterize an MMS system. Based on his or her own experience, the human being executes and implements regulatory interventions, whose quality depends on the operator's knowledge and practice. In the described context, the human is a self-learning adaptive regulator with properties analogical to commercial regulators, which normally perform interventions based on mathematical description and analysis. However, there is a significant aspect of difference between the human operator and a commercial regulator: the human mind. This specific property enables the operator to integrate their brain, whose functions cannot be later excluded or overridden, into the regulation process.

2. Man-Machine Systems

Most machines are designed to be manipulated, managed, and controlled by humans. The man and the machine (a technical system) together form a specific higher system in which both these basic subjects cooperate and interact. In scientific literature, these systems are referred to as MMS. There are various types of human - machine interacting systems; from the simplest MMS such as hand tools, we can continue to mention the more complicated or specific ones represented by the car or the computer. An example of highly complex MMS systems can be seen in a nuclear plant or aircraft control.

In machines and devices, technical and economic parameters such as cost or sales prices are currently preferred to the significant aspects of safety and reliability. Thus, it is not possible to expect that an MMS system will operate and without any problems. From the perspective of safety, the system has to be designed in such a manner as to prevent major failures that could cause health risk, property loss, or ecological damage. Failures can be caused by either the machine or the human, who is not able to work and remain alert for an unlimited period of time; therefore, in the latter case, it is necessary to avoid an incorrect operation or a wrong decision.

A multitude of mathematical methods are employed to analyse human reliability, and many sources are focused on this issue [1], [2], [3].

2.1. Operator Activity Levels in MMS System

The human operator in an MMS system performs working and controlling operations at various stages of difficulty. The knowledge and description of the operator are among the necessary preconditions for the creation of accurate MMS models, which facilitate the analysis of critical points and the detection of hazardous system states as well as wrong operator actions.

An understanding of the overall human role and the operating principles related to human activity within a system is required for the successful evaluation of the safety and reliability aspects and enables further advancement in the communication between a human being and a machine [3].

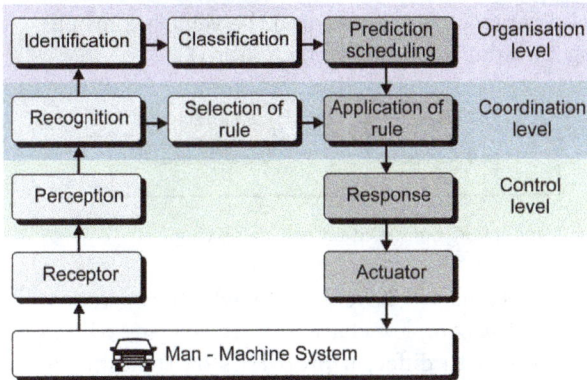

Fig. 1: Three levels of the human-based control according to Rasmussen's model.

Human activities in an MMS system depend on the specific difficulty aspects characterizing a system. The activities can be classified into categories (Fig. 1) based on these different difficulty aspects, which include elements such as the time and the function. The most widely applied classification approach was proposed by Professor J. Rasmussen, who categorized the human activities by difficulty criteria and the operator body parts used to carry out a given activity [4]. The discussed approach comprises the following levels:

- The control level, where a human being assumes the role of the regulator to perform regulatory activities and machine-controlling interventions. The active and executive elements of the human regulator are his or her kinetic devices such as the upper and lower limbs. This activity level is characterized by the smallest intellectual requirements; the human being performs his/her role to function as an executive element.

- The coordinating level, which comprises activities based on controlling a specific machine. The human operator must recognize several states of the controlled system, analyze the situation, and select a relevant activity to make the actual state of the system conform to specific rules, standards and techniques. The human is required to learn these activities in advance. The operator employs his or her brain to conduct a large number of trained states including specific and task-related activities, procedures, or methods. After many repetitions of one activity, the human being adopts the stereotype and learned practice; this is the stage when the operator "disconnects" his/her brain from the activity.

- The cognitive level, also known as the tactic level, includes activities related to decisions or analyses of unexpected and abnormal situations (system states) to which no specific action has been assigned yet. Other similar activities comprise the processes optimizing the selected human-preferred criteria, rush decisions (such as the reaction to an unexpected situation where the solution depends on the human experience, knowledge, and abilities). Within the described activity level, the human brain is activated: the operator incorporates his/her own mind into the system control procedures [5].

2.2.   The Reliability of an MMS

Previous evaluations of MMS reliability focused principally on technical subsystems, and the influence of the human factor on the system reliability was not quantitatively monitored. The necessity to start new research in the field of the human factor and reliability arose from failures in nuclear power plants, chemical factories, and frequently repeated air and sea disasters. The scientific research has shown that it is very difficult to design universal evaluation procedures, mainly because human activities are markedly diversified. Each such activity comprises specific working methods which cannot be unified or merged, and thus it is not possible to label them with corresponding tabular values. Technical and human reliability are two aspects incompatible especially as regards data processing and the procedures of achieving a goal.

Humans actively use their brains and mind to set or complete certain aims and objectives, and their behaviour is generally directed towards achieving a goal. Based on the analysis of the current state, the human operator may choose tools and methods other than those recommended or ordered. The operator is capable of permanent monitoring and modification of his or her behaviour; thus, he or she can effectively correct wrong steps performed earlier within the system operation procedure. At this point, let us note that although the probability of an error occurring in certain human activities can be very high, the completion of the given aim is usually not jeopardized. The references [4], [5], [6] define human reliability as the ability to perform a task flawlessly under certain conditions and within the stipulated time.

Quantitative evaluation of human reliability is based on the total probabilistic safety analysis (PSA) of the whole MMS [5]. This analysis also comprises human reliability assessment (HRA), which carries information regarding the following elements: the safety and readiness of the technical system with respect to human interventions; human faults in comparison with technical faults; and the possibility of increasing the reliability and safety of the system.

Human Reliability Assessment (HRA) is the part of the reliability discipline where the human performance in operating actions is studied. Human reliability is usually defined as the probability that a person will correctly perform some system-required activity during a given time period (if time is a limiting factor) without performing any extraneous activity that can degrade the system.

Human Error Probability (HEP) expresses erroneous performance of an action during the observation period. The determination of HEP is based in particular on research studies of comparable activities (generic data) and on the observation of incorrect actions in the analyzed or similar MMS.

There exist many methods for human reliability probabilistic assessment [6], [7] that pursue identical goals. These techniques are as follows: quantitative analysis of human behaviour; identification of erroneous activities; and identification of weak points of the system carried out together with the formation of preconditions for suitable remedial steps. The best known HRA methods are THERP (Technique for Human Error Rate Prediction), SLIM (Success Likelihood Index Method), HEART (Human Error Assessment and Reduction Technique), ATHENA (A Technique for Human Error Analysis), and CREAM (Cognitive Reliability and Error Analysis Method) [8].

3.   The Human Driver as a Regulator

Driving a vehicle is a complex activity. However, currently there does not exist any universal driver model capable of simulating the total of driving activities across all control levels (feedback control, coordination level based on the application of rules, knowledge-based cognitive level). Driver simulation models can be classified into two basic categories that result from the description of driving-related activities. These two classes based on the mode of vehicle driving are as follows:

- Transverse driving, which is defined by both the quality of road holding and the car position inside the traffic lane.

- Longitudinal driving, which is determined by the control of the car speed and acceleration in a linear direction.

The basic control circuit for the transverse compensation vehicle driving is shown in Fig. 2. The eye perceives the control process, and the information from the visual field is transferred to the central nervous system by back coupling. The vehicle dynamics are represented by the transmission function $Y_M(p)$, and the dynamics of the human regulator are expressed by the transmission function $Y_H(p)$. The driver executes feedback control of the momentary transverse car location $y(t)$; the aim is to achieve a situation when the control divergence $e(t)$ is zero and the vehicle continues moving towards the desired position $y_z(t)$.

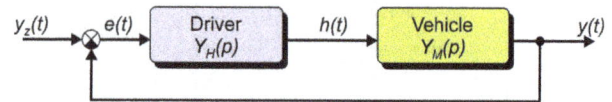

Fig. 2: A model of compensation vehicle driving.

In practice, as we have mentioned above, vehicle driving is of a complex character: It is a set of partial activities with different properties on the different control levels (Fig. 3). The memorized stereotypes and routine manoeuvres are realized by the R_{pg} precognitive controller based on knowledge, qualifications and idea processes. The ability of prediction, which facilitates the estimation of the future trajectory and situation on the roadway. The predictive controller R_ψ participates in the vehicle control. By this controller, the driver holds his car in the required direction $\psi_r(t)$.

The compensation controller R_y is used for the minimization of the control error $e(t)$. With this controller, the action interferences are controlled based on the visual information about the required location $y_z(t)$ and the actual location $y(t)$.

In feedback compensation vehicle control, the control circuit has the structure of eye – brain – hand and is defined by permanent feedback. The information is obtained predominantly from visual sensation, and its processing is performed in the corresponding centres of the gray cerebral cortex (ectocinerea). The functions of the feedback predictive controller R_ψ and the precognitive controller R_{pg} are suppressed; their action interferences are not a priority, and they participate in the control only minimally [9].

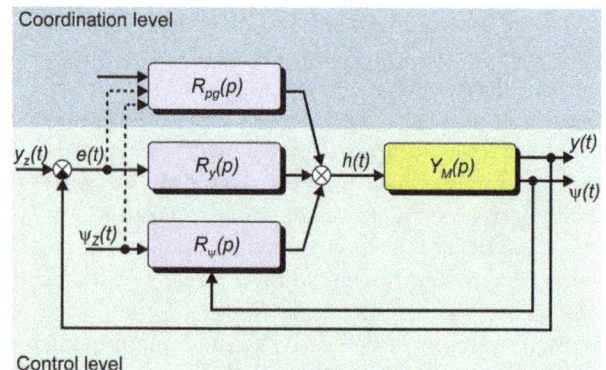

Fig. 3: Types of the driver controller [9].

3.1. Transfer Functions of the Human Driver

The transfer function of the driver may assume different structures depending on either the driver's control abilities or the simplifying hypothesis to be applied.

The model of compensation vehicle control derived from the transfer of the open loop of the control circuit F_0 is denoted as the Crossover Model [12]. The different transfer functions $Y_H(p)$, $Y_M(p)$ are validated for every driver, vehicle and ambient conditions, but certain properties are invariable in different types of the control task. These properties can be defined as follows:

- The closed loop of the feedback control is stable if the driver manages his work. The unstable state occurs only in the case of control ability failure on the part of the driver.

- The frequency characteristic of the open loop crosses the frequency axis with slope of approximately –20 dB.

These transfer functions of the driver models are correct for small changes of the input signals and drive with low dynamics. The variables of the activator (driver $Y_H(p)$) are set to ensure convenient conditions for the open loop transmission $F_0(p)$ in equation Eq. (4); thus, the stability of the control system will be secured.

$$F_0(p) = Y_H(p) \cdot Y_M(p) = \frac{\omega_C e^{-T_D p}}{p}, \qquad (1)$$

where F_0 - the open loop transmission, $Y_H(p)$ - transmission function of driver, $Y_M(p)$ - transmission function of vehicle, ω_C - cut frequency given by the product of the amplification of the system and the intensification of the action element, T_D - the driver response delay.

The simple form of the transfer function of the driver $Y_H(p)$ is expressed by formula Eq. (2) and applies to the compensation type of control [13], [14]. The action interferences of the driver $h(t)$ assume only an integrating role, and they are performed with a certain time delay given by the quantity of the driver delay T_d:

$$Y_H(p) = \frac{K}{p} \cdot e^{-T_D p}, \qquad (2)$$

where T_d - the constant describing the delay of the driver response, K- the constant determining the size of the driver intensification.

The transfer function type shown below is presented very frequently [10], [11].

$$Y_H(p) = \frac{K \cdot e^{-T_D p} \cdot (T_a p + 1)}{(T_n p + 1) \cdot (T_I p + 1)}, \qquad (3)$$

where T_D - the constant describing the delay of the driver brain response to visual sensation, T_n - the inertia constant determining the driver's delay with respect to the activity of the neuromuscular system, T_a - the predictive constant associated with the driver's practice, T_I - the counterproductive inertia delay associated with the learned stereotype and routine process, K - the attenuation describing the driver's custom.

The form of the transfer function Eq. (3) results from the hypothesis that the driver behaves like a linear component. Some nonlinear components always occur in the real control circuit. The extended form of the transfer function of the driver $Y_H(p)$ is presented within equation Eq. (4) [14], and the nonlinearity of the action component is implicated in the remnant factor:

$$Y_H(p) = \frac{K \cdot e^{-T_D p} \cdot (T_a p + 1)}{(T_n p + 1) \cdot (T_I p + 1)} + remnant. \qquad (4)$$

A very interesting form of the transfer function $Y_H(p)$ is shown in [15], where the driver model consists of two parts that contain two information inputs. The transfer function of the driver comprises two controllers (Fig. 4). The controller with the transfer function $Y_{Hy}(p)$ reacts to the respective divergence $e(t)$ of the vehicle location $y(t)$ and the second controller $Y_{H\psi}(p)$ reacts to the vehicle direction $\psi(t)$.

At this point, we may conclude that the complex forms of the transfer function faithfully represent the real behaviour of the driver or operator in MMS systems; it is also possible to say that the identification of the parameters of the transfer function forms is a difficult and problematic task.

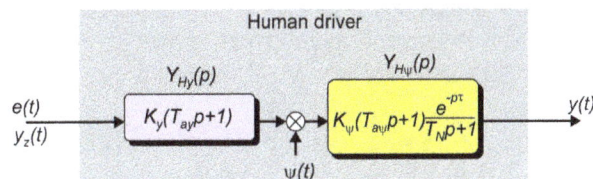

Fig. 4: A driver model with more inputs and transfer functions.

3.2. Simulations of a Nonlinear, Driver - Vehicle Model

The driver does not execute the control functions according to the linear model; rather than that, his or her actions are invariably encumbered with negative effects of nonlinear components, such as hysteresis, insensitivity, saturation, or nonlinear amplification.

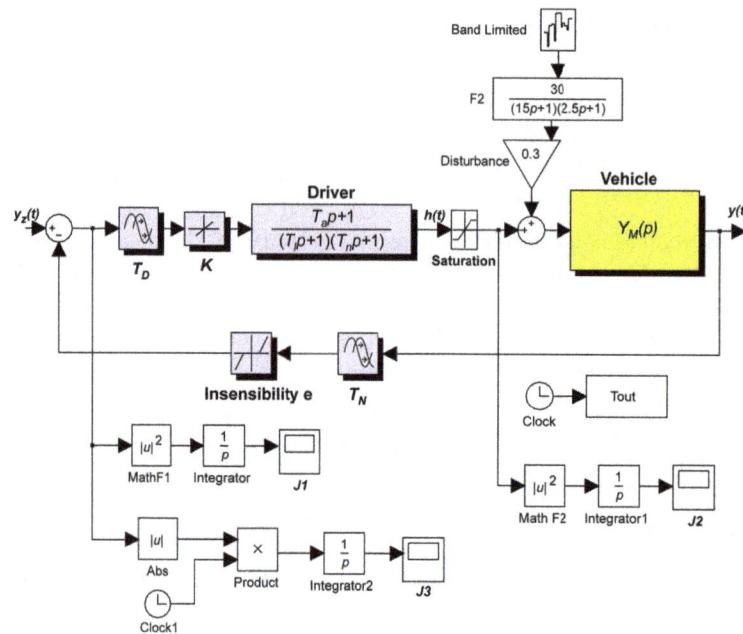

Fig. 5: The nonlinearity components in the driver - vehicle simulation model.

Compensation vehicle control is characterized by single feedback, where the driver feedback compensation controller is active. The nonlinear components can then be located in either the direct branch or the feedback, and they affect the vehicle control process in the manner described below.

- Insensitivity e is associated with the decision of the driver to respond to a situation by performing an intervention $h(t)$ depending on the actual car position $y(t)$.

- Saturation is located explicitly on the output of the action element in the direct branch, and it can be defined as the undesired limitation restricting the driver's intervention $h(t)$.

- Amplification K expresses such driver's response where the intensity of the intervention corresponds to the magnitude of the control deviation $e(t)$.

An example of possible location of nonlinearities in the control circuit of the driver – vehicle simulation model is shown in Fig. 5. The variant amplification K and the saturation of the driver's intervention are located in the direct branch; the nonlinearity of the insensitivity e is contained in the feedback.

3.3.   Quality Criteria of the Control Process

The quality of the control processes in the driver – vehicle simulation models is determined by means of the following iterative criteria:

- Quadratic integral criterion used to classify the control deviation value $e(t)$ (marked by the symbol *J1* in the simulation models).

- Quadratic integral criterion applied to classify the action intervention value $h(t)$ (marked by the symbol *J2* in the simulation models).

- Integral criterion ITAE for the classification of the control deviation value $e(t)$ (marked by the symbol *J3* in the simulation models).

The block diagram of the components for the calculation of the quadratic criteria in the simulation model is presented in Fig. 5. The numerical values of the integral criteria *J1*, *J2* and *J3* are generated by program algorithms based on circuit solution of the simulation model.

3.4.   Simulation of the Parameters and Time Constants of the Human Driver Model

The simulations for different models of the **human driver – vehicle** system are implemented in Matlab 7.9.0. The aim of the simulations is to determine the influence of the parameter changes and time constants of the model driver transfer function $Y_H(p)$ on both the quality of the regulatory process. In the given context, it is also necessary to define what values can be used to preserve the stability of the control loop.

Thus, we specify the intervals of the values of the parameters monitored in the **human driver – vehicle** simulation models; the aim of the simulation was to secure a stable regulatory process by obtaining the lowest possible values of the quadratic integral criteria $J1$ and $J2$, see Fig. 6.

3.5. Results of the Simulations

Changes of the time constants T_a and T_l, the transport delay T_D, and the value of the variable amplification K in the transfer functions of the driver $Y_H(p)$ are given in Tab. 1. Similar interval values of the analyzed parameters are reported in references [14], [12], [9].

The effectivity of the driver's intervention significantly depends on the amplification parameter K. If the intervention is performed with an insufficiently small or excessively large force, it will destabilize the entire regulatory system. This resulting effect is vital for the monitoring of the fatigue factor.

Tab. 1: The range of the simulated parameter values and time constants in a closed loop the with insensitivity e.

| Parameter | Insensitivity e | |
	Direct branch	Feedback
T_a [s]	1.0–2.0	1.0–2.2
T_D [s]	0.1–0.2	0.1–0.2
T_l [s]	0.20–0.41	0.2–0.5
e [m]	0.1–0.4	0.1–0.4
K [rad·m^{-1}]	0.010–0.015	0.010–0.013

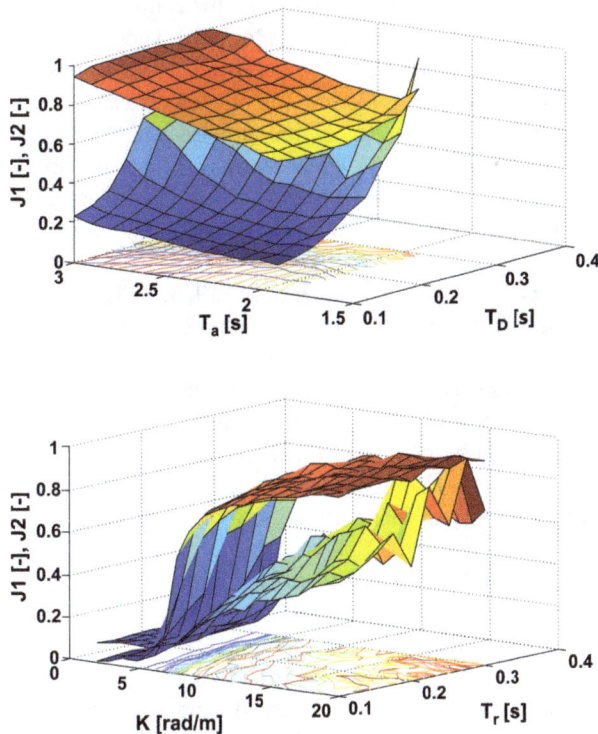

Fig. 6: The values of the integral criteria $J1$, $J2$ in the simulations of the time constants T_a, T_D, T_r and the variable amplification K.

4. Conclusion

The growing number of aeroplane crashes, chemical plant accidents, and disasters related to dangerous operations has led to increased interest in human functions within MMS systems. It follows from various detailed analyses of accident causes that most negative events of this type are based on a human factor failure.

The reliability of MMS depends largely on the human factor, and therefore it is very important to monitor any human activity in the system. The performed simulations of the **driver (man) – vehicle** nonlinear models showed that even small changes in the parameters of the transfer function of the human operator may destabilize the whole regulatory system and jeopardize its overall safety.

Acknowledgment

The research was financially supported by Brno University of Technology and the European Regional Development Fund under project No. CZ.1.05/2.1.00/01.0014. The above-mentioned funds and institutions facilitated efficient performance of the presented research and associated tasks. This work was supported also by the project "TA02010864 - Research and development of motorized ventilation for the human protection against chemical agents, dust and biological agents" and project "TA03020907 - REVYT - Recuperation of the lift loss energy for the lift idle consumption" granted by Technology Agency of the Czech Republic (TACR). Part of the work was supported by project "FR-TI4/642 - MISE - Employment of Modern Intelligent MEMS Sensors for Buildings Automation and Security" granted by Ministry of Industry and Trade of Czech Republic (MPO).

Part of the work was carried out with the support of core facilities of CEITEC – Central European Institute of Technology under CEITEC – open access project, ID number LM2011020, funded by the Ministry of Education, Youth and Sports of the Czech Republic under the activity „Projects of major infrastructures for research, development and innovations".

Part of this paper was made possible by grant No. FEKT-S-14-2429 - "The research of new control methods, measurement procedures and intelligent instruments in automation", and the related financial assistance was provided from the internal science fund of Brno University of Technology.

References

[1] OLSON, J. R. and G. M. OLSON. The growth of cognitive modelling in human computer interaction since GOMS. *Human Computer Interaction.* 1997, vol. 5, iss. 2–3, pp. 221–265. ISSN 0737-0024.

[2] HELD, J. and H. KRUGER. FIT: Ein Mensch-Maschine System zur Analyse von Mensch-Maschine Interaktionen. In: *Schriftenreihe Berufliche Bildung.* Bremen: Donat, 2000. pp. 263–279.

[3] BUBB, H. *Menschliche Zuverlaessigkeit.* Landsberg: Huthig Jehle Rehm, 1992. ISBN 978-36-096-9000-1.

[4] RASMUSSEN, J. *Information processing and human-machine interaction: an approach to cognitive engineering.* New York: Elsevier, 1986. ISBN 04-440-0987-6.

[5] HOLLNAGEL, E. *Cognitive reliability and error analysis method: CREAM.* New York: Elsevier, 1998. ISBN 00-804-2848-7.

[6] REASON, J. *Human Error.* Cambridge: University Press, 1999. ISBN 05-213-1419-4.

[7] COOPER, S., A. RAMEY-SMITH, J. WREATHALL, G. W. PERRY, D. C. BLEY, W. J. LUCKAS, J. H. TAYLOR and M. T. BARRIERE. *A Technique for Human Error Analysis (ATHEANA) - Technical Basis and Methodology.* Description. NRC. Washington DC, 1996.

[8] FORESTER, J., D. BLEY, S. COOPER, E. LOIS, N. SIU, A. KOLACZKOWSKI and J. WREATHALL. Expert elicitation approach for performing ATHEANA quantification. *Reliability Engineering.* 2004, vol. 83, iss. 2, pp. 207–220. ISSN 0951-8320. DOI: 10.1016/j.ress.2003.09.011.

[9] VYSOKY, P. Dynamic properties of human operator as a driver (in czech). *Automatizace.* 2003, vol. 46, iss. 12, pp. 796–800. ISSN 0005-125X.

[10] MCRUER, D., W. F. CLEMENT, P. M. THOMSON and R. E. MAGDALENO. *Pilot Modelling for Flying Quantities Applications.* System Technology, Technical Report, Hawtborne. 1989.

[11] BORIL, J. and R. JALOVECKY. Response of the Mechatronic System, Pilot-Aircraft, on Incurred Step Disturbance. In: *Proceedings ELMAR-2011: 53rd International Symposium.* Zagreb: ITG, 2011, pp. 261–264. ISBN 978-1-61284-949-2.

[12] MCRUER, D. Human Dynamics in Man Machine Systems. *Automatica.* 1980, vol. 16. iss. 3, pp. 237–253. ISSN 0005-1098. DOI: 10.1016/0005-1098(80)90034-5.

[13] HESS, R. A. and A. MODJTAHEDZADEH. A control theoretic model of driver steering behavior. *IEEE Control Systems Magazine.* 1990, vol. 10, iss. 5, pp. 3–8. ISSN 0272-1708. DOI: 10.1109/37.60415.

[14] MCRUER, D. and D. K. SCHMIDT. Pilot-vehicle analysis of multi axis task. *Journal of Guidance, Control and Dynamics.* 1990, vol. 13, iss. 2, pp. 348–355. ISSN 0731-5090. DOI: 10.2514/3.20556.

[15] OSCARSON, M. *Variable Vehicle Dynamics Design.* Master Thesis. Reg. Nr: LiTH-ISY-EX-3348-2003. 2003.

About Authors

Marie HAVLIKOVA was born in 1957. She received her Ph.D. in technical cybernetics s in 2009 at the Brno University of Technology, doctoral thesis *Diagnostic of Systems with a Human Operator.* Her research interests include electrical measurement, human reliability and signal processing. She is an assistant professor at Department of Control and Instrumentation, Faculty of Electrical Engineering and Communication, Brno University of Technology.

Sona SEDIVA was born in 1975. She received her Ph.D. in technical cybernetics in 2003 at the Brno University of Technology, doctoral thesis *Parameter Optimization of the Multiport Averaging Probe.* Her research interests include electrical and nonelectrical measurement, signal processing. She is an assistant professor at Department of Control and Instrumentation, Faculty of Electrical Engineering and Communication, Brno University of Technology.

Zdenek BRADAC was born in 1973. He received his Ph.D. in technical cybernetics in 2004 at the Brno University of Technology. His research interests include HMI systems, fault-tolerant systems, information systems safety and security. He is an associated professor at Department of Control and Instrumentation, Faculty of Electrical Engineering and Communication, Brno University of Technology.

Miroslav JIRGL was born in 1988. He received his M.Sc. in Electrical Engineering in 2012 at the Brno University of Technology. Currently he is a Ph.D. student at Department of Control and Instrumentation, Faculty of Electrical Engineering and Communication, Brno University of Technology and his research interests include Man-Machine Systems and their safety.

15

Analysis and Control of STATCOM/SMES Compensator in a Load Variation Conditions

Mahmoud Reza SHAKARAMI[1], *Reza SEDAGHATI*[1], *Mohammad Bagher HADDADI*[2]

[1]Department of Electrical and Electronic Engineering, Engineering Faculty, Lorestan University,
Daneshgat Street, 71234-98653 Khoramabad, Lorestan, Iran
[2]Department of Electrical Engineering, School of Engineering, Shiraz University,
Bahar Azadi Street, 73159-88553 Kazeroon, Shiraz, Iran

shakarami@iust.ac.ir, reza_sedaghati@yahoo.com, haddadimohammadbagher@gmail.com

Abstract. *The utilization of Flexible AC Transmission System (FACTS) devices in a power system can potentially overcome limitations of the present mechanically controlled transmission system. Also, the advanced technology makes it possible to include new energy storage devices in the electrical power system. The integration of Superconducting Magnetic Energy Storage (SMES) into Static Synchronous Compensator (STATCOM) can lead to increase their flexibility to improve power system dynamic behavior by exchanging both active and reactive powers with power grids. This paper describes structure and behavior of STATCOM/SMES compensator in power systems. A control strategy based on direct Lyapanov method for compensator is used. Moreover, the performance of the STATCOM/SMES compensator in a load variation condition is evaluated by PSCAD/EMTDC software in test system. Also, SMES capacity effects on integrated compensator are investigated.*

Keywords

Direct Lyapanov method, load variation condition, SMES capacity, STATCOm/SMES compensator.

1. Introduction

In recent years, by ongoing growth in electric power demand and deregulation in the electrical power industry, numerous changes have been introduced to modern electricity industry. One of the most significant problems in power systems is that its power swings between synchronous generators and subsystems damped weakly and it must be controlled in appropriate way, otherwise the power system will encounter a significant problem and lose the conventional operation. Due to recent advances in high power semiconductor technology, FACTS technology has been proposed to solve this problem [1], [2].

Furthermore, as a typical FACTS device, STATCOM have been developed and utilized to improve transient stability margin, power quality improvement and damping power system oscillations by controlling reactive power [3], [4], [5]. Whereas significant increase in energy storage capacity for STATCOM lead to increase in degree of freedom and as a result its reliability and flexibility, therefore an energy storage system (ESS) for integration of STATCOM is proposed.

There are different technologies for energy storage such as ultra-capacitors, batteries, flywheels and SMES which the SMES system for power utility applications have received considerable attention due to rapid response, high power, high efficiency and four quadrant control [6], [7]. STATCOM and SMES are considered to cooperate and emerge as a compensator with prominent capability in power swings damping improvement. In [7] the SSSC/SMES application for frequency stabilization is examined and in [8] the experimental system integration of a battery energy storage system (BESS) into a STATCOM is discussed.

Specifically, this paper will present:

- Specifications and performance principles of the STATCOM/SMES compensator.

- Behavior of STATCOM/SMES in a load variation condition.

- SMES capacity effects on integrated compensator.

2.   Proposed Model of Integrated STATCOM/SMES Compensator

A STATCOM can only absorb/inject reactive power, and consequently is limited in the degree of freedom. The STATCOM/SMES combination can provide a better dynamic performance than a stand alone STAT-COM [7], [9]. A functional model of a STATCOM integrated with a SMES coil is shown in Fig. 1. This model consists main parts of the STATCOM controller, the SMES coil and the interface between both devices. The inclusion of a SMES in the dc bus of the STAT-COM requires to adapt the voltage and current levels of both devices by utilizing an interface. In this case, a two-quadrant three-phase dc-dc converter is chosen as interface.

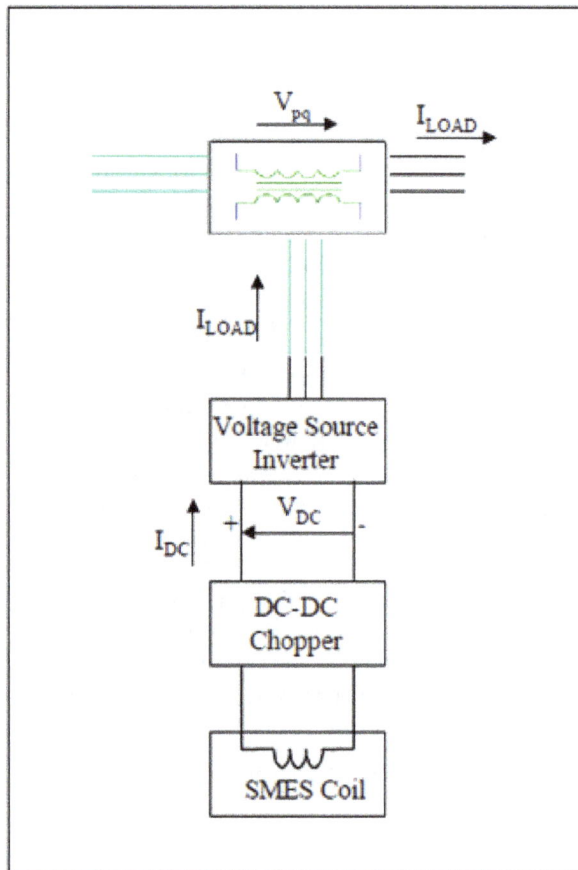

Fig. 1: General model of the integrated STATCOM/SMES compensator.

The chopper changes the dc current from the SMES coil to dc voltage, and a VSC changes the voltage into a three-phase ac current. Both the chopper and VSC need GTOs. The control of active and reactive powers was accomplished by controlling firing angles of the GTOs and the dc voltage that is determined by

the duty ratio of the chopper. The angle and voltage differences between the utility line and the VSC output built current through the leakage inductance of the transformer that became the utility line current. The dc-dc chopper play in role of energy flow controller through the SMES coil [10]. When the SMES needs to be charged, the chopper connects the dc link voltage to the SMES so that the current inside the SMES increases and make a power flow from the dc link to the SMES coil. When the SMES needs to be discharged, the chopper connects the opposite voltage. The rate of charge/discharge is controlled by the voltage magnitude of the SMES coil. In other words, the dc-dc chopper changes the constant dc link voltage into a variable voltage required by the SMES coil to make the desired energy flow. Figure 2 shows a detailed configuration of the chopper. As shown in Fig. 2, the coil can be charged when the two GTOs are fired simultaneously and the diodes become reverse-biased. When both of GTOs are turned off, the coil discharges and the diodes become forward biased. At a duty cycle of 0.5, the SMES coil's average voltage and the VSC's average dc current are both zero, and no net power is transferred throughout one switching cycle. At a duty cycle larger than 0.5, the coil is charged; while at less than 0.5, the coil is discharged . Therefore, the control of charge/discharge process is accomplished by controlling the duty cycle [11].

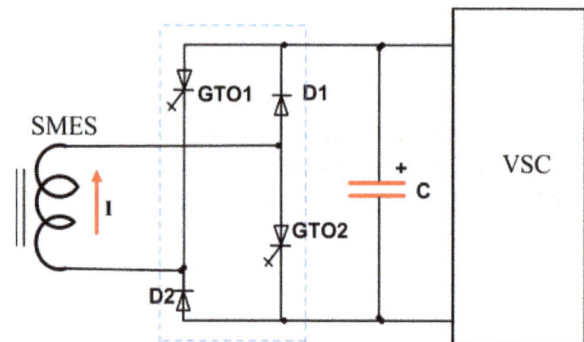

Fig. 2: Structure of a two-quadrant chopper.

3.   Control Strategy for Integrated Compensator

3.1.   Power System Model

Structure preserving model (SPM) of power systems has been presented to improve the modeling of generators and load representations such that system parts represent more realistic behavior [9]. Generators are modeled as one-axis generator model which includes one circuit for the field winding and also loads are modeled by constant active power and reactive power with following equations:

$$P_{Lk} = P_{Lk0} \left(\frac{V_k}{V_{k0}} \right)^{mp},$$

$$Q_{Lk} = Q_{Lk0} \left(\frac{V_k}{V_{k0}} \right)^{mq},$$

$$(1)$$

where P_{Lk} and Q_{Lk} will be the active and reactive powers at the nominal voltage Vo, respectively the P_{Lk} and Q_{Lk} real and reactive power of k-th load. mp and mq can be an arbitrary integer from 0 to 3. A power system with N buses and M generators without exciter and governor is considered. The system is assumed to be lossless. The governing equations of the system are:

$$\dot{\delta}_i = \omega_i,$$

$$M_i \dot{\omega}_i = P_{mi} - P_{Gi} - D_i \omega_i$$

$$i = 1...M,$$

$$(2)$$

$$T'_{doi} \dot{E}'_{qi} = \frac{x_{di} - x'_{di}}{x'_{di}} V_{M+i} \cos(\delta_i - \theta_{M+i} +$$

$$+ E_{fdi} - \frac{x_{di}}{x'_{di}} E'_{qi}),$$

where ω_i and δ_i are velocity and mechanical angle of the ith generator. M_i and D_i are inertia and damping factors for the ith generator, x_{di} and x_{qi} are d and q-axis synchronous reactance of the ith generator. T'_{doi} is the d-axis transient open circuit time constant of the ith machine. E_{fdi} is the ith generator exciter voltage, P_{mi} and P_{Gi} are the ith generator mechanical and electrical power, respectively. Generated active and reactive electric powers are shown by:

$$P_{Gi} = \frac{1}{X'_{di}} E'_{qi} V_{M+i} \sin(\delta_i - \theta_{M+i}) -$$

$$- \frac{x'_{di} - x_{qi}}{2 x'_{di} x_{qi}} V^2_{M+i} \sin(2(\delta_i - \theta_{M+i})),$$

$$(3)$$

$$Q_{Gi} = \frac{1}{X'_{di}} \left[E'_{qi} V_{M+i} \cos(\theta_{M+i} - \delta_i) - V^2_{M+i} \right] +$$

$$+ \frac{x'_{di} - x_{qi}}{2 x'_{di} x_{qi}} V^2_{M+i} [\cos(2(\theta_{M+i} - \delta_i) - 1)].$$

The injected real and reactive powers into k^{th} node are:

$$P_k = \sum_{i=M+1}^{M+N} B_{ki} V_k V_i \sin(\theta_k - \theta_i),$$

$$Q_k = - \sum_{i=M+1}^{M+N} B_{ki} V_k V_i \cos(\theta_k - \theta_i),$$

$$(4)$$

where V_i and θ_i are the magnitude and phase voltage of the ith bus, B_{ki} is the susceptance of the k-i branch. P_k and Q_k are the active and reactive power injected into k^h node. P_{Lk} and Q_{Lk} are active and reactive powers of k^h load.

Therefore, the equilibrium of powers at load buses offers the load flow equations as:

$$P_k + P_{Lk} - P_{Gk} = 0,$$

$$Q_k + Q_{Lk} - Q_{Gk} = 0.$$

$$(5)$$

3.2. Direct Lyapanov Method

Let $w(x)$ be the direct Lyapunov function or energy function explained for the power system model described by Eq. (2) through Eq. (5). Any disturbance in power system involves a power imbalance that moves the system trajectory from the pre-fault stable equilibrium point to a transient point $x_i(t)$ that has a higher energy level than post-fault equilibrium point. If $\dot{w} = dw/dt$ is negative, direct Lyapunov function $w(x)$ decreases with time and tends towards its minimum value which appears at the post-fault equilibrium point $\hat{x}_i$. The more negative value of $\dot{w}$ means the system returns to the equilibrium point $\hat{x}_i$ quickly (i.e. the better damping in power system) [12], [13]. With assumption of $\hat{x}_i = 0$, the direct Lyapunov function for SPM power system without any control is written by:

$$w(\omega, \delta, E'_q, V, \theta) = w_1 + w_2 + C_0,$$

$$w_1 = \frac{1}{2} \sum_{k=1}^{M} M_k \omega_k^2,$$

$$(6)$$

$$w_2 = \sum_{i=1}^{8} w_{2i},$$

where w_1 is kinetic energy and w_2 is potential energy that will be described as:

$$w_{21} = -\sum_{k=1}^{M} P_{mk}\delta_{mk},$$

$$w_{22} = \sum_{k=M+1}^{M+N} P_{Lk}\theta_k,$$

$$w_{23} = \sum_{k=M+1}^{M+N} \int \frac{Q_{Lk}}{V_k} dV_k,$$

$$w_{24} = \sum_{k=M+1}^{2M} \frac{1}{2x'_{dk-M}}[E'^2_{qk-M} + V_k^2 -$$

$$-2E'_{qk}V_k\cos(\delta_{k-M} - \theta_k)],$$

$$w_{25} = -\frac{1}{2}\sum_{k=M+1}^{M+N}\sum_{l=M+1}^{M+N} B_{kl}V_kV_l \quad (7)$$

$$\cos(\theta_k - \theta_l),$$

$$w_{26} = \sum_{k=M+1}^{M} \frac{x'_{dk-M} - x_{qk-M}}{4x'_{dk-M}x_{qk-M}}$$

$$\left[V_k^2 - V_k^2\cos(2(\delta_{k-M} - \theta_k))\right],$$

$$w_{27} = -\sum_{k=1}^{M} \frac{E_{fdk}E'_{qk}}{x_{dk} - x'_{dk}},$$

$$w_{28} = -\sum_{k=1}^{M} \frac{E'^2_{qk}}{2(x_{dk} - x'_{dk})}.$$

C_0 is a constant, such that at post-fault equilibrium point, the total energy, Eq. (7), is equal to zero. More details about energy function are given in [12], [13]. As it could be shown, the time derivative of the direct Lyapunov function Eq. (7) across the trajectories of the uncontrolled system is written by:

$$\dot{w}(x) = -\sum_{k=1}^{M} D_k\omega_k^2 -$$

$$-\sum_{k}^{M} \frac{T'_{dok}}{x_{dk} - x'_{dk}}(\dot{E}'_{qk})^2 \le 0. \quad (8)$$

STATCOM supports grid voltage by exchanging reactive power with power system, which means the output voltage of VSI ($\bar{V}_{sh}$) is in phase with linked bus voltage ($\bar{V}_i$). But if dc-bus voltage (V_{dc}) could be supported by an energy storage system, the VSI voltage angle (θ_{dc}) can varies from 0 to 2π and satisfies the condition of active power exchange.

Therefore STATCOM could be modeled as ideal voltage source and transformer leakage reactance x_{sh} in series (Fig. 3). Assume STATCOM/SMES is connected to bus i and its voltage amplitude relates to V_i by r_{sh}. Thus, VSI voltage is described as in Eq. (9).

$$\bar{V}_{sh} = r_{sh}V_ie^{j\theta_{sh}} = V_i[r_{sh}\cos(\theta_{sh}) +$$

$$jr_{sh}\sin(\theta_{sh})]. \quad (9)$$

This voltage in synchronous reference frame has been explained as direct and quadrature component by Eq. (10):

$$\bar{V}_{sh} = r_{sh}V_ie^{j\theta_{sh}} = V_i[u_d + ju_q]. \quad (10)$$

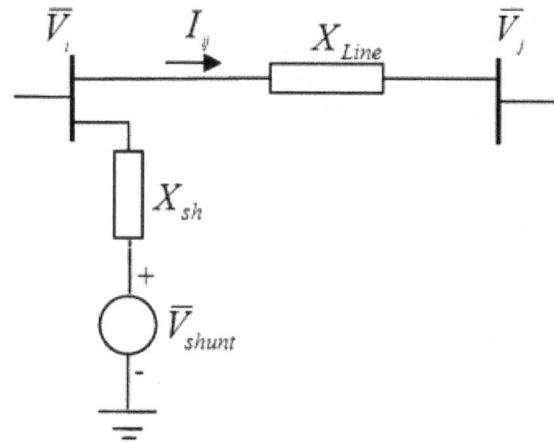

Fig. 3: Equivalent circuit of STATCOM.

STATCOM/SMES existence into the power system will modify the load flow Eq. (5) and also time derivative of Lyapannov function Eq. (5) will be changed. The powers equilibrium in i^{th} bus is based on Eq. (11):

$$P_i + P_{Li} - P_{Gi} + P_{shi} = 0,$$

$$Q_i + Q_{Li} - Q_{Gi} + Q_{shi} = 0, \quad (11)$$

where P_{shi} and Q_{shi} are interchanging active and reactive powers Between i^{th} bus and STATCOM/SMES consequently, which have been stated by Eq. (12):

$$P_{shi} = b_{sh}V_i^2[u_d\sin(\theta_i) - u_q\cos(\theta_i)],$$

$$Q_{shi} = b_{sh}V_i^2[1 - u_q\sin(\theta_i) - u_d\cos(\theta_i)], \quad (12)$$

$$b_{sh} = 1/X_{sh}.$$

Time derivative of Lyapunov function with STATCOM/SMES existence is related to Eq. (13):

$$\dot{w} = \dot{w}_{uncontrol} - P_{shi}\dot{\theta}_i - Q_{shi}\frac{\dot{V}_i}{V_i}. \quad (13)$$

Based on Eq. (13), it is obvious that by controlling the inputs the negative rate of Lyapanov function will be increased. Controlling parameters (u_d, u_q) consist in transient (u_d'', u_q'') and steady-state controlling components (u_d', u_q'), which have large time constant and improving system dynamic behavior extremely effective, Thus just single transient components are considered for power oscillation damping. By replacing Eq. (12) in Eq. (13) and set them based on controlling parameters, Eq. (14) will be gained:

$$\dot{w} = \dot{w}_{uncontrol} - b_{sh}V_i\frac{d}{dt}[-V_i\cos(\theta_i)]u_d''-$$

$$-b_{sh}V_i\frac{d}{dt}[-V_{i\sin(\theta_i)}]u_g''. \tag{14}$$

$\bar{V}_i$ is written into the synchronous reference frame as:

$$\bar{V}_i = V_i\cos(\theta_i) + jV_i\sin(\theta_i) = V_{di} + jV_{qi}. \tag{15}$$

By replacing Eq. (15) in Eq. (14), the Eq. (16) will be gained:

$$\dot{w} = \dot{w}_{uncontrol} - b_{sh}V_i\frac{d}{dt}(-V_{di})u_d''-$$

$$-b_{sh}V_i\frac{d}{dt}(-V_{qi})u_g''. \tag{16}$$

In order to keep negative sign in Eq. (16), transient controlling components (u_d'', u_q'') should have a similar sign with their sign of coefficients. Therefore STATCOM controlling law that has been derived from Eq. (16) will be as in Eq. (17):

$$u_d'' = k_1\frac{d}{dt}(-V_{di}),$$

$$u_g'' = k_2\frac{d}{dt}(-V_{gi}), \tag{17}$$

where k_1 and k_2 are positive coefficients and their value depends on oscillation damping time. In order to decentralize control strategy, all of the measurement signals must be local. On the other hand due to obtained control strategy from Eq. (17) a reference frame signal, that should be a local signal is required. If V_i consider in a reference frame, u_q'' will became zero in Eq. (17) and in this case Lyapanov method wouldn't be an optimum strategy, so we consider I_{ij} to provide synchronous reference frame in order to calculate the direct-axis and quadrature-axis components of ac voltage bus and converter voltage and provides a complete decentralized control law.

$$P_{shi} = b_{sh}V_i(u_d'V_{qi} - u_q'V_{di}) = 0,$$

$$u_d'V_{qi} - u_q'V_{di}) = 0. \tag{18}$$

The voltage error value $(V_{refi} - V_i)$ is given to PI-controller to generate control signal (λ_{qsh}') which is related to STATCOM reactive power. Thus, following equations could be defined:

$$V_i - u_d'V_{di} - u_q'V_{qi} = \lambda_q',$$

$$u_d'V_{di} + u_q'V_{qi} = \lambda_q,$$

$$\lambda_q = V_i - \lambda_q'. \tag{19}$$

From Eq. (18) to Eq. (19) the control parameters for steady-state could be present as Eq. (20):

$$u_d' = \frac{\lambda_q V_{di}}{(V_{qi})^2 + (V_{di})^2} = \frac{\lambda_q V_{di}}{V_i^2},$$

$$u_q' = \frac{\lambda_q V_{qi}}{(V_{qi})^2 + (V_{di})^2} = \frac{\lambda_q V_{qi}}{V_i^2}. \tag{20}$$

4. Simulation Results

The effectiveness of the proposed control strategy will be illustrated using a three-machine test system in PSCAD/EMTDC software [14]. A single line of a sample system diagram has been demonstrated in figure 4. In this system, Generator G_3 has high power and consider as an infinite bus. Generators G_1 and G_2 have different inertial constant, hence system reveal similar behavior like actual power systems. A temporary short circuit occurred in bus 5 (B_5) while a STATCOM/SMES compensator placed in line 4 (L_4). In this simulation a unit SMES (100MJ/10H) have been utilized. The System responses with and without STATCOM/SMES presence are shown.

4.1. Without Compensation

The characteristics related to variation in speed, active and reactive power of generator G_1 have been represented in Fig. 5, Fig. 6 and Fig. 7, respectively. According to these characteristics, it can be seen that during sampling, all characteristics were in oscillation states and they are not damped.

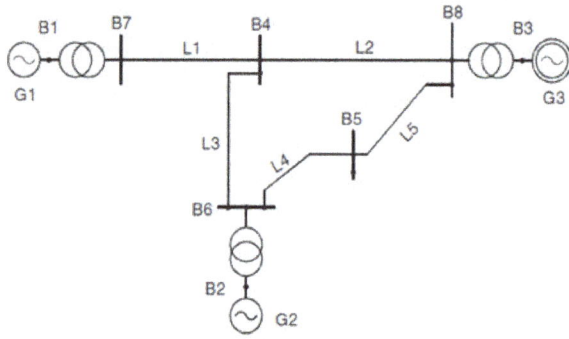

Fig. 4: Three-machine test system.

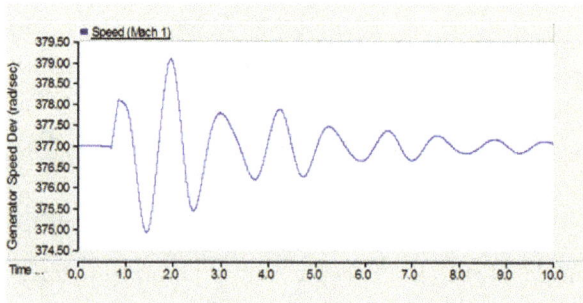

Fig. 5: Generator No.1 speed without compensation presence.

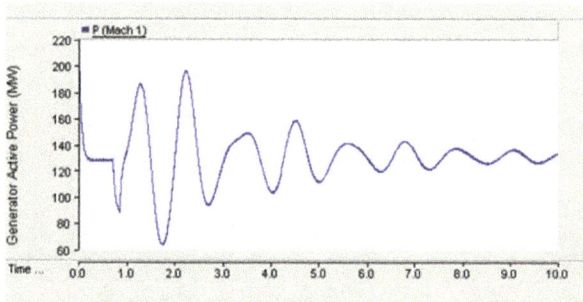

Fig. 6: Generator No.1 active power without compensation presence.

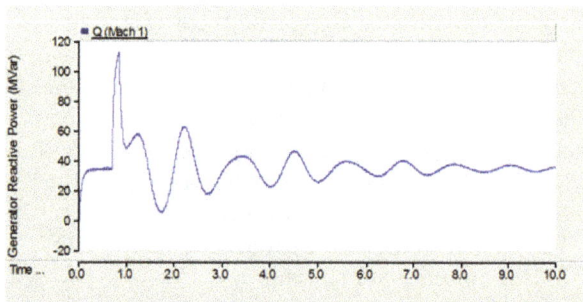

Fig. 7: Generator No.1 reactive power without compensation presence.

G_1 active power deviations have been represented in Fig. 9, as it can be seen under this conditions, oscillations amplitude of power in comparison to previous state decrease rapidly and this fluctuation have been damped and also these oscillations are damped in less time. Generator G_1 speed and reactive power variation characteristics have been shown in Fig. 8 and Fig. 10 respectively, that express desirable performance of STATCOM/SMES combinational compensation is in damping oscillations and increasing system safety and improvement in dynamic performance of power system.

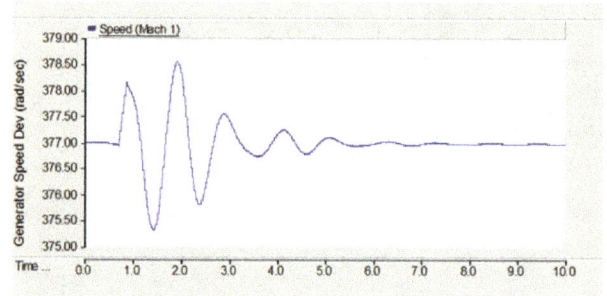

Fig. 8: Generator No.1 speed with STATCOM/SMES compensator presence.

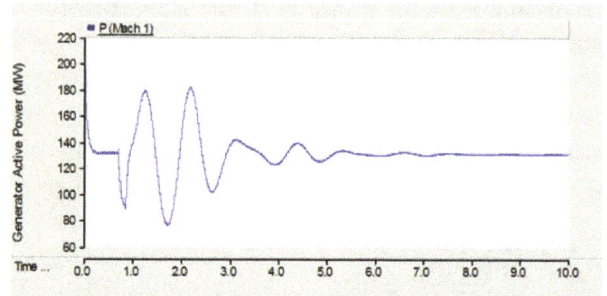

Fig. 9: Generator No.1 active power with STATCOM/SMES compensator presence.

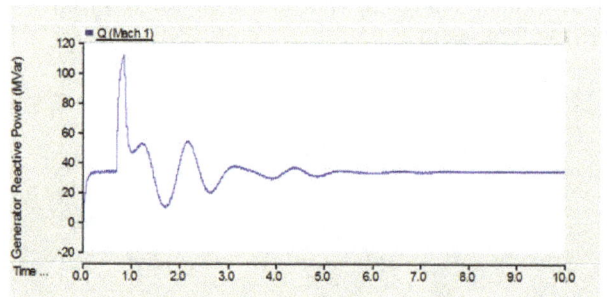

Fig. 10: Generator No.1 reactive power with STATCOM/SMES compensator presence.

4.2.   With STATCOM/SMES Compensator

Now in this condition, STATCOM/SMES compensator has been put in the bus5 B_5 of test system. Generator

4.3.   Overload in Bus 6 (B_6)

In this case, regarding to short circuit which occurred in system, a 100 MW and 6.6 MVAR load have been place in bus 6 B_6. Simulation results related to speed and active power deviations of generator G_1 have been represented in Fig. 11 and Fig. 12, respectively. Now STATCOM/SMES compensator have been placed in the bus5 B_5 of test system, characteristics of generator G_1 have been represented in Fig. 13 and Fig. 14, respectively. By comparison between specification in two state of with and without compensator it can be seen that integrated compensator have a significant effect on power oscillation damping, while the first oscillation damping have been decreased by presence of STATCOM/SMES compensator.

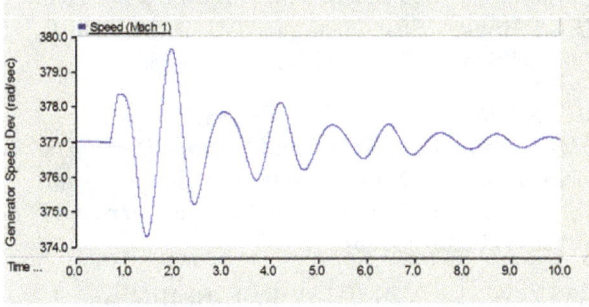

Fig. 11: Generator No.1 speed in state of overload in system and without compensation.

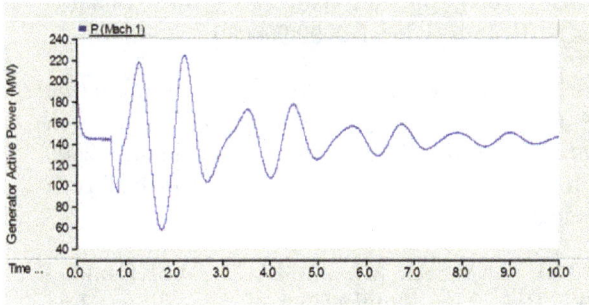

Fig. 12: Generator No.1 active power in state of overload in system and without compensation.

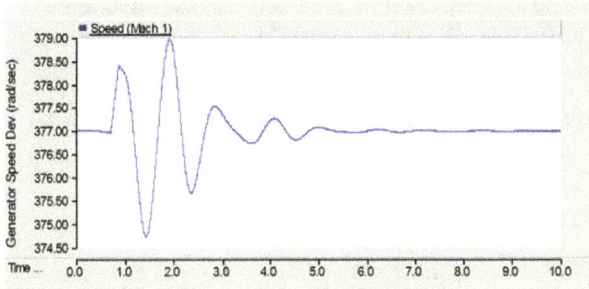

Fig. 13: Generator No.1 speed in state of overload in system and with STATCOM/SMES compensator presence.

4.4. Change in SMES Capacity

In the simulation which have been done in sections 4.1, 4.2 and 4.3 a unit SMES (10H) coil have been used, now we put a unit SMES (12H) coil and we expect that by

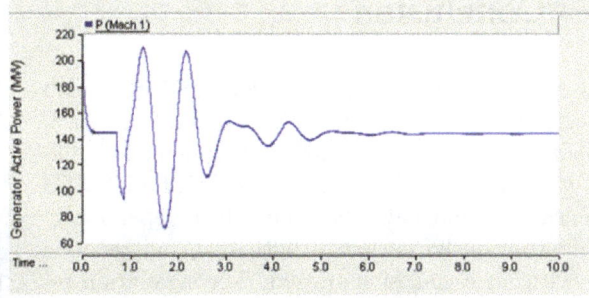

Fig. 14: Generator No.1 active power in state of overload in system and with STATCOM/SMES compensator presence.

increasing SMES coil capacity according to equation $E = \frac{1}{2}LI^2$ the stored energy in coil increase and as a result more energy transfer will occurr in system. On the other hand, since transient and dynamic stability are significantly control by active power and SMES storage transfers active power therefore it expects that increasing in storage capacity leads to improvement of power system dynamic operation. Simulation results related to generator G_1, active and reactive power characteristics have been demonstrated in Fig. 15 and Fig. 16, respectively.

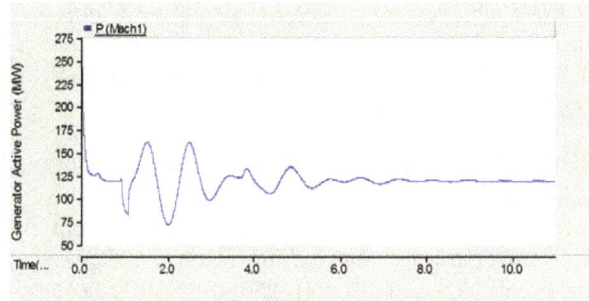

Fig. 15: Generator No.1 active power in state of SMES capacity variation.

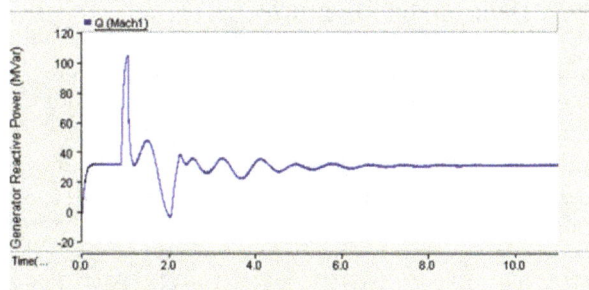

Fig. 16: Generator No.1 reactive power in state of SMES capacity variation.

5.   Conclusion

This paper presents the specifications and operation of integrated STATCOM/SMES compensator. Also, a benefit of integration of SMES system into STATCOM was presented. Integrated STATCOM/SMES compensator would have the ability to independently and simultaneously, exchange both real and reactive power with a transmission system. The simulation results show that STATCOM/SMES compensator has a more significant effect on dynamic performance improvement.

References

[1] UZUNOVIC, E. *Transient Stability and Power Flow Models of VSC FACTS controllers.* Waterloo, 2001. Ph.D. dissertation thesis. University of Waterloo.

[2] GLANZMANN, G. *FACTS, Flexible Alternating Current Transmission Systems.* Zurich: EEH-Power Systems Laboratory, 2005. DOI: 10.3929/ethz-a-004891251.

[3] TYLL, H. K. FACTS technology for reactive power compensation and system control. In: *2004 IEEE/PES Transmision and Distribution Conference and Exposition: Latin America.* Sao Paulo: IEEE, 2004, pp. 976–980. ISBN 0-7803-8775-9. DOI: 10.1109/TDC.2004.1432515.

[4] PULESTON, P. F., S. A. GONZALEZ and F. VALENCIAGA. A STATCOM based variable structure control for power system oscillations damping. *International Journal of Electrical Power.* 2007, vol. 29, iss. 3, pp. 241–250. ISSN 0142-0615. DOI: 10.1016/j.ijepes.2006.07.003.

[5] RAO, P., M. L. CROW and Z. YANG. STATCOM control for power system voltage control applications. *IEEE Transactions on Power Delivery.* 2000, vol. 15, iss. 4, pp. 1311–1317. ISSN 0885-8977. DOI: 10.1109/61.891520.

[6] TAN, Y. L. and Y. WANG. A robust nonlinear excitation and SMES controller for transient stabilization. *International Journal of Electrical Power.* 2004, vol. 26, iss. 5, pp. 325–332. ISSN 0142-0615. DOI: 10.1016/j.ijepes.2003.10.017.

[7] ZHANG, L., C. SHEN, M. L. CROW, L. DONG, S. PEKAREK and S. ATCITTY. Performance Indices for the Dynamic Performance of FACTS and FACTS with Energy Storage. *Electric Power Components and Systems.* 2004, vol. 33, iss. 3, pp. 299–314. ISSN 1532-5016. DOI: 10.1080/15325000590474438.

[8] YANG, Z., C. SHEN, L. ZHANG, M.L. CROW and S. ATCITTY. Integration of a StatCom and battery energy storage. *IEEE Transactions on Power Systems.* 2001, vol. 16, iss. 2, pp. 254–260. ISSN 0885-8950. DOI: 10.1109/59.918295.

[9] ARSOY, A., Y. LIU, P. F. RIBEIRO and F. WANG. Power converter and SMES in controlling power system dynamics. In: *Industry Applications Conference, 2000. Conference Record of the 2000 IEEE.* Rome: IEEE, 2000, pp. 2051–2057. ISBN 0-7803-6401-5. DOI: 10.1109/IAS.2000.883109.

[10] JOHNSON, B. K. and H. L. HESS. *Incorporating SMES Coils into FACTS and Custom Power Devices.* Moscow, 2000. University of Idaho. Available at: http://www.ee.uidaho.edu/ee/power/brian/PAPERS/smesapp.pdf.

[11] GYUGYI, L., Y. LIU, P. F. RIBEIRO and F. WANG. Application characteristics of converter-based FACTS controllers. In: *2000 International Conference on Power System Technology.* Perth: IEEE, 2000, pp. 391–396. ISBN 0-7803-6338-8. DOI: 10.1109/ICPST.2000.900089.

[12] GHANDHARI, M.. *Control Lyapunov Function: A Control Strategy for Damping of Power Oscillations in Large Power Systems.* Stockholm, 2000. Ph.D. Dissertation thesis. Royal Institute of Technology.

[13] JANUSZEWSKI, M., J. MACHOWSKI and J. W. BIALEK. Application of the direct Lyapunov method to improve damping of power swings by control of UPFC. *IEE Proceedings - Generation, Transmission and Distribution.* 2004, vol. 151, iss. 2, pp. 252–260. ISSN 1350-2360. DOI: 10.1049/ip-gtd:20040054.

[14] Manitoba HVDC Research Center, *PSCAD/EMTDC User's Manual.* Manitoba, 1988. Available at: https://hvdc.ca/uploads/ck/files/reference_material/EMTDC_User_Guide_v4_2_1.pdf.

About Authors

Mahmoud Reza SHAKARAMI was born in Khorramabad, Iran, in 1972. He received his M.Sc. and Ph.D. degree in Electrical Engineering from Iran University of Science and Technology of Tehran, Tehran, Iran, in 2000 and 2010 and is currently an Assistant Professor in Electrical Engineering Department of Lorestan University, Lorestan, Iran. His current research interests are: power system dynamics and stability and FACTS devices.

Reza SEDAGHATI was born in Kazeroon, Iran, on September 21, 1983. He received his M.Sc. degree in Electrical Engineering in 2009. He is currently as a Ph.D. student in Department of Electrical Engineering of Lorestan, Lorestan, Iran. His research interests include renewable energies, optimization, FACTS devices and power system dynamics.

Mohammad Bagher HADDADI was born in Shiraz, on September 1, 1988. He received his B.Sc. degree in Electrical Engineering from Shahid Beheshti University, Tehran, Iran, in 2011 and M.Sc. degree in Electrical Engineering from the university of Shiraz, Shiraz, Iran in 2013. His research interests also include power electronic, digital signal processing in optoelectronics and radiation protection.

16

Resistorless Cascadable Current-Mode Filter Using CCCFTAS

Witthaya MEKHUM[1], *Winai JAIKLA*[2]

[1]Department of Industrial Management, Faculty of Industrial Technology, Suan Sunandha Rajabhat University, 103 00 Bangkok, Thailand
[2]Department of Engineering Education, Faculty of Industrial Education, King Mongkut's Institute of Technology Ladkrabang, 105 20 Bangkok, Thailand

witthaya.me@ssru.ac.th, kawinai@kmitl.ac.th

Abstract. *The realization of current-mode biquad filter using current controlled current follower transconductance amplifiers (CCCFTAs) and grounded capacitors is presented. The proposed filter can simultaneously provide three standard transfer functions (low pass, high pass and band pass filter). The tuning of quality factor can be done without affecting the pole frequency. Low input and high output impedances of the configurations enable the circuit to be cascaded without additional current buffers. The use of only grounded capacitors is ideal for integration. The circuit performances are depicted through PSpice simulations, they show good agreement to theoretical anticipation.*

Keywords

CCCFTA, current-mode, filter, integrated circuit.

1. Introduction

Active filter is important in electrical and electronic applications, widely used for continuous-time signal processing. It can be found in many fields, including, communications, measurement, and instrumentation, and control systems [1], [2], [3]. The universal filter with one input and multiple output can be found in many applications, for example in touch-tone telephone tone decoder, in phase-locked loop FM stereo demodulator, or in the crossover network as a part of the three-way high-fidelity loudspeaker [2]. With growing interest in design of current-mode filters, more attention is being paid to the filters which have the high-output impedance because they make them easy to drive loads and they facilitate cascading without using a buffering device [4], [5].

The synthesis and design of analog filters using modern electronically controllable active building blocks (ABBs) give flexibility and convenience for designer. These filters can be easily controlled by microcomputer or microcontroller. Also some filter circuits which use active building block can avoid the use of the external resistors. This will reduce the cost and chip area. The design of analog circuits using active building blocks, taking into account several various criteria such as the minimum number of active elements or others, has been receiving considerable attention. Biolek et al. [6] proposed several circuit ideas of building blocks for voltage-, current- and mixed mode applications. One of them is the current follower transconductance amplifier (CFTA). This device allows applications with interesting features, especially those providing the electronic controllability. Later, Herencsar et al [7] introduced the modification of CFTA, called current controlled current follower transconductance amplifier (CCCFTA) which the parasitic resistance at input terminal is electronically tuned. It seems to be a versatile component in the realization of a class of analog signal processing circuits, especially analog frequency filters. It is really current-mode element whose input and output signals are currents.

It is obvious from the literature survey that a few current-mode filter circuits using CCCFTA have been hitherto published [7], [8], [9], [10]. These filters are focused on the use of single CCCFTA with grounded capacitors. This is ideal for IC implementation. However, the pole frequency and quality factor of filter in [7], [8], [9], [10] cannot be independently controlled. The high pass function of filter in [7] is not easy to cascade in current-mode circuit because this output current flows through the grounded capacitor.

This contribution presents a single-input three-output current-mode filter with low input and high impedance, employing CCCFTAs. It is suitable for fabricating as a monolithic chip or also for off-the-shelf implementation, consisting of 2 active elements and 2 grounded capacitors. The proposed filter can provide three standard functions (low-pass, high-pass and band-pass). The tuning of quality factor can be done without affecting the pole frequency.

2.   Circuit Configuration

There are two topics in this section as follows:

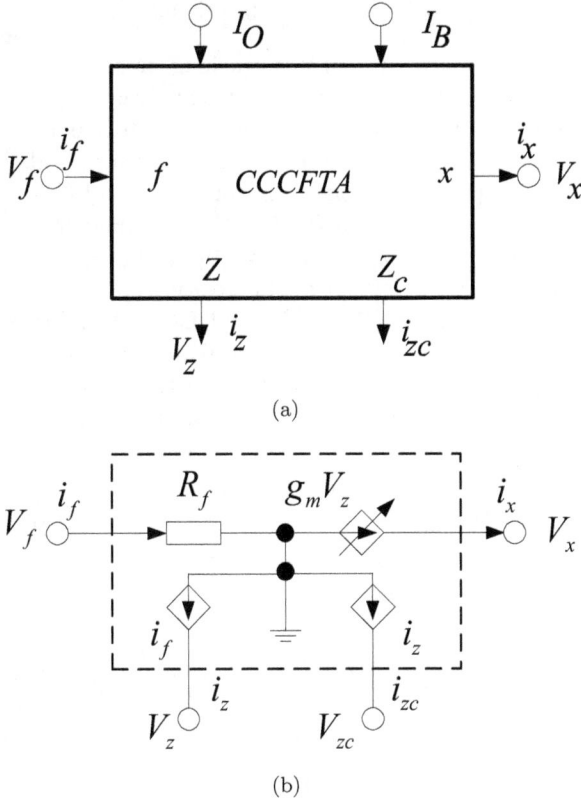

(a)

(b)

Fig. 1: Circuit symbol and equivalent circuit of CCCFTA.

2.1.   Basic Concept of CCCFTA

The circuit symbol and the ideal equivalent circuit model of the CCCFTA are shown in Fig. 1(a) and Fig. 1(b), respectively. It has finite input resistance R_f at f terminal. This parasitic resistance can be controlled by the bias current I_O. The input current i_f flows into f terminal which will be sent to z terminal. In some applications, to utilize the current through z terminal, an auxiliary z_c (z-copy) terminal is used [7], [8]. The internal current mirror provides a copy of the current flowing out of the z terminal to the z_c terminal. The voltage v_z on z terminal is converted into current using transconductance g_m, which flows into output terminal x. The g_m is tuned by I_B. In general,

CCCFTA can contain an arbitrary number of x terminals, providing currents i_x of both directions. The characteristics of the CCCFTA are represented by the following hybrid matrix:

$$\begin{bmatrix} V_f \\ i_{z,zc} \\ i_x \end{bmatrix} = \begin{bmatrix} R_f & 0 & 0 \\ 1 & 0 & 0 \\ 0 & 0 & \pm g_m \end{bmatrix} \begin{bmatrix} i_f \\ V_x \\ V_z \end{bmatrix}. \quad (1)$$

If the CCCFTA is realized using CMOS technology, R_f and g_m can be respectively written as

$$R_f = \frac{1}{\sqrt{8k_R I_o}}, \quad (2)$$

and

$$g_m = \sqrt{k_g I_B}. \quad (3)$$

For the internal construction of CMOS CCCFTA in Fig. 2, the parameters $k_R = \mu_n C_{ox}(W/L)_{1,2} = \mu_n C_{ox}(W/L)_{3,4}$ and $k_g = \mu_n C_{ox}(W/L)_{25,26}$. I_O and I_B are input bias current to control R_f and g_m, respectively.

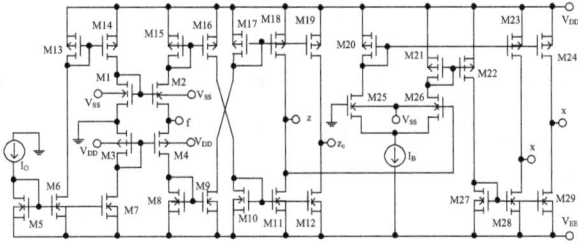

Fig. 2: Internal construction of CMOS CCCFTA.

2.2.   Proposed Current-Mode Filter

The proposed current-mode biquad filter is illustrated in Fig. 3. It is found that the filter consists of two CCCFTAs, and two grounded capacitors. The output current terminals are high impedance. Moreover, the low input impedance terminal can be achieved by setting IO1 as high as possible. With this configuration, the proposed filter is easy to cascade in the current-mode system. Considering the circuit in Fig. 3 and using CCCFTA properties in section 2.1, the current transfer functions for high pass, low pass and band pass filter are respectively shown as

$$\frac{I_{HP}}{I_{in}} = \frac{s^2}{s^2 + \dfrac{g_{m1}s}{C_1} + \dfrac{g_{m2}}{C_1 C_2 R_{f2}}}, \quad (4)$$

$$\frac{I_{LP}}{I_{in}} = \frac{\dfrac{g_{m2}}{C_1 C_2 R_{f2}}}{s^2 + \dfrac{g_{m1}s}{C_1} + \dfrac{g_{m2}}{C_1 C_2 R_{f2}}}, \quad (5)$$

$$\frac{I_{BP}}{I_{in}} = \frac{\dfrac{sg_{m1}}{C_1}}{s^2 + \dfrac{g_{m1}s}{C_1} + \dfrac{g_{m2}}{C_1 C_2 R_f 2}}. \tag{6}$$

The following relations are valid for the pole frequency and the quality factor:

$$\omega_0 = \sqrt{\frac{g_{m2}}{C_1 C_2 R_{f2}}}, \tag{7}$$

and

$$Q = \frac{1}{g_{m1}} \sqrt{\frac{C_1 g_{m2}}{C_2 R_{f2}}}. \tag{8}$$

From Eq. (7) and Eq. (8), if the R_f and g_m are equal to Eq. (2) and Eq. (3), the pole frequency and quality factor are re-written as

$$\omega_0 = \sqrt{\frac{(8k_g k_R I_{B2} I_{O2})^{1/2}}{C_1 C_2}}, \tag{9}$$

and

$$Q = \sqrt{\frac{(8k_g k_R I_{B2} I_{O2})^{1/2}}{k_g I_{B1} C_2}}. \tag{10}$$

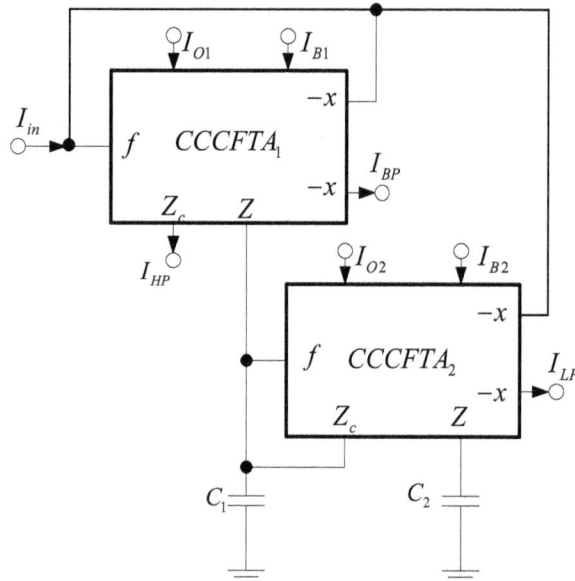

Fig. 3: Proposed current-mode filter.

It is apparent from Eq. (10) that the quality factor could be controlled by I_{B1} without affecting the pole frequency. It should remark that the parameters k_g and k_R are proportional to mobility and mobility falls with increasing temperature. Then the temperature variation will affect the ω_0 and Q [11]. From Eq. (7), the ω_0-sensitivity analysis with respect to the parameters of the active and passive element used can be given by:

$$S_{C_1, C_2, R_{f2}}^{\omega_0} = -\frac{1}{2}, \quad S_{g_{m2}}^{\omega_0} = \frac{1}{2}. \tag{11}$$

3. Analysis of Non-Ideal Case

In practice, the influences of voltage and current tracking errors and also the parasitic terminal impedances of CCCFTA will affect the filter performance. In this Section, these parameters will be taken into account. For non-ideal case, the CCCFTA can be respectively characterized with the following equations:

$$\begin{bmatrix} V_f \\ i_z \\ i_{zc} \\ i_x \end{bmatrix} = \begin{bmatrix} R_f & 0 & 0 & 0 \\ \alpha & 0 & 0 & 0 \\ 0 & 0 & \gamma & 0 \\ 0 & 0 & \pm\beta g_m & 0 \end{bmatrix} \begin{bmatrix} i_f \\ i_z \\ V_z \\ V_x \end{bmatrix}, \tag{12}$$

where α and γ are the current tracking errors from f and z ports to z port z_c port, respectively. β is the transconductance error gain from z port to x port. The influences of parasitic impedances are resistive and capacitive parts affecting the z, z_c, and x ports of CC-CFTA. Let us denote them R_z, C_z, R_{zc}, C_{zc}, and R_X, C_X, respectively. If R_{f1} is very low (by setting I_{O1} as high as possible), the influences of parasitic impedances of the x terminals of CCCFTA1 and CCCFTA2 are negligible because of their connection to low-impedance input f (CCCFTA1). Considering into these effects, the current transfer functions will be modified to the more general forms:

$$\frac{I_{HP}^*}{I_{in}} = \frac{\alpha_1 \gamma_1 s^2 + s\dfrac{\omega_z^*}{Q_z^*} + \omega_z^*}{s^2 + s\dfrac{\omega_0^*}{Q^*} + \omega_0^{*2}}, \tag{13}$$

$$\frac{I_{BP}^*}{I_{in}} = \frac{\alpha_1 \beta_1 g_{m1}\left(s\dfrac{1}{C_1^*} + \dfrac{1}{C_1^* C_2^* R_{z2}}\right)}{s^2 + s\dfrac{\omega_0^*}{Q^*} + \omega_0^{*2}}, \tag{14}$$

$$\frac{I_{LP}^*}{I_{in}} = \frac{\dfrac{\alpha_1 \alpha_2 \beta_2 g_{m2}}{C_1^* C_2^* R_{f2}}}{s^2 + s\dfrac{\omega_0^*}{Q^*} + \omega_0^{*2}}, \tag{15}$$

where parameters:

$$C_1^* = C_1 + C_{z1} + C_{zC2},$$

$$C_2^* = C_2 + C_{z2},$$

$$\frac{\omega_z^*}{Q_z^*} = \frac{\alpha_1 \gamma_1}{C_2^* R_{z2}} + \frac{\alpha_1 \gamma_1}{C_1^*}\left(\frac{1}{R_{z1}} + \frac{1}{R_{zc2}}\right) + \frac{\alpha_1 \gamma_1}{C_1^* R_{f2}}(1 - \alpha_2 \gamma_2),$$

$$\omega_z^* = \frac{\alpha_1 \gamma_1}{C_1^* C_2^* R_{z2}} + \left(\frac{1}{R_{z1}} + \frac{1}{R_{zc2}}\right) + \frac{\alpha_1 \gamma_1}{C_1^* C_2^* R_{f2} R_{z2}}(1 - \alpha_2 \gamma_2).$$

In this case, the pole frequency and quality factor is modified to

$$\omega_0^* = \sqrt{\frac{1}{C_1^* C_2^* R_{z2}} \left(\frac{1}{R_{z1}} + \frac{1}{R_{zc2}} + \frac{1 - \alpha_2 \gamma_2}{R_{f2}} + \alpha_1 \beta_1 g_{m1} \right) + \frac{\alpha_1 \alpha_2 \beta_2 g_{m2}}{C_1^* C_2^* R_{f2}}}, \tag{16}$$

and

$$Q^* = \frac{R_{f2} \sqrt{\dfrac{C_1^*}{C_2^* R_{z2}} \left(\dfrac{1}{R_{z1}} + \dfrac{1}{R_{zc2}} + \dfrac{1 - \alpha_2 \gamma_2}{R_{f2}} + \alpha_1 \beta_1 g_{m1} \right) + \dfrac{\alpha_1 \alpha_2 \beta_2 C_1^* g_{m2}}{C_2^* R_{f2}}}}{\dfrac{C_1^* R_{f2}}{C_2^* R_{z2}} + \left(\dfrac{R_{f2}}{R_{z1}} + \dfrac{R_{f2}}{R_{zc2}} + 1 - \alpha_2 \gamma_2 + \alpha_1 \beta_1 g_{m1} R_{f2} \right)}. \tag{17}$$

It should be mentioned that the stray/parasitic z-terminal capacitances are absorbed by C_1 and C_2 as it appears in shunt with them. However, the parasitic resistance R_{z1}, R_{zc2} and R_{z2} not only affect the ω_0 and Q by they also add parasitic zeros to the HP and BP transfer functions. The parameters α, γ and β of the CCCFTA affect the gain of all filter responses.

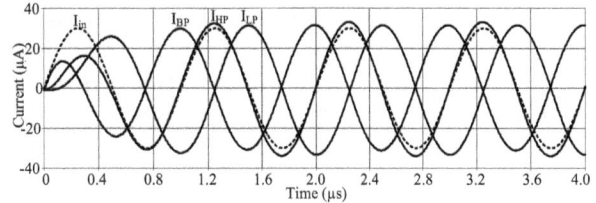

Fig. 4: Gain responses.

4. Simulation Results

To verify the theoretical analyses, the proposed CMOS CCCFTA implementation in Fig. 2 is examined using the PSPICE simulation program. The PMOS and NMOS transistors have been simulated by respectively using the parameters of a 0,25 μm TSMC CMOS technology [12]. The DC power supply voltages are ±1,25 V. The aspect ratios (W/L) of transistors are as follows: M1-M2 is 2 μm/0,5 μm; M3-M4 is 7,5 μm/0,5 μm; M25-M26 is 24 μm/0,5 μm; other PMOSs are 2 μm/0,25 μm and other NMOSs are 3 μm/0,25 μm. The filter was simulated with the following parameters of its components: $C_1 = C_2 = 5$ pF, $I_{O1} = 400$ μA, $I_{O2} = 40$ μA, $I_{B1} = 100$ μA, and $I_{B2} = 270$ μA. According to these conditions, the tracking errors and parasitic components are as follows: $\alpha_1 = \gamma_1 = 1,1$, $\alpha_2 = \gamma_2 = 1,15$, $\beta_1 = 1,3$, $\beta_2 = 1,42$, $R_{z1} = R_{zC1} = 55,46$ kΩ, $C_{z1} = C_{zC1} = 54$ fF, $R_{x1} = 43,56$ kΩ, $C_{x1} = 5,6$ fF, $R_{z2} = R_{zC2} = 141,58$ kΩ, $C_{z2} = C_{zC2} = 53$ fF, $R_{x2} = 32,62$ kΩ, and $C_{x2} = 5,5$ fF. It is found that the current transfers α and γ have an error of 10 % and 15 %, respectively. The errors in the transadmittance g_m represented by coefficients β are 30 % and 42 %. These errors, mainly in case of the transadmittance g_m, are still very high. Anyway, these errors do not seem to be that significant from the view point of final behavior of the proposed structure. Also, using the tuning feature, the real behavior of the active elements can be further suppressed by proper adjustment of the bias currents. The simulated gain responses of the proposed filter are shown Fig. 4. It is clearly seen that the filter can simultaneously provide low-pass, high-pass and band-pass functions without modifying the circuit topology. The simulations yield the pole frequency of 30 MHz and

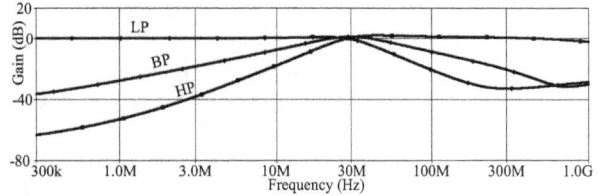

Fig. 5: Time domain responses.

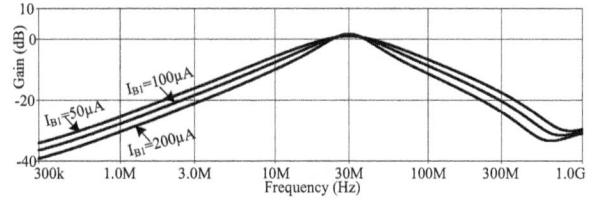

Fig. 6: Band-pass response for different values of I_{B1}.

the quality factor of 0,95. The expected value of pole frequency from Eq. (9) is 32,13 MHz (thus the deviation is 6.63 %). The expected value of quality factor from Eq. (10) is 0,92 (thus the deviation is 3,26 %). This error is from the influences of current tracking errors and parasitic impedances of CCCFTA as analyzed in section 4. The proposed filter was excited by 25 μA/30 MHz sinusoidal signal. The transient responses are shown in Fig. 5. The total harmonic distortions (THD) for IHP, IBP and ILP are 1,26 %, 0,73 % and 0,52 %, respectively. Considering Eq. (10), the Q can be controlled by I_{B1} without affecting the ω_0. The Q tuning is confirmed via the BP response in Fig. 6. By varying I_{B1} with different values of 50 μA, 100 μA and 200 μA, the simulated quality factors are 1,39, 0,95 and 0,704, respectively. The magnitudes of output impedance for HP, BP and LP are 52,14 kΩ,

112,78 kΩ and 32,62 kΩ, respectively. The power consumption is about 8 mW.

5.   Conclusion

The current-mode biquad filter based-on CCCFTA has been presented. The proposed filter can provide three standard transfer functions (LP, BP and HP) with low input and high output impedances. It consists of two CCCFTAs and two grounded capacitors. The orthogonal current control of the quality factor and pole frequency is achieved. As mentioned advantages, the proposed circuit is convenient to fabricate in integrated circuit (IC). The PSPICE simulation results agree well with the theoretical anticipation.

Acknowledgment

The authors would like to thank the anonymous reviewers for providing valuable comments which helped improve the paper substantially. Research described in this paper was financially supported from Suan Sunandha Rajabhat University and King Mongkut's Institute of Technology Ladkrabang (KMITL).

References

[1] SEDRA, A. S. and K. C. SMITH. *Microelectronic circuits*. 3rd ed. Philadelphia: Oxford University Press, 1991. ISBN 9780030516481.

[2] IBRAHIM, M. A., S. MINAEI, and H. A. KUNTMAN. A 22.5 MHz current-mode KHN-biquad using differential voltage current conveyor and grounded passive elements. *International Journal of Electronics and Communication (AEU)*. 2005, vol. 59, iss. 5, pp. 311–318. ISSN 1434-8411. DOI: 10.1016/j.aeue.2004.11.027.

[3] PANDEY, R., N. PANDEY, S. K. PAUL, A. SINGH, B. SRIRAM, and K. TRIVEDI. Voltage mode OTRA MOS-C single input multi output biquadratic universal filter. *Advances in Electrical and Electronic Engineering*. 2012, vol. 10, no. 5, pp. 337–344. ISSN 1336-1376.

[4] HORNG, J.-W. High output impedance current-mode universal biquadratic filters with five inputs using multi-output CCIIs. *Microelectronics Journal*. 2011, vol. 42, iss. 5, pp. 693–700. ISSN 0026-2692. DOI: 10.1016/j.mejo.2011.02.007.

[5] SOLIMAN, A. M. New current mode filters using current conveyors. *International Journal of Electronics and Communication (AEU)*. 1997, vol. 51, iss. 5, pp. 275–278. ISSN 1434-8411.

[6] BIOLEK, D., R. SENANI, V. BIOLKOVA and Z. KOLKA. Active elements for analog signal processing, classification, review and new proposals. *Radioengineering*. 2008, vol. 17, no. 4, pp. 15–32. ISSN 1210-2512.

[7] HERENCSAR, N., J. KOTON, K. VRBA, A. LAHIRI and O. CICEKOGLU. Current-controlled CFTA-based current-mode SITO universal filter and quadrature oscillator. In: *International Conference on Applied Electronics (AE), 2010*. Pilsen: IEEE, 2010, pp. 1–4. ISBN 978-80-7043-865-7.

[8] JAIKLA, W., S. LAWANWISUT, M. SIRIPRUCHYANUN and P. PROMMEE. A four-inputs single-output current-mode biquad filter using a minimum number of active and passive components. In: *35th International Conference on Telecommunications and Signal Processing (TSP2012)*. Prague: IEEE, 2012, pp. 378–381. ISBN 978-1-4673-1117-5. DOI: 10.1109/TSP.2012.6256319.

[9] KUMNGERN, M. Electronically tunable current-mode universal biquadratic filter using a single CCCFTA. In: *IEEE International Symposium on Circuits and Systems (ISCAS), 2012*. Seoul: IEEE, 2012, pp. 1175–1178. ISBN 978-1-4673-0218-0. DOI: 10.1109/ISCAS.2012.6271443.

[10] DUANGMALAI, D., P. SILAPAN and W. JAIKLA. A synthesis of electronically controllable current-mode universal biquadratic: filter based on CCCFTA. In: *The 2nd International Conference on Electrical Engineering and Applied Sciences*. Tokyo: Emerlad, 2013, pp. 1740–1746. ISBN 978-986-87417-1-3.

[11] LAHIRI, A. New CMOS-based resistor-less current-mode first-order all-pass filter using only ten Transistors and one external capacitor. *Radioengineering*. 2011, vol. 20, no. 3, pp. 638–643. ISSN 1210-2512.

[12] PROMMEE, P., K. ANGKEAW, M. SOMDUNYAKANOK, and K. DEJHAN. CMOS-based near zero-offset multiple inputs max-min circuits and its applications. *Analog Integrated Circuits Signal Processing*. 2009, vol. 61, iss. 1, pp. 93–105. ISSN 0925-1030. DOI: 10.1007/s10470-009-9281-2

About Authors

Witthaya MEKHUM was born in Nonthaburi, Thailand. He received Ph.D. in Technology Management from Phranakhon Rajabhat University in 2009. He has been with Department of Industrial Management, Faculty of Industrial Technology, Suan Sunandha Rajabhat University, Bangkok, Thailand since 1995. His research interests include technology management and communication system.

Winai JAIKLA was born in Buriram, Thailand. He received the B.Sc. I. Ed. degree in telecommunication engineering from King Mongkut's Institute of Technology Ladkrabang (KMITL), Thailand in 2002, M. Tech. Ed. in electrical technology and Ph.D.

in electrical education from King Mongkut's University of Technology North Bangkok (KMUTNB) in 2004 and 2010, respectively. From 2004 to 2011 he was with Electric and Electronic Program, Faculty of Industrial Technology, Suan Sunandha Rajabhat University, Bangkok, Thailand. He has been with Department of Engineering Education, Faculty of Industrial Education, King Mongkut's Institute of Technology Ladkrabang, Bangkok, Thailand since 2012. His research interests include electronic communications, analog signal processing and analog integrated circuit. He is a member of ECTI, Thailand.

17

On Multidimensional Linear Modelling Including Real Uncertainty

Jana NOWAKOVA, Miroslav POKORNY

[1]Department of Cybernetics and Biomedical Engineering, Faculty of Electrical Engineering and Computer Science, VSB–Technical University of Ostrava, 17. listopadu 15/2172, 708 33 Ostrava-Poruba, Czech Republic

jana.nowakova@vsb.cz, miroslav.pokorny@vsb.cz

Abstract. *The theoretical background for abstract formalization of vague phenomenon of complex systems is fuzzy set theory. In the paper are defined vague data as specialized fuzzy sets - fuzzy numbers and there is described a fuzzy linear regression model as a fuzzy function with fuzzy numbers as vague regression parameters. To identify the fuzzy coefficients of model the genetic algorithm is used. The linear approximation of vague function together with its possibility area are analytically and graphically expressed. The suitable numerical experiments are performed namely in the task of two-dimensional fuzzy function modelling and the time series fuzzy regression analysis as well.*

Keywords

Complex systems, fuzzy linear regression, fuzzy number, fuzzy set, genetic algorithms, possibility area, vague property.

1. Introduction

Regression models are often used in engineering practice wherever there is a need to reflect more independent variables together with the effects of other unmeasured disturbances and influences. In classical regression, we assume that the relationship between dependent variables and independent variables of the model is well-defined and sharp. In the real world, however, hampered by the fact that this relationship is more or less non-specific and vague. This is particularly true when modelling complex systems which are difficult to define, difficult to measure or in cases where it is incorporated into the human element [8].

The suitable theoretical background for abstract formalization of vague phenomenon of complex systems is fuzzy set theory. In the paper are defined vague data as specialized fuzzy sets - fuzzy numbers. Next, a fuzzy

linear regression model as a fuzzy function with fuzzy numbers as vague parameters is identified using the genetic algorithms.

2. Model Definition

2.1. Ordinary Lienar Model

The ordinary linear regression model of the investigated system [11] is given by a linear combination of values of its input variables:

$$Y = A_0 + A_1 x_1 + \ldots + A_n x_n = A_0 + \sum_{i=1}^{n} A_i x_i, \quad (1)$$

where $(x_1, \ldots, x_n)$ are input variables and $(A_0, A_1, \ldots, A_n)$ are ordinary regression coefficients.

The conventional regression model is based on the assumption that the system characteristics is defined as sharp, precise and deviations between the observed and estimated values of dependent variables are results of observation errors. The origin of a deviation between the observed and estimated values of dependent variables may not be of significant extent caused by poor local variables of the system structure. The causes of these variations are in a not very sharp nature of the system parameters. Such fuzzy phenomenon must also be reflected in the fuzziness of the corresponding parameters of the model.

2.2. Uncertainty Interval Linear Model

The development of the indeterminate regression model is the development of the model of vagueness, using the formalization of uncertainty rather than nu-

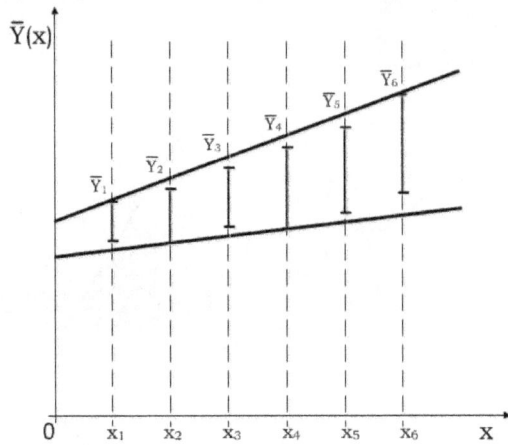

Fig. 1: Ordinary one-dimensional linear interval regression model.

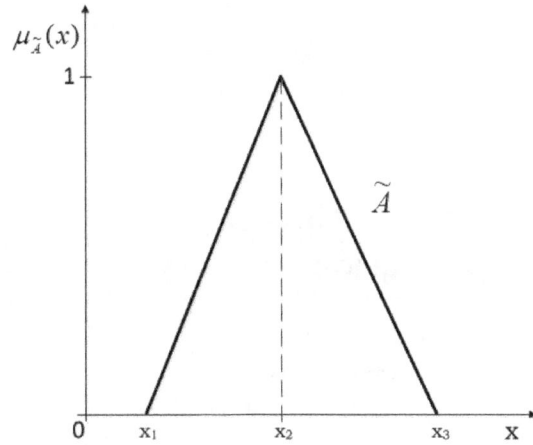

Fig. 2: Triangular membership function of fuzzy number $\widetilde{A}$.

merical intervals [3], [9]:

$$\overline{Y} = \overline{A}_0 + \overline{A}_1 x_1 + \ldots + \overline{A}_n x_n = A_0 + \sum_{i=1}^{n} \overline{A}_i x_i. \quad (2)$$

Regression coefficients are numeric interval:

$$\overline{A}\left(a_i + e_i, a_i - e_i\right) \equiv \overline{A}_i\left(a_i, e_i\right), \quad (3)$$

where a is the middle of the interval is and e is half of its width. For one dimensional function:

$$\overline{Y}(x) = \overline{A}_0 + \overline{A}_1 x = (a_0, e_0) + (a_1, e_1) x, \quad (4)$$

the interval regression model is depicted in Fig. 1.

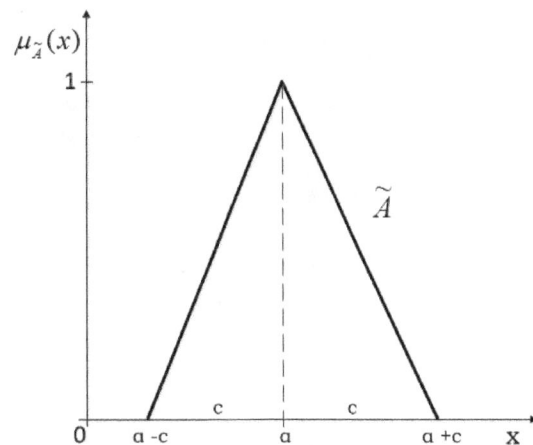

Fig. 3: Membership functions of fuzzy regression coefficients.

2.3. Interval Model Fuzzification

Regression models reflecting the vagueness of the modelled systems are called fuzzy regression models [10], [1] and [8]. The indeterminate nature of the fuzzy regression model is represented by the fuzzy output values and the fuzzy regression coefficients in the form of specialized fuzzy sets - fuzzy numbers. The shape of fuzzy linear regression model is given by:

$$\widetilde{Y} = \widetilde{A}_0 x_0 + \widetilde{A}_1 x_1 + \ldots + \widetilde{A}_n x_n = \sum_{i=0}^{n} \widetilde{A}_i x_i, \quad (5)$$

where $x_0 = 1$ and $\left(\widetilde{A}_0, \widetilde{A}_1, \ldots, \widetilde{A}_n\right)$ are fuzzy regression coefficients - fuzzy sets. The fuzzy set $\widetilde{A}$ is defined as image, which assigns to every element x of universum X number $\mu_{\widetilde{A}}(x) \in \langle 0, 1 \rangle$ as a degree of its membership function $\widetilde{A}$ [6], [7]:

$$\widetilde{A} = \left\{ x, \mu_{\widetilde{A}}(x) \mid x \in X \right\}; \mu_{\widetilde{A}}(x) \geq 0, \forall x \in X. \quad (6)$$

At least piecewise continuous function $\mu_{\widetilde{A}}(x) = f(x)$ is called membership function, which defined

fuzzy set $\widetilde{A}$ conclusively. The membership function is usually in engineering praxis approximated by broken line (Fig. 2).

Triangular fuzzy set $\widetilde{A}$ then formalizes uncertain number (fuzzy number) "about x_2". The degree of uncertainty of number x_2 is defined as the width of the carrier bearing of the fuzzy set $\widetilde{A}$ as closed interval $\langle x_1, x_3 \rangle$ (Fig. 2). Parameters of that fuzzy sets constitute structured vector of values of breaking point $[x_1, x_2, x_3]$. Using this vector fuzzy sets are computer-formalized.

In the fuzzy regression model, the fuzzy regression coefficients (fuzzy numbers) $\widetilde{A}$ are defined using its triangular shape membership function $\mu_{\widetilde{A}}(x)$ (Fig. 3), where α is the mean value (core) of fuzzy number $\widetilde{A}$ and c is a half of the width of the carrier bearing $\widetilde{A}_i = \{\alpha, c\}$. The output variable $\widetilde{Y}$ of fuzzy regression model (Eq. (5)) is fuzzy number defined using the triangular membership function (similar Fig. 3), where β is the mean value (core) of fuzzy number $\widetilde{Y}$ and d is a half of the width of the carrier bearing $\widetilde{Y} = \{\beta, d\}$.

Fuzzy regression modelling (Eq. (5)) requires operation with fuzzy numbers. For this types of operations it is needed to use relations of fuzzy arithmetic with usage of extensional principle.

3.   Fuzzy Arithmetic Application

Extensional principle (principle of extension) allows to transfer operation over ordinary numbers to operation over fuzzy numbers. It allows to create fuzzy arithmetic for computing with imprecise (fuzzy) numbers [2].

Let consider universum U and V and function f, which maps U to V, i.e.:

$$f : U \to V \qquad (7)$$

and fuzzy set $A \subseteq U$. Fuzzy set A then in V induces fuzzy set, whose membership function is defined by relation:

$$\mu_f(v) =$$
$$= \begin{cases} sup_{f(u)=v} \, \mu_A(u) & \text{if } \exists \, u \in U \text{ such that} \\ & v = f(u), \\ 0 & \text{elsewhere} \end{cases} \cdot \quad (8)$$

Using the extension principle fuzzy numbers arithmetic can be defined [6]. Take the case of the sum of two fuzzy numbers m ("about m") and n ("about n"). These relations are needed for calculation of output value $\widetilde{Y}$ (Eq. (5)):

$$\mu_{\widetilde{m} \otimes \widetilde{n}} = sup_{x,y/z=x \cdot y} min \left(\mu_{\widetilde{m}}(x), \mu_{\widetilde{n}}(x) \right), \quad (9)$$

$$\mu_{\widetilde{m} \oplus \widetilde{n}} = sup_{x,y/z=x+y} min \left(\mu_{\widetilde{m}}(x), \mu_{\widetilde{n}}(x) \right). \quad (10)$$

4.   Fuzzy Model Identification

4.1.   Identification Method Description

Fuzzy number Y_j^0 is mentioned of a triangular type. The values d_j can be calculate by the formula:

$$d_j = \frac{1}{2} \left| y_{j+1}^0 - y_{j-1}^0 \right|, \qquad (11)$$

where $j = 1, 2, \ldots, m$ is the number of observations. Finding values α_i and c_i as searched parameters of fuzzy regression coefficients $\widetilde{A}_i$ (Fig. 3) is defined as an optimization issue.

Fitness of the linear regression fuzzy model to the given data is measured through the Bass-Kwakernaaks's index H, Fig. 4, [4]. Adequacy of the

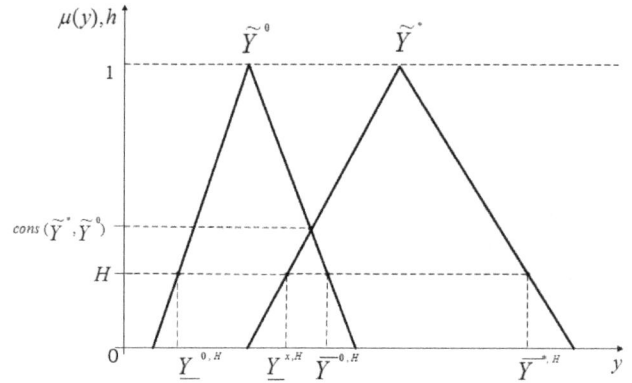

Fig. 4: Adequacy of linear regression model.

observed and estimated values is conditioned by the relation (Eq. (12)), the maximum intersection (consistency) of two fuzzy sets, the estimated $\widetilde{Y}_j^*$ and the examined $\widetilde{Y}_j^0$; must be higher than the set value H:

$$max_y \left\{ \mu_{\widetilde{y}^0}(y) \wedge \mu_{\widetilde{y}^*}(y) \right\} =$$
$$= Cons \left(\widetilde{Y}^0, \widetilde{Y}^* \right) \geq H. \quad (12)$$

Only if the condition (Eq. (12)) is fulfilled we assume good estimation of the observed output value $\widetilde{Y}_j^0$. The relation (Eq. (13)) is satisfied under the condition (Fig. 4):

$$\underline{Y}^{*,H} \leq \overline{Y}^{0,H}, \qquad (13)$$

$$\underline{Y}^{0,H} \leq \overline{Y}_j^{*,H}. \qquad (14)$$

Consider the determined level H the boundary of intervals $Y^{*,H}$ and relations (Eq. (13), Eq. (14)) can be expressed:

$$\underline{Y}^{*,H} = -(1-H) \sum_{i=0}^{n} c_i |x_i| + \alpha x_i, \qquad (15)$$

$$\overline{Y}^{*,H} = (1-H) \sum_{i=0}^{n} c_i |x_i| + \alpha x_i. \qquad (16)$$

According to the Fig. 3, it can be written:

$$\underline{Y}^{*,0} = y_j^0 (1-H) d^0, \qquad (17)$$

$$\overline{Y}^{*,0} = -y_j^0 (1-H) d^0. \qquad (18)$$

The conditions (Eq. (13), Eq. (14)) can be written in the shape:

$$\sum_{i=0}^{n} \sum_{j=1}^{m} a_{i,j} x_{i,j} +$$
$$+ (1-H) \sum_{i=0}^{n} \sum_{j=1}^{m} c_{i,j} |x_{i,j}| \geq$$
$$\geq y_j^0 + (1-H) d^0, \qquad (19)$$

$$-\sum_{i=0}^{n}\sum_{j=1}^{m} a_{i,j} x_{i,j} +$$
$$+ (1-H)\sum_{i=0}^{n}\sum_{j=1}^{m} c_{i,j} |x_{i,j}| \geq$$
$$\geq -y_j^0 + (1-H) d^0,$$
$$c_{i,j} \geq 0. \qquad (20)$$

The requirement on adequacy of the estimated and observed values will be complemented by the requirement on minimum possible total uncertainty of the identified fuzzy regression function:

$$\sum_{i=0}^{n}\sum_{j=1}^{m} c_{i,j} \to min, i = 0, 1, \ldots, n,$$
$$j = 1, 2, \ldots, m, \qquad (21)$$

where $i = 0, 1, \ldots, n$ is the number of input values of the regression function and $j = 1, 2, \ldots, m$ is the number of observations.

4.2. Model Parameters Optimization

Then we can set the optimization problem:

- minimization of fuzzy model vagueness (Eq. (21)),

- under the condition (Eq. (12)).

To solve the minimization problem under the condition, many authors use the linear programming method. Nevertheless, in this paper we use the genetic algorithm method to solve this problem [4]. Mainly, the reason is that the authors are oriented to use unconventional methods of artificial intelligence in order to prove their quality and efficiency in solving complex tasks. Genetic algorithms are a representative of evolutionary methods; their higher computational complexity is nowadays eliminated by high-performance computing. They are widely used in the search for optimal solutions. They can be well used for the identification of fuzzy regression models where they deal with the task of finding the optimal fuzzy regression coefficients as triangular fuzzy numbers. The identification of fuzzy regression coefficients – fuzzy numbers: $\widetilde{A}_0, \widetilde{A}_1, \ldots, \widetilde{A}_n$, was divided into two tasks:

- the identification of the mean value (core) α_i of fuzzy number $\widetilde{A}_i$ and

- the identification of c_i as a half of the width of the carrier bearing $\widetilde{A}_i = \{\alpha_i, c_i\}$.

The tasks are solved by using the genetic algorithm in series. First the identification of α_i and then the identification of c_i are done. Thus, the optimization of the fuzzy linear regression model is a two-step process when two genetic algorithms, designated G1 and G2,

are used. For the identification of the mean value (core) α_i of fuzzy number $\widetilde{A}_i$ the minimization of the fitness function J_1 is defined in the form:

$$min\, J_1 = min\frac{1}{m}\sum_{j=1}^{m}\left(y_j^0 - \beta_j\right)^2, \qquad (22)$$

and the genetic algorithm GA1 is used. For the identification of as a half of the width of the carrier bearing $\widetilde{A}_i$ the minimization of the fitness function J_2 is defined in the form:

$$min\, J_2 = min\sum_{j=1}^{m}\sum_{i=0}^{n} |c_{j,i}|, \qquad (23)$$

and the genetic algorithm GA2 with two constraints (Eq. (21)) is used. Minimization of the fitness function J_2 is based on the previous identification of the role of the mean value (core) α_i and uses the already identified values of α_i for determining the width of the carrier bearing α_i. The value of $H = 0.5$ is expertly determined in the next part of paper.

4.3. Genetic Algorithms Utilization

As mentioned before, the classical method of linear programming used for the identification of fuzzy regression coefficients [11] was substituted by using a genetic algorithm (GA) [4]. Mainly, the reason is that the authors are oriented to use unconventional methods of artificial intelligence in order to prove their quality and efficiency in solving complex tasks. Genetic algorithms are a representative of evolutionary methods; their higher computational complexity is nowadays eliminated by high-performance computing. They are widely used in the search for optimal solutions. They can be well used for the identification of fuzzy regression models where they deal with the task of finding the optimal fuzzy regression coefficients as triangular fuzzy numbers.

The identification of fuzzy regression coefficients – fuzzy numbers: $\widetilde{A}_0, \widetilde{A}_1, \ldots, \widetilde{A}_n$, was divided into two tasks:

- the identification of the mean value (core) α_i of fuzzy number $\widetilde{A}_i$ and

- the identification of c_i as a half of the width of the carrier bearing $\widetilde{A}_i = \{\alpha_i, c_i\}$.

The tasks are solved by using the genetic algorithm in series. First the identification of α_i and then the identification of c_i are done.

As it was mentioned before, the genetic algorithm is an unconventional optimisation method, which is used for minimization of the target optimization function

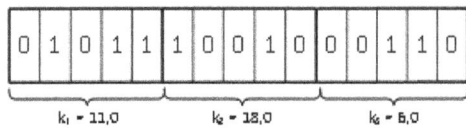

Fig. 5: Binary coded parameters in structure of chromosome.

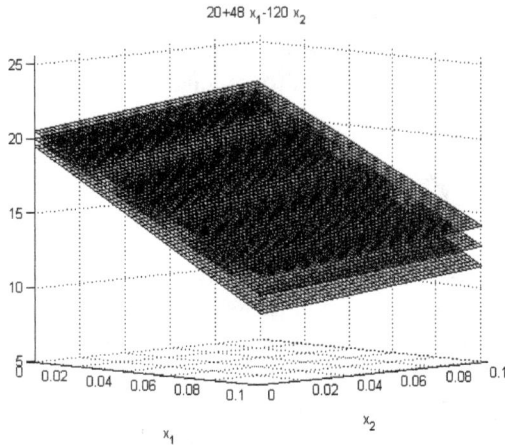

Fig. 6: Adequacy of linear regression model.

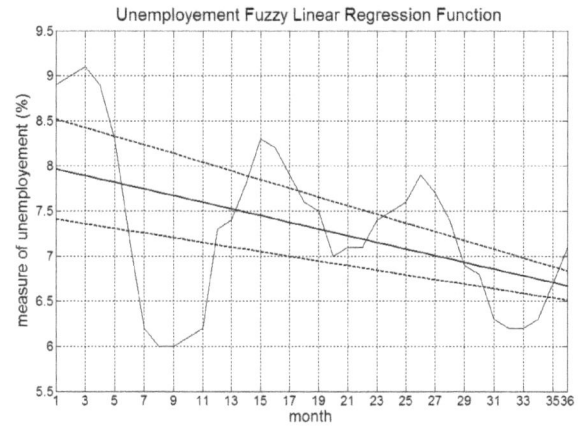

Fig. 7: Fuzzy linear regression function for unemployment in the Czech Republic.

(fitness function). It is used instead of conventional methods, such as the linear programming method.

GA is a seeking procedure which looks for the best solutions according to the fitness function based on the processes observed in Nature, on the principle of natural selection and genetic laws, i.e. selection, crossover and mutation. The basis of GA is to use a character string, also called a chromosome, in which parameters of an optimized model are stored. An example of a chromosome which is composed of three parameters k_1, k_2 and k_3 expressed by three 5-bit binary words is shown in Fig. 5.

Individual bits represent the string of chromosome genes; at the particular optimization step their specific values represent binary codes of three parameters of the model. Each chromosome is evaluated by the size of its fitness function, the value of which determines the distance of a solution (which is represented by a particular chromosome) from the optimal solution.

The set of evaluated n-chromosomes represents one population, the best individuals (solutions) of which are genetic operations of selection and are picked out for follow-up populations. Selected individuals are subjected to genetic operations of crossover, in which two individuals (parents) interchange gene circuits and generate two new chromosomes - offspring with different combinations of k_1, k_2 and k_3. The descendants, who were generated this way, then form a new population where individuals (solutions) appear to have better characteristics (better fitness function value) than the best individual in the population of parents. Then, an

appropriate follow-up offspring population is created (solution step, iteration) and the genetic crossing procedure is repeated. Good convergence for finding an optimal individual (solution) is supported by a genetic operation - mutation. The features of genetic operations as selection, crossover and mutation are defined by setting the internal parameters of the genetic algorithm in the way that the convergence of a solution to optimum is favorable.

The procedure of the genetic algorithm is usually finished by a solution step (population) in which the values of the fitness function of the best current individual and the best individual in the last step vary less than the specified limit (stop-criterion). As an optimal solution is then determined the best chromosome of the last population. Corresponding (coded) parameter values are used in the optimal model.

The main tasks while designing a genetic algorithm are the method of encoding the optimized parameters to a chromosome string and the definition of its fitness function. Optimization of the fuzzy linear regression model is a two-step process when two genetic algorithms, designated G1 and G2, are used.

5. Fuzzy Model Verification

For proving of efficiency of proposed method, the two dimensional linear function in form:

$$Y^0 = 20 + 48x_1 - 120x_2, \qquad (24)$$

was chosen. The set of Y^0 with ten members using (Eq. (19), Eq. (20)) was created. For creating the set of Y^0 the values of x_1 and x_2 were chosen randomly from the standard uniform distribution on the open interval $(0, 1)$ but multiplied by random integer. For fuzzification of observed value $a = 0.1$ was used. The result can be seen in Fig. 6, [5].

The usage in economic area is depicted in Fig. 7. As an input data the unemployment rate in the Czech republic for years 2009 to 2011 was used.

6. Conclusion

Abstract mathematical models of complex systems are often not very adequate because they do not accurately reflect the natural uncertainty and vagueness of the real world. The suitable theoretical background for abstract formalization of vague phenomenon of complex systems could be fuzzy set theory, which was shortly described. In the paper vague data as specialized fuzzy sets - fuzzy numbers are defined and it is described a fuzzy linear regression model as a fuzzy function with fuzzy numbers as vague parameters. Interval and fuzzy regression technology are discussed, the linear fuzzy regression model is proposed. It is used the effective genetic algorithm instead of commonly used linear programming method for identification of fuzzy regression coefficients of the model. The two-dimensional numerical example and practical economic usage are presented and the possibility area of vague model is graphically illustrated. Next research will be focused on model vague non-linear systems.

Acknowledgment

This work has been supported by Project SP2014/156, "Microprocessor based systems for control and measurement applications", of the Student Grant System, VSB–Technical University of Ostrava.

References

[1] ABDALLA, A. and J, BUCKLEY. Monte Carlo Methods in Fuzzy Non-Linear Regression. *New Mathematics and Natural Computation.* 2008, vol. 4, iss. 2, pp. 123–141. ISSN 1793-0057. DOI: 10.1142/S1793005708000982.

[2] HANSS, M. *Applied fuzzy arithmetic: an introduction with engineering applications.* New York: Springer-Verlag, 2005. ISBN 35-402-4201-5.

[3] ISHIBUCHI, H. and H. TANAKA. Identification of fuzzy parameters by interval regression models. *Electronics and Communications in Japan (Part III Fundamental Electronic Science).* 1990, vol. 73, iss. 12, pp. 19–27. ISSN 1520-6440. DOI: 10.1002/ecjc.4430731203.

[4] REEVES, C. R. *Genetic algorithms: principles and perspectives: a guide to GA theory.* Boston: Kluwer Academic, 2003. ISBN 14-020-7240-6.

[5] MATHWORKS. *MATLAB and Simulink for Technical Computing* [online]. Available at: http://www.mathworks.com.

[6] MOHAN, C. *Fuzzy set theory and fuzzy logic: an introduction.* Tunbridge Wells: Anshan, 2009. ISBN 18-482-9025-X.

[7] NOVAK, V. and I. PERFILIEVA. *Mathematical Principles of Fuzzy Logic.* Boston: Springer US, 1999. ISBN 978-146-1552-178.

[8] NOWAKOVA, J. and M. POKORNY. Fuzzy Linear Regression Analysis. In: *12th IFAC Conference on Programmable Devices and Embedded Systems.* Velke Karlovice: VSB–Technical University of Ostrava, 2013, pp. 245–249. ISBN 978-3-902823-53-3. DOI: 10.3182/20130925-3-CZ-3023.00079.

[9] POLESHCHUK, O. and E. KOMAROV. A fuzzy linear regression model for interval type-2 fuzzy sets. In: *Annual Meeting of the North American Fuzzy Information Processing Society (NAFIPS).* Berkeley: IEEE, 2012, pp. 1–5. ISBN 978-1-4673-2336-9. DOI: 10.1109/NAFIPS.2012.6290970.

[10] SHAPIRO, A. F. *Fuzzy Linear Regression.* Penn State University, 2005.

[11] A. TANSU. *Fuzzy Linear Regression: Fuzzy Regression.* New York: LAP LAMBERT Academic Publishing, 2012. ISBN 978-38-44-38442-0.

About Authors

Jana NOWAKOVA was born in 1987 in Trinec. She received her M.Sc. in Measurement and Control from VSB–Technical University of Ostrava, Faculty of Electrical Engineering and Computer Science in 2012. Nowadays she continues her studies in Technical Cybernetics. She is interested in addition to fuzzy modelling, also in statistical data processing in cooperation with University Hospital Ostrava.

Miroslav POKORNY was born in 1941. He received his M.Sc. in High-frequency communication technologies from Technical University Brno, Faculty of Electrical Engineering in 1963. His Ph.D. he received in 1994 in Cybernetics and Informatics from Technical University Brno, Faculty of Electrical Engineering and Informatics. Since 1993 he worked as associate professor and since 1998 he has worked as professor at VSB–Technical University of Ostrava,

Faculty of Electrical Engineering and Computer Science, Department of Cybernetics and Biomedical Engineering. Between 1964 and 1971 he worked as researcher in Research institute of metallurgy VZKG Ostrava-Vitkovice and between 1972 and 1992 as researcher and head of development in VUHZ-Research Institute of metallurgy.

18

Modelling of Data Stream Time Characteristics for Use of Inverse Multiplexer

Petr JARES, Jiri VODRAZKA

Department of Telecommunication Engineering, Faculty of Electrical Engineering,
Czech Technical University in Prague, Technicka 2, 166 27 Prague 6, Czech Republic

petr.jares@fel.cvut.cz, jiri.vodrazka@fel.cvut.cz

Abstract. *Today, increasing of the transmission rate in a telecommunications network is possible in various ways. One of them is inverse multiplexing. The inverse multiplexer divides a data stream to multiple parallel channels. This principle not only allows to increase the total available transmission rate, but also allows to reduce the error rate and interruption in data stream. The digital subscriber line may be used for the implementation of the inverse multiplex. Accurate knowledge of the transmission parameters of a digital subscriber line and the entire infrastructure of the network provider is necessary for the effective functioning of the terminal device with inverse multiplexing. It is necessary to know not only the parameters related to the transmission rate, but above all the parameters relevant to the time characteristics of data transmission. This paper describes how to obtain the transmission parameters of real digital subscriber lines and their modelling.*

Keywords

Digital subscriber line, inverse multiplexing, packet delay variation, transmission delay.

1. Introduction

The increase of a transmission rate or the error-free data transmission in telecommunication networks can be done with the use of different methods. The inverse multiplexer on transmit side divides a single data stream into multiple parallel data channels and the multiplexer on receive side puts together data from all channels to single data stream [1].

The inverse multiplexing requires installation of the special network element - multiplexer/demultiplexer (the Muldex) - one on the end user side and one on the service provider side. On the transmitting side, the Muldex is responsible for the proper packets dis-tribution in the particular channels with respect to the current transfer conditions and the selected transmission mode (speed increase, higher reliability). Therefore, on the transmitting side the Muldex must have the accurate information on the characteristics of data channels.

On the receiving side the Muldex is responsible for the proper arrangement of received packets, so they are set up into the original sequence (packet on transmitting side are numbered sequentially), and for the next hop. In the transmission mode, which increases fault tolerance, the Muldex must decide which of the received packet is error-free and which has to be discarded. The transmitted data are encapsulated into proprietary communication protocol. The described principle of inverse multiplex is shown on the Fig. 1.

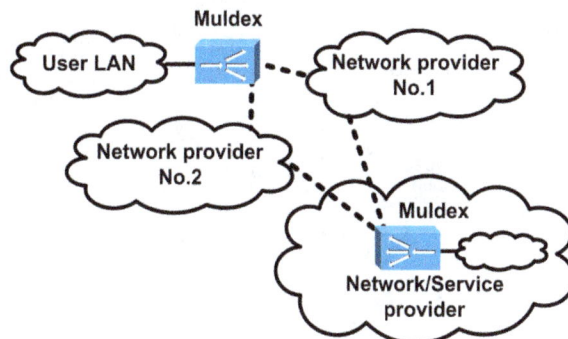

Fig. 1: The principle of inverse multiplexing.

The Muldex must be fully transparent for the TCP/IP protocol family. The reason is, the transmission channel is formed by networks of multiple network providers. The picture shows the parallel paths through two independent network providers presented by a dashed line. Two Muldex devices are placed on the user's end and the service provider side [2].

2.   Experiments with Inverse Multiplexer over DSL

Today, realistically achieved transmission rate in the digital subscriber lines ADSL2+ and VDSL2 (in general xDSL) in the downstream direction is in the order of few to tens of Mbps. Reached speeds in the upstream direction are significantly lower. Usually there are hundreds of kbps to few Mbps. The resultant value of the transmission rate is affected by the type of digital line, the actual conditions of the transmission, a network infrastructure of the provider and concentration (aggregate) point.

Development of the digital subscriber lines has not finished yet, and VDSL2 will not be the last one. The new ITU-T G.9701 recommendation allows further use of a metallic line in the last mile of the access network. When using the twisted pairs for data transmission with gigabit rate is expected [3].

2.1.   Network Infrastructure of the Digital Subscriber Lines

The network infrastructure is illustrated on Fig. 2. The diagram shows the xDSL line, aggregation network infrastructure including aggregation point and backbone networks of the multiple network providers. The aggregation point combines data streams from end users [4]. At this point, data packets are classified in the priority queues. When these queues are full, next incoming packets are dropped. Therefore, increased the delay packet transmission and packet loss may occur due to the existence of priority queues. Features of the aggregation point may also cause packet delay variation during the transmission [6], [8].

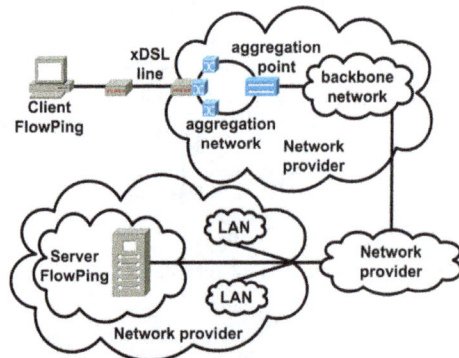

Fig. 2: Typical xDSL network infrastructure.

For testing use, at both ends of the transmission circuit a personal computer must be present. The software application FlowPing has been used for analysis of transmission parameters. The FlowPing server has sufficiently sized bandwidth (Ethernet 1 Gbps), so it will not affect the results of measurements of the digital subscriber line parameters. As already mentioned, for xDSL lines data transmission in the upstream direction is slower than in the downstream direction. Therefore, it is assumed that inverse multiplexer will enforce mainly the weaker direction of data transfer.

2.2.   Measuring Data Transmission Time Characteristics of the Digital Subscriber Lines

Currently, a large number of diagnostic tools for the analysis of the data channels transmission parameters or bi-directional data links are available. In most cases, these applications are able to measure only the value of the maximum achievable transmission rate and the transmission Round Trip Time (RTT).

However, these parameters are insufficient for implementing inverse multiplexer. During the transmission, the Muldex must have available not only the value of the current transmission rate, but also information about the current Packet Loss (PL), Packet One Way Delay (OWD) and Packet Delay Variation (PDV).

Custom developed application FlowPing [5] is able to accurately test the parameters of the data channel. On the transmitting side (FlowPing client) application generates precisely defined UDP (User Datagram Protocol) stream and creates log files. On the receiving end (FlowPing server), incoming packets are processed, and activity is logged. The FlowPing client and server are synchronized using the NTP protocol.

Thanks to log files post-processing, it is possible to obtain information about the parameters of the data channel. For example, due to internal clock synchronization of the testing PC running client of FlowPing application and testing server running server of FlowPing application, it is possible to calculate OWD parameter as time difference between receiving time and transmitting time for each packet. PDV parameter is calculated as the difference between OWD mean value and OWD value of a particular packet.

Analysis and theoretical modelling of these and other results are needed to control the Muldex. Knowledge of the behaviour characteristics of the data channel in time can help to prevent abnormal and unwanted states during data transmission.

2.3.   Transmission Parameters of the Experimental Digital Subscriber Lines

For the Muldex management requirements the transmission characteristics of two real lines, ADSL2+ from first provider and VDSL2 from the second provider, were measured. Basic parameters of the lines are shown in Tab. 1.

Tab. 1: Basic xDSL lines parameters.

Type of xDSL	Actual net data rate [kbps]	
	Upstream	Downstream
ADSL2+	515	8199
VDSL2	1415	24127

Parameters of the tested data stream are listed in the Tab. 2.

Payload filed size of UDP protocol packet was always 1464 B. Profile of the test data stream gradually loads the data channel with the increasing transmission rate. With the gradual increase of the transmission rate to a maximum value, an influence of the aggregation point and the network infrastructure can occur. The results of previous tests have shown that the data stream of ADSL2+ or VDSL2 is not affected by the value of the transmission rate to 75 % or 90 % of Actual Net Data Rate. The maximum value of 100 % of Actual Net Data Rate cannot be achieved with regard to the existing protocol structure PPPoEoA (ADSL2+) and PPPoE (VDSL2). Actual Net Data Rate is the transmission rate provided to higher communication layers by xDSL physical layer.

Tab. 2: Parameters of the testing data stream.

Time [s]	Test profile ADSL2+	Test profile VDSL2
	Percent of actual net data rate	Percent of actual net data rate
0	75	90
900	85	96.5
1800	85	96.5
2700	87	97.2
3600	87	97.2
4500	89	98
5400	89	98

2.4. Modelling of the Subscriber Line Delay

For each of the lines, a total of 40 tests were performed. The tests were carried out at different times of the day and different days of the week.

For ADSL2+ line it was found that the value of the maximum transmission rate V_{pmax} is equal to 89 % of the Actual Net Data Rate. This fact corresponds to the used PPPoEoA protocol structure, because this protocol structure requires approximately 11 % for the service communication (overhead). Before reaching V_{pmax} rate, OWD value equals 28.4 ± 1.2 ms (level of significance 95 %).

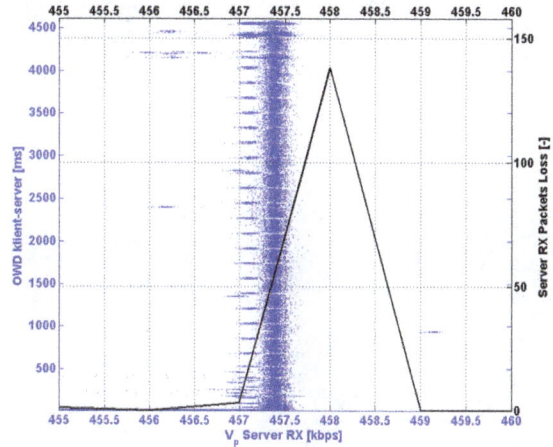

Fig. 3: The measured values of OWD and packet loss.

The graph on Fig. 3 shows the packet OWD (blue points) and packet loss data stream (black curve) dependency to its bit rate on the receiver side. The packet loss is calculated with 1 kbps interval, and it is always carried by a higher value in the interval. The graph evidently shows that up to V_{pmax} value, data transmission was error free. With transmission rate of V_{pmax} stream the OWD value is increased approximately 140 times and then PL starts to increase as well. In general, increase of PL occurs, when buffers of network elements are full, so transmission of other incoming packets is not possible.

Possible increase of PL before reaching V_{pmax} value can have several reasons. In general, in xDSL technology, it is possible that due to influence of external disturbance, during transmission, packets will be lost even if the V_{pmax} value was not reached. It is also possible that aggregation point on purpose denied specific packets of particular data flow to decrease transmission rate by using principles of TCP protocol (Transmission Control Protocol). The data flow tested in FlowPing application uses UDP protocol, where transmission rate is not affected by packet-loss of data channel.

The Goodness-of-Fit test (Kolmogorov-Smirnov) confirmed that it was possible to model the packet delay variation parameters of ADSL2+ lines with the logistic distribution with Probability Distribution Functions (PDF) according to Eq. (1):

$$f(x) = \frac{e^{\frac{x-\mu}{\sigma}}}{\sigma \left(1 + e^{\frac{x-\mu}{\sigma}}\right)^2}. \tag{1}$$

Distribution parameters are listed in Tab. 3. These values were obtained using Maximum-Likelihood Estimation (MLE). The following graph shows the PDV histogram and modelled theoretical (red curve) PDF logistic distribution.

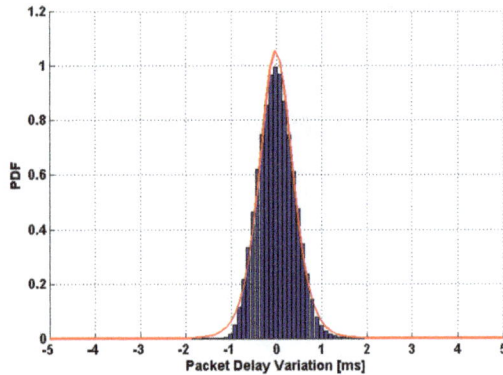

Fig. 4: Comparison of ADSL2+ PDV histogram and PDF logistic distribution.

Tab. 3: Parameters of the PDF logistic distribution.

Parameter	Value	Confidence interval	
		Low 5 %	High 95 %
Location μ [s]	−0.006	−0.00827	−0.00416
Scale σ [s]	0.2372	0.2372	0.2382

In the same way, the measured parameters from the VDSL2 line from the second provider were analysed. For this line it was found, that the maximum possible value of the transmission rate Vpmax is approximately 98 % of Actual Net Data Rate. PPPoE protocol structure requires overhead of around 2 %. Relation of OWD value to V_{pmax} is approximately 19.4 ± 9 ms (level of significance 95 %).

The Goodness-of-Fit test confirmed that PDV parameters can be statistically modelled by a normal distribution, which has the values listed in Tab. 4.

Tab. 4: Parameters of the PDF normal distribution.

Parameter	Value	Confidence interval	
		Low 5 %	High 95 %
Mean [s]	$-2 \cdot 10^{-5}$	$-1 \cdot 10^{-3}$	$9 \cdot 10^{-4}$
Variance [s²]	0.340	0.339	0.341

The following graph shows the PDV histogram and compliance with the theoretical Probability Distribution Functions of the normal distribution.

3. Conclusion

From a technical point of view, ADSL2+ and VDSL2 lines are very similar, both use the same modulation principles and the same method of the data protection during data transmission.

With regard to the same concept of the aggregation network infrastructure of network provider (in-

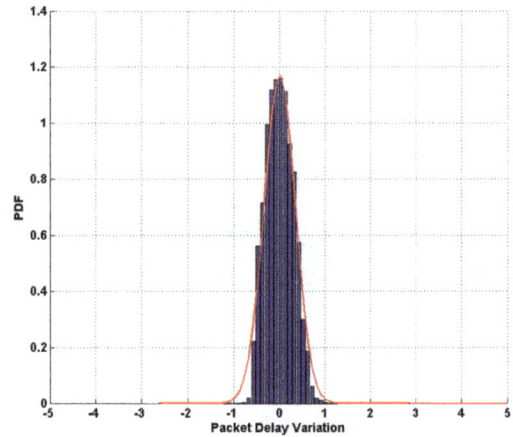

Fig. 5: Comparison of VDSL2 PDV histogram and PDF normal distribution.

cumbent's network in the Czech Republic), the same function of the aggregation point for both lines can be assumed. Generally, it can be said, that the aggregation point significantly does not affect the parameters of the data stream in the upstream direction. With respect to the existing protocol structure, both lines achieved almost theoretical maximum transmission speed values. In addition, the achieved value did not change during the day or in different days of the week.

Different time characteristics of the data channel which implement both lines are probably influenced with the next network infrastructure provider. The OWD values for ADSL2+ are higher than for VDSL2. The OWD in VDSL2 is less stable with a higher variance. With regard to the limit parameters of nowadays usually used services (e.g. limit value of 20 ms for VoIP or TDM over IP [7]) delay variation for both lines is generally very small. Packet delay variation for these two lines has different statistical dependencies for modelling. Although the PDV has similar nature for both xDSL systems, for ADSL+ it correlates empirically determined histogram of PDV amplitudes with the logistic distribution. During data transmission, VDSL2 technology has less stable time characteristics, for example, with bigger value of PDV variance.

For theoretical modelling, better correspondence between empirical PDV histogram and the distribution function of the normal distribution is visible. As already mentioned, ADSL2+ and VDSL2 use the same principle on the physical layer of OSI model, but implementation on higher communication layers is different. Also, the network infrastructure behind aggregation point has different parameters. All this is obviously visible in distinctive time characteristics of transmitted data flows in both technologies.

To control network element with the function of inverse multiplexing, the results of behaviour of both lines are important. This means value of the maximum achievable transmission rate, a way of OWD increase and the corresponding parameter of packet delay variation during the transmission will be used for set of the packet regulator in the inverse multiplexer [2].

Acknowledgment

This work was supported by the Grant of the Technology Agency of the Czech Republic, No. TA02011015, „Research and development of a new communication system with multi-channel approach and multi-layer co-operation for industrial applications", and was researched in cooperation with CERTICON.

References

[1] CHLUMSKY, P., Z. KOCUR and J. VODRAZKA. Comparison of Different Scenarios for Path Diversity Packet Wireless Networks. *Advances in Electrical and Electronic Engineering*. 2012, vol. 10, no. 4, pp. 199–203. ISSN 1336-1376. DOI: 10.15598/aeee.v10i4.713.

[2] KOCUR, Z., P. MACEJKO, P. CHLUMSKY, J. VODRAZKA and O. VONDROUS. Adaptable System Increasing the Transmission Speed and Reliability in Packet Network by Optimizing Delay. *Advances in Electrical and Electronic Engineering*. 2014, vol. 12, no. 1, pp. 13–19. ISSN 1336-1376. DOI: 10.15598/aeee.v12i1.878.

[3] NEVOSAD, M., P. LAFATA and P. JARES. Modeling of Telecommunication Cables for Gigabit DSL Application. *Advances in Electrical and Electronic Engineering*. 2013, vol. 11, no. 5, pp. 336–341. ISSN 1336-1376. DOI: 10.3403/30169077.

[4] STARR, T., M. SORBARA, J. M. CIOFFI and P. J. SILVERMAN. *DSL Advances*. Upper Saddle River, USA: Prentice Hall, 2002. ISBN 9780130938107.

[5] VONDROUS, O. *New UDP Based Methodology for Throughput and Stress Tests*. Prague, 2013. Available at: http://www.flowping.comtel.cz.

[6] VOZNAK, M. and M. HALAS. Delay Variation Model with RTP Flows Behavior in Accordance with M/D/1 Kendall's Notation. *Advances in Electrical and Electronic Engineering*. 2010, vol. 8, no. 5, pp. 124–129. ISSN 1336-1376.

[7] VODRAZKA, J., P. JARES and P. LAFATA. Time Division Multiplex over Packet Oriented Digital Subscriber Line. In: *Proceedings of 15th Mechatronika 2012*. Prague: Czech Technical University in Prague, 2012, pp. 177–181. ISBN 978-80-01-04985-3.

[8] VODRAZKA, J. and P. LAFATA. Transmission Delay Modeling of Packet Communication over Digital Subscriber Line. *Advances in Electrical and Electronic Engineering*. 2013, vol. 11, no. 4, pp. 260–265. ISSN 1336-1376.

About Authors

Petr JARES is an assistant professor at the Department of Telecommunication Engineering, Faculty of Electrical Engineering (FEE), Czech Technical University in Prague (CTU in Prague). He received his doctor (Ph.D.) degree in 2008 at FEE, CTU in Prague, specializing in Telecommunication Engineering. For past few years he has worked on various projects in the transmission systems. His current focus of interest is on data transmission in the metallic and optical access networks.

Jiri VODRAZKA was born in Prague, Czech Republic in 1966. He joined the Department of Telecommunication Engineering, FEE, CTU in Prague in 1996 as a research assistant and received his Ph.D. degree in electrical engineering in 2001. He has been the head of the Transmission Media and Systems scientific group since 2005 and became Associate Professor in 2008. He participates in numerous projects in cooperation with external bodies. Currently he acts also as vice-head of the Department.

19

REDUCTION OF PACKET LOSS BY OPTIMIZING THE ANTENNA SYSTEM AND LAYER 3 CODING

Petr CHLUMSKY[1], *Zbynek KOCUR*[1], *Jiri VODRAZKA*[1], *Tomas KORINEK*[2]

[1]Department of Telecommunication Engineering, Faculty of Electrical Engineering,
Czech Technical University in Prague, Technicka 2, 166 27 Prague, Czech Republic
[2]Department of Electromagnetic Field, Faculty of Electrical Engineering,
Czech Technical University in Prague, Technicka 2, 166 27 Prague, Czech Republic

petr.chlumsky@fel.cvut.cz, zbynek.kocur@fel.cvut.cz, vodrazka@fel.cvut.cz, korinet@fel.cvut.cz

Abstract. *The article describes the verification of the application of methods of network coding in the mobile wireless network environment with specific antenna system in use. Network coding principles are used for increase the reliability of data transmission. The proposed principles are demonstrated by modelling of the transmission system using data from real mobile networks. The effectiveness of network coding for deployment in wireless mobile networks is demonstrated on the basis of simulation in the simulation environment Omnet++. The input data of the simulation model are obtained by measuring in real mobile network of 2nd generation.*

This measurement, during which values of packet error rate (PER), round trip time (RTT) and data throughput were obtained, was performed as a result of parallel data communication over data networks of Czech mobile operators Telefonica O2 CZ and T-Mobile CZ on a route by car. Antennas system for the measurement was designed as a multiple band and multi-polarization. We want to use these data for more accurate results how will operate the proposed scheme under realistic conditions with focus on the data resiliency.

Keywords

Data transmission, measurement, mobile network, simulation.

1. Introduction

We developed the innovative data transmission scheme used in this article during work on a project that addressed the problem of data transmission over multiple transmission channels with an emphasis on the wireless environment [1]. We optimized common backup path diversity transmission scheme [2] in order to reach a better resiliency of transmission. Results from the performance analysis of the proposed scheme were beneficial; improved resiliency of the transmission was confirmed [3]. In this mentioned work, transmission channels error rate was modeled as a theoretical packet error rate steady on a certain value. In this article, the channels packet error rate is modeled based on data from a real measurement.

2. System Design

Innovative data transmission scheme, which analysis on real data is presented, is intended for use in systems with two separate data channels. It is developed as a part of the system under project "Development of adaptable and data processing systems for high-speed, secure and reliable communication in extreme conditions". It is designed to be more robust compared to the common backup data transmission scheme. Therefore, the overall transmission packet loss is the most important value. This scheme has its basis in the network coding theory which was firstly introduced in [4]. The key feature is in the idea of not only moving packets to different paths but also modifying packet content. We realized this modification using bitwise exclusive disjunction (XOR) for manipulation with the content of packets. The benefit of this data operation is in the possibility of decoding missing data from correctly received packets without any retransmissions. The scheme is designed as end-to-end rather than point-to-point correction system, therefore, it brings the possibility to overcome network nodes in the transmission path.

2.1. Transmission System Design

Principle of the scheme is shown in Fig. 1. Packet from the data source on the left side is sent directly to channel 0, whereas its copy is combined with the previous packet which is saved in the queue. This combination is done by exclusive-or (XOR) operation. After the operation is executed, the resulting packet is sent through channel 1. The packet that was previously saved in the queue is replaced by the current one (the copied packet used for a combination). Therefore, the size of the queue in the coding part of the scheme is just the length of two consecutive packets.

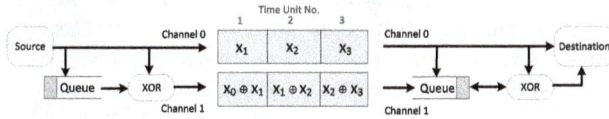

Fig. 1: Transmission scheme principle.

When a packet from channel 0 is lost, or corrupted, there is still possibility of its recovery. Moreover, the recovery of the original packet from channel 0 is still feasible even in a situation where the corresponding packet from the redundant channel is also lost. This is possible because of the characteristics of the XOR operation and the decoder function. This recovery is an advantage compare to the common backup scheme where both channels transmit the same data, so it is impossible to recover the missing packet. Proposed scheme is able to reconstruct up to three consecutively missing packets. For this reason, the decoder part (on the right side of Fig. 1) requires a queue for saving previously, correctly received or reconstructed packets. Example of this situation is shown in Eq. (1). The packet from channel 0 (x_1) and its correspondent packet from channel 1 ($x_0 \oplus x_1$) are lost and, moreover, the following packet from channel 0 is also lost (x_2). The proposed transmission scheme is capable of reconstructing both of the lost packets from channel 0 without any retransmissions. Packets from channel 1 are no need to recover because this channel is intended as a redundant one. For a reconstruction of the original data it is necessary to receive the redundant packet for the second lost packet from channel 0 ($x_1 \oplus x_2$) and also receive the following pair of original (x_3) and its corresponding redundant packet ($x_2 \oplus x_3$).

$$x_1, (x_0 \oplus x_1), x_2 \ldots lost$$
$$x_3 \oplus (x_2 \oplus x_3) \Rightarrow x_2 \qquad (1)$$
$$x_2 \oplus (x_1 \oplus x_2) \Rightarrow x_1$$

Process of reconstruction in the decoder relies on the number of consecutively missing packets. The example in Eq. (1) describes situations when two original packets in a row and a single redundant packet are lost.

When the number of consecutively missing packets is lower, then the process of decoding is less complicated and it has lower demands on the length of the decoding queue. Complete description of the scheme and its performance analysis is in [3].

2.2. Antenna System Design

Instead of previous configuration of measurement system described in [5], the current version used traffic generator powered by FlowPing application [6]. This application is highly configurable UDP measurement tool with possibility of round trip time measurement. The other difference besides previous is in use of the antenna system. The previous system used simple dipole antennas. Now, we used multiple band and multi-polarization antennas which are more suitable for the mobile environment in urban and suburban areas. The designed antenna stabilizes received signal level during the movement and significantly reduces packet losses. Comparison of this antenna with different types is shown in Fig. 2. The detail description of used antennas is in the Tab. 2 and in Fig. 3. Reflection coefficient (S11 parameter) for both frequency bands shows the graph in the Fig. 4.

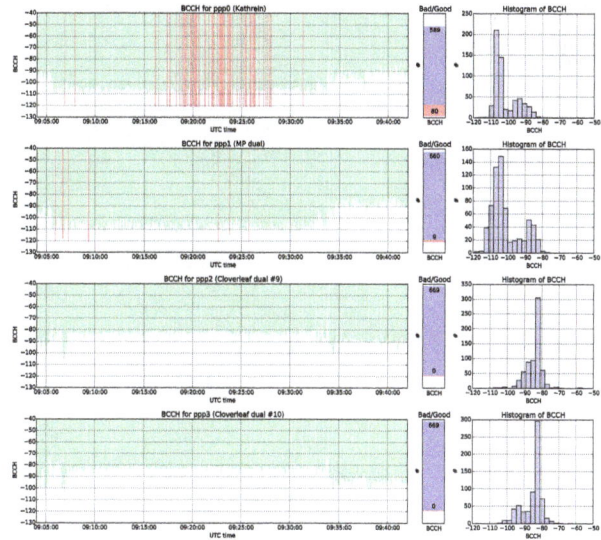

Fig. 2: Comparison of antennas signal level.

Tab. 1: Selected antenna parameters.

Type	Cloverleaf omnidirectional
Polarization	Circular
Frequency Band	900, 1800 MHz
Gain	3 dBi (900 MHz) 5 dBi (1800 MHz)

Fig. 3: Antenna testbed at the roof of the car.

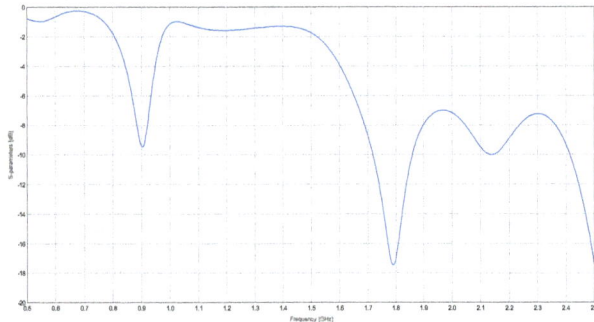

Fig. 4: S11 parameter of the dual band cloverleaf antenna.

Tab. 2: Setting of the measurement equipment.

Parameter	T-Mobile	Telefonica O2
Channel number	0	1
Technology	EDGE	EDGE
Traffic generator	Flowping	
Traffic type	UDP	
Traffic throughput	45 kb$\cdot$s^{-1}	
Packet size	1000 Bytes	
Measured parameters	PER, RTT, Throughput	
Measurement time	140 s	
Velocity of car	~ 20 km$\cdot$h^{-1}	
Measurement date	27.5.2014	
Measurement location	Dejvice, Prague, Czech Republic	

3. Data Transmission Modeling with Layer 3 Coding

The proposed scheme was completely implemented in C++ programming language in the simulation framework OMNeT++. OMNeT++ is an object oriented modular system which is based on the processing of discrete events; more information about this framework are in [7]. It was necessary to simulate packet losses for the testing of the transmission scheme recovery capabilities. We decided to model the channel packet error rate from the data from real measurement. Obtained data allowed us to model the behavior of the transmission scheme more accurately under realistic conditions of mobile networks. For this purpose, it was necessary to implement a new channel for the simulation; this

channel is able to read parameters obtained during the measurement and adjust the simulation based on them. Implementation is based on the reading of the input file and matching actual channel parameters based on the simulation time.

3.1. Input Data from Real Measurement

Real measurement was done by a special testbed based on system which was presented in [5]. The measurement system is able to measure the actual delay and throughput of passing packets in each of the communication interfaces. System is able, on the interface of mobile networks, to determine the current physical layer parameters, such as the type of the mobile network (2G/3G), level of the received signal (RSSI), or cell ID (Cell ID) to which is a data terminal connected. A key part of that system is a unit of time synchronization that allows to perform the correlation of measured quantities.

The purpose of the measurement was to obtain data about transmission channels behavior under real conditions. The measurement was based on the packet error rate on separate communication interfaces during movement of the testbed. A constant data flow was generated between the mobile testbed and the Internet server part in the uplink direction from the testbed. Data were received and analyzed on the server part. Among the measured parameters were also throughput and round trip time. Subsequently, these values were adapted for use in the simulation environment Omnet++.

3.2. Simulation Results

The obtained parameters of packet round trip time were used as delay values for the transmission channels, the throughput parameters were set as a maximum throughput for the channels. The most important parameter for our purpose, the packet error rate, was also set, as previous parameters, specifically for every specific time based on the measured values. Table 3 shows average values of mentioned parameters. The simulation was performed with the same settings also for the common backup scheme. Table 4 shows the channel error rates from the measurement and, in the *Resulting PER column*, there is an overall packet error rate for the data transmission in simulation. The advantage of using two channels, in terms of communication robustness, is obvious. The common backup scheme was able to reduce the loss ratio almost by an order of magnitude. Nevertheless, the impact of the recovery abilities of the proposed scheme is clear. The possibility of the recovery of two packets in a row from one of the channels results in the packet error rate of 0.003.

Tab. 3: Measurement results.

Parameter	T-Mobile	Telefonica O2
RTT	4904 ms	3946 ms
Throughput	6.2 kb·s^{-1}	7.9 kb·s^{-1}
PER	13.5 %	11.9 %

Tab. 4: Simulation results.

	Channels PER		Resulting PER
Proposed scheme	13.5 %	11.9 %	0.3 %
Backup scheme	13.5 %	11.9 %	2.6 %

It is important to mention that the character of measured data has inevitable impact on the resulting packet error rate. This impact is based on the spread of the lost packets and based on the number of these lost packets in a row. Long bursts of errors that occur only at one of the channels favors the common backup system where the second channel continuously transmits the same original data as on the erring channel. Short transmission interrupts that occur while roaming between mobile operators cells or when the antenna is shaded has similar impact on both schemes in case that the second channel continues in transmission without error. The backup scheme has full redundancy for these cases, the proposed scheme is able of packet reconstruction. Nevertheless, in case that both channels are influenced with this short packets outage, then the proposed scheme gains considerable advantage in the possibility of recovery of missing packets.

4.   Conclusion

This article presents verification of innovative data transmission scheme in simulation where transmission channel parameters were modeled by data from measurement. The proposed transmission scheme operates with two separate transmission channels. The first channel is used for original data and the second channel is designed as a redundant one. An important part of this scheme is in the packets combinations that are transmitted through the second channel. These combinations are realized by the exclusive-or operation. Because of the possibilities given by the combined packets from the second channel is the decoding part of the scheme capable of reconstruction of missing packets without any retransmissions.

Advantage of this scheme in transmission resiliency was already confirmed [3] with channel packet error rate modeled as a theoretical value, now, we wanted to verify this scheme with transmission channel modeled based on the data from measurement. Measurement was performed by a special testbed with specific multiple band and multi-polarization antennas. The parameters of packet error rate, round trip time and throughput were obtained and used for modeling the channel parameters in dependency on the time. Simulation was also performed for the common backup scheme for the possibility of comparison. Results proved that the proposed scheme is able to be more robust compared to the backup scheme that uses both channels for transmission of the same data. The difference between these schemes depends on the actual transmission channels conditions, whether packet losses occurs in long bursts or as an individual interrupts.

Acknowledgment

This work was supported by Ministry of the Interior of the Czech Republic Grant no. VG20122014095, "Development of adaptable and data processing systems for high-speed, secure and reliable communication in extreme conditions", and the research was carried out in cooperation with CERTICON and the Student's grant of Czech Technical University in Prague, No.SGS12/186/OHK3/3T/13. This work was also supported by the Grant of the Ministry of the Interior of the Czech Republic, No. VG20132015104, "Research and development of secure and reliable communications network equipments to support the distribution of electric energy and other critical infrastructures", and was researched in cooperation with TTC Telecomunikace.

References

[1] KOCUR, Z., P. MACEJKO, P. CHLUMSKY, J. VODRAZKA and O. VONDROUS. Adaptable System Increasing the Transmission Speed and Reliability in Packet Network by Optimizing Delay. *Advances in Electrical and Electronic Engineering*. 2014, vol. 12, no. 1, pp. 13–19. ISSN 1336-1376. DOI: 10.15598/aeee.v12i1.878.

[2] CHLUMSKY, P., Z. KOCUR and J. VODRAZKA. Comparison of Different Scenarios for Path Diversity Packet Wireless Networks. *Advances in Electrical and Electronic Engineering*. 2012, vol. 10, no. 4, pp. 199–203. ISSN 1336-1376. DOI: 10.15598/aeee.v10i4.713.

[3] CHLUMSKY, P. and J. VODRAZKA. Delay analysis of data transmission system with channel coding. In: *Proceedings of the 10th International Conference Elektro*. Rajecke Teplice: IEEE, 2014, pp. 31–35. ISBN 978-1-4799-3720-2.

[4] AHLSWEDE, R., N. CAI, S. R. LI and R. W. YE-
    UNG. Network information flow. *IEEE Transac-
    tions on Information Theory*. 2000, vol. 46, iss. 4,
    pp. 1204–1216. ISSN 0018-9448.

[5] KOCUR, Z., P. CHLUMSKY, P. MACEJKO,
    M. KOZAK, L. VOJTECH and M. NERUDA.
    Measurement of Mobile Communication Devices
    on the Testing Railway Ring. In: *15th Interna-
    tional Conference on Research in Telecommuni-
    cation Technologies*. Bratislava: Slovak Univer-
    sity of Technology in Bratislava, 2013, pp. 34–37.
    ISBN 978-80-227-4026-5..

[6] VONDROUS, O. *FlowPing - UDP based ping ap-
    plication*. Prague, 2013. Available at: http://
    www.flowping.comtel.cz/.

[7] OMNET++. *Framework homepage*. Available at:
    http://www.omnetpp.org/.

About Authors

Petr CHLUMSKY was born in 1985. He received
his M.Sc. from the Czech Technical University in
Prague in 2010. Since 2010 he has been studying
Ph.D. degree. His research interests include wireless
transmission, network coding and network simulation.

Zbynek KOCUR was born in 1982. He received his
M.Sc. degree in electrical engineering from the Czech
Technical University in Prague in 2008 and Ph.D.
degree in electrical engineering in 2014. He is teaching
communication in data networks and networking
technologies. His research is focused on wireless
transmission and data flow analysis, simulation and
optimization. He is currently actively involved in
projects focused on high speed data transmission from
fast moving objects and data optimization via satellite
network.

Jiri VODRAZKA was born in Prague, Czech
Republic in 1966. He joined the Department of
Telecommunication Engineering, Faculty of Electrical
Engineering, Czech Technical University in Prague in
1996 as a research assistant and received his Ph.D.
degree in electrical engineering in 2001. He has been
the head of the Transmission Media and Systems
scientific group since 2005 and became Associate Professor in 2008. He participates in numerous projects
in cooperation with external bodies. Currently he acts
also as vice-head of the Department.

Tomas KORINEK was born in Jicin, the Czech
Republic, in 1979. He received the M.Sc. degree and
the Ph.D. degree in radio electronics from the Czech
Technical University in Prague, the Czech Republic,
in 2005 and 2012 respectively. From 2007 to 2008, he
was a research and designer engineer at RFspin s.r.o.,
where he was engaged in antennas and microwave
circuits. He is currently an Assistant Professor and
the head of laboratories at the Department of Electromagnetic Field at the Czech Technical University
in Prague. His research interests include the area of
measurements in EMC and antennas.

20

Model-Based Attitude Estimation for Multicopters

Radek BARANEK, Frantisek SOLC

Department of Control and Instrumentation, Faculty of Electrical Engineering and Communication,
Brno University of Technology, Technicka 12, 616 00 Brno, Czech Republic

xbaran10@stud.feec.vutbr.cz, solc@feec.vutbr.cz

Abstract. *The paper deals with model-based attitude estimation for multicopters and is mainly focused on investigation of accuracy degradation due to wind and inaccurate model parameters which are conditions always present when using in real world. At first the need for model-base estimation is motivated. Then the multicopter model is described. Based on the mathematical model of multicopter, the estimation algorithm utilizing the extended Kalman filter is constructed. The main contribution of the paper is the investigation of the negative impact of the wind and of inaccurate knowledge of the model parameters.*

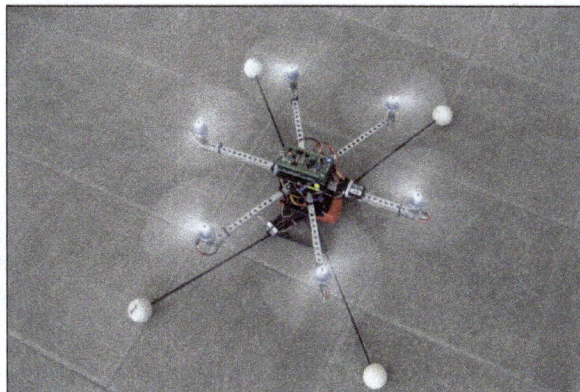

Fig. 1: The experimental multicopter: hexacopter.

Keywords

Attitude estimation, extended Kalman filter, multicopter.

1. Introduction

Multicopters are VTOL (Vertical Take-Off and Landing) aerial vehicles (AVs) included within the class of multirotor helicopters. These vehicles differ from standard helicopters in that they use rotors with fixed-pitch blades, and thus their rotor pitch does not vary as the blades rotate. Currently existing versions comprise 4 or more rotors (x-copters); the hexacopter uses 6 rotors (Fig. 1).

There also exist versions with fewer rotors, but they require additional moving components for stabilization, thus they do not possess the mechanical simplicity feature.

At present, these types of AVs are used as a standard platform for robotics research. The first objective in the construction of automatic control systems for these AVs is to ensure stable flight at low velocities, particularly during the hovering phase. The design of x copter control elements has hitherto been discussed by a large number of authors. Generally, the controllers are designed as linear SISO systems [1], PID systems [2], special nonlinear controllers [3] or [4], and even neural net controllers [5]. The most important component of the control loop is a good feedback signal providing correct information about the AV's attitude. Although the majority of the above-mentioned systems use on-board sensors, the necessary feedback can be also provided by off-board sensing elements [6]. The most widely used devices for attitude estimation are inertial sensors and magnetic field sensors based on MEMS technology, namely MEMS accelerometers, gyroscopes and magnetometers. In fact, these devices form a strap-down inertial measurement unit (IMU), and they have become the main attitude sensors due to their low cost. Other sensors, such as sonar range finders, cameras, lasers, or GPS, are used especially for position feedback.

A significant factor in the design of a control system is a good mathematical model which describes both dynamics of the AV itself and dynamics of the IMU as well. In most papers, the presented models of controlled AVs consider only the forces and moments caused by propellers in hovering and neglect all other aerodynamic effects because of the low linear velocities. Such models do not produce accelerometer data usable for good feedback ensuring accurate information on the attitude of an AV [7].

The basic sensors of IMU do not provide direct information about the attitude of the AV. To provide the right information about the attitude of the AV, the sensor signals must be appropriately handled. At present, the processing of such signals is performed by the algorithm of Kalman filtering. This algorithm is used in the calculation of the attitude in form of the combination of prediction and correction. For the prediction a suitable dynamic model is used, the correction is carried out by direct measurement. Thus well chosen model of prediction can improve information about the attitude of the AV. The following article shows and discusses characteristics of such a model and its use for estimation.

2. Model Based Attitude Estimation

The typical attitude estimation algorithms [10] use no assumptions about dynamics of the examined object. They rely only on the used sensors (typically accelerometers, gyroscopes, magnetometers, GPS etc.). With good model including the knowledge of the dynamic properties can provide more accurate results. But any unmodeled effects or inaccurate model parameters can degrade the accuracy and even make the results worse than without the dynamic model.

Multicopters are obviously good adepts (inputs to the system are known, several dynamics models were constructed) where including the dynamic model should improve the accuracy. The analysis of inaccurate model parameters and unmodeled effects (namely wind) is subject of this paper.

In the following section multicopter dynamic model based on the recent papers is described. In the fourth section the dynamic model based attitude estimation algorithm is constructed. The results of simulations, where mainly the effects of wind and inaccurate model parameters are studied, are in section 5.

Finally in the section 6 the wind effect mitigation is described.

3. Multicopter Model

In this chapter the mathematical model of multicopter is mentioned. This model was used for generation of input data for simulation of attitude estimation algorithm. Additionally the model was also used for designing the estimation algorithm itself. The biggest advantage of using mathematical model for generation of testing data is the knowledge of the true values of all states. This feature enables direct comparison of the estimated state with the true state.

Mathematical model of multicopter is studied in many recent publication focused mainly on attitude and position control [1], [2], [3], [4], [5] but also for attitude estimation [10]. They differ in level of precision and in number of modeled effects. Here mentioned model is based on all cited works, and the main effects of interest are aerodynamic drag, wind and precise sensor models.

This particular model assumes the multicopter has 6 motors with propellers arranged equidistantly on a circle with radius L. The scheme of the multicopter is in Fig. 2.

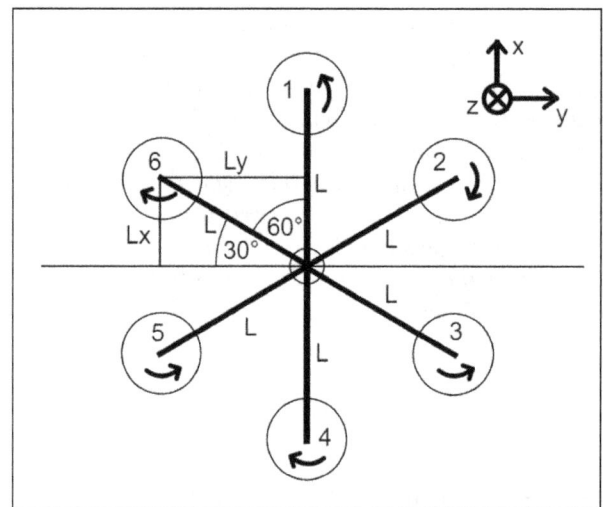

Fig. 2: Scheme of multicopter with 6 motors (top view).

Two coordinate frames are used throughout the description of the multicopter model. Body frame is the coordinate system rigidly linked with the multicopter. The orientation of body frame axes with respect to multicopter is depicted in Fig. 2. Reference frame is a coordinate system with respect to which the attitude of body frame is investigated. Usually this frame is linked to local vertical direction and to true north direction. In this work this frame is assumed to be inertial frame. It means that Coriolis force, earth rotation and transport rate, which are common for frames linked with earth surface, are neglected. This approximation is valid when the usage of low-cost sensors is expected since these effects are beyond the resolution of the sensors.

3.1. Forces and Torques

Forces and torques are driving the motion of the multicopter. The total force and torque are composed of different sources: thrust of individual propellers, aerodynamic drag and gravity.

1) Motors with Propellers

Thrusts of individual propellers are related to the control signal. This relation is modeled by this equation.

$$T_i = c_T S_i^2, \qquad (1)$$

where T_i is thrust of i-th motor, c_T is a positive constant parameter and S_i is control value for i-th motor. Each motor also generates reactive torque, which has opposite direction than the angular rate vector of the motor shaft. The magnitude of the reactive torque is assumed to be proportional to motor thrust.

$$\vec{m}_i = -c_R T_i \frac{\vec{\omega}_i}{\|\vec{\omega}_i\|}, \qquad (2)$$

where $\vec{m}_i$ is the reactive torque vector of i-th motor, c_R is a positive constant parameter and $\vec{\omega}_i$ is the angular rate vector of i-th motor shaft. The resulting force and torque vector generated by thrusts of motors are computed using.

$$\begin{bmatrix} F_z \\ M_x \\ M_y \\ M_z \end{bmatrix} = \mathbf{A} \begin{bmatrix} T_1 & T_2 & T_3 & T_4 & T_5 & T_6 \end{bmatrix}^T,$$

$$\mathbf{A} = \begin{bmatrix} -1 & -1 & -1 & -1 & -1 & -1 \\ 0 & -L\frac{\sqrt{3}}{2} & -L\frac{\sqrt{3}}{2} & 0 & L\frac{\sqrt{3}}{2} & L\frac{\sqrt{3}}{2} \\ L & \frac{L}{2} & -\frac{L}{2} & -L & -\frac{L}{2} & \frac{L}{2} \\ c_R & -c_R & c_R & -c_R & c_R & -c_R \end{bmatrix}, \quad (3)$$

where F_z is the z-component of force vector, $\vec{F}_{BT}$ expressed in body frame respecting the coordinate system defined in Fig. 2, M_x, M_y, M_z are components of torque $\vec{M}_{BT}$ expressed again in body frame, L is length of the multicopter arm and c_R is constant parameter defined in Eq. (2). The matrix $\mathbf{A}$ can be derived using simple mechanics and Fig. 2.

2) Aerodynamic Drag

Aerodynamic drag is very often neglected in multicopter models. However, this phenomenon enables use of the accelerometer measurements in the attitude estimation. This is the main reason why this drag is implemented in this multicopter model.

The aerodynamic drag force is modeled as linearly dependent on the relative velocity with respect to wind.

$$\vec{F}_{BA} = \begin{bmatrix} -k_x v_{xb}^w \\ -k_y v_{yb}^w \\ -k_z v_{zb}^w \end{bmatrix}, \qquad (4)$$

where $\vec{F}_{BA}$ is drag force vector expressed in body frame, k_i are positive constant parameters and v_{ib}^w are

components of relative velocity with respect to wind expressed in body frame. This is a simplified version of propeller induced aerodynamic drag from [7]. The reason why the parameters k_i are constant in body frame comes from the constant rotation axis of propellers with respect to this frame. The parameter k_z is assumed to be of order smaller than the k_x and k_y, because the z body axis is parallel to the axis of rotation of the propellers and the main contributor to the aerodynamic drag is the blade-flapping phenomenon which occurs only in the axis perpendicular to the propeller rotation [7].

The aerodynamic drag torque is similarly modeled as linearly dependent on angular rate.

$$\vec{M}_{BA} = -k_M \vec{\omega}_B, \qquad (5)$$

where $\vec{M}_{BA}$ is drag torque vector expressed in body frame, k_M is positive constant parameter and $\vec{\omega}_B$ is angular rate vector expressed in body frame. The aerodynamic drag torque is based on the experience that angular rate settles on some finite value when constant torque is applied. In many recent multicopter models this effect is omitted.

3) Total Acceleration and Angular Acceleration

All the mentioned forces and torques are summed and divided by mechanical properties to form the final acceleration $\vec{a}_I$ and angular acceleration $\vec{\varepsilon}_B$.

$$\vec{a}_I = \vec{g}_I + \frac{\mathbf{R}_{\mathrm{BI}}(\vec{F}_{BT} + \vec{F}_{BA})}{m}, \qquad (6)$$

where $\vec{a}_I$ is acceleration expressed in inertial frame, $\vec{g}_I$ is gravitational acceleration expressed in inertial frame, $\mathbf{R}_{\mathrm{BI}}$ is transformation matrix from body to inertial frame and m is the mass of the multicopter.

$$\vec{\varepsilon}_B = [(\vec{M}_{BT} + \vec{M}_{BA}) - (\vec{\omega}_B \times \mathbf{I}_{\mathrm{B}}\vec{\omega}_B)] \cdot \mathbf{I}_{\mathrm{B}}^{-1}, \qquad (7)$$

where $\vec{\varepsilon}_B$ is angular acceleration, $\vec{\omega}_B$ is angular rate and $\mathbf{I}_{\mathrm{B}}$ is inertia matrix of the multicopter, all variables expressed in body frame.

3.2. Multicopter Motion

The motion of multicopter is modeled as the 6–DoF rigid body motion driven by forces and torques described above. The motion state is described by velocity and position for translational motion and by angular rate and attitude quaternion for rotational motion. This part is valid for any aerial vehicle and is expressed by the following differential equations.

1)   Position and Velocity

$$\vec{p} = \vec{v}_I, \ \vec{v}_I = \vec{a}_I, \tag{8}$$

where $\vec{p}$ is position vector, $\vec{v}_I$ is velocity vector expressed in inertial frame and $\vec{a}_I$ is acceleration vector expressed in inertial frame.

2)   Attitude and Angular Rate

$$
\vec{\dot{q}} = \begin{bmatrix} \dot{q}_1 \\ \dot{q}_2 \\ \dot{q}_3 \\ \dot{q}_4 \end{bmatrix} =
$$

$$
= 0.5 \begin{bmatrix} q_1 & -q_2 & -q_3 & -q_4 \\ q_2 & q_1 & -q_4 & q_3 \\ q_3 & q_4 & q_1 & -q_2 \\ q_4 & -q_3 & q_2 & q_1 \end{bmatrix} \begin{bmatrix} 0 \\ \omega_x \\ \omega_y \\ \omega_z \end{bmatrix}, \tag{9}
$$

where $\vec{q}$ is attitude quaternion and ω_x, ω_y, ω_z are components of angular rate vector expressed in body frame.

$$\vec{\dot{\omega}}_B = \vec{\varepsilon}_B, \tag{10}$$

where $\vec{\omega}_B$ is angular rate vector expressed in body frame and $\vec{\varepsilon}_B$ is angular acceleration vector expressed in body frame.

3.3.   Wind

Since the aerodynamic drag force is dependent on velocity with respect to wind and the acceleration measured by accelerometers (used for attitude estimation) is the time derivative of the inertial velocity, it is reasonable to model the wind to be able to reveal possible negative impact of this fact on the accuracy of attitude estimation.

The wind is specified by instant velocity vector expressed in inertial frame and in this work is assumed to be only function of time. The total wind velocity consists of static and dynamic part. Static part is a constant velocity vector expressed in inertial frame and corresponds to dominant constant wind experienced in real condition. Dynamic part is time dependent and each component of the dynamic velocity is modeled as a first order Gauss Markov process [11]. The dynamic component corresponds to short time variation of the wind (wind gust). The behavior of the dynamic part can be adjusted by steady state variance and by time constant of Gauss Markov process.

In Fig. 3 there is the output of the wind model simulation with time constant $\tau = 7$ s, constant speed vector $\vec{b} = [2, 0, -2]$ m·s^{-1} and steady state deviation of Gauss-Markov process $\sigma = 0.3$ m·s^{-1}.

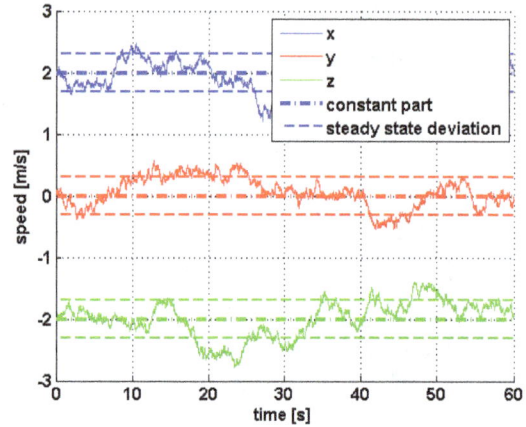

Fig. 3: Wind speed simulation.

3.4.   Sensors

With respect to attitude estimation these sensors are of interest:

- Gyroscope – measuring angular rate.

- Accelerometer – measuring specific force.

- Magnetometer – measuring magnetic field.

Each of this sensor senses the physical quantity along three perpendicular axes parallel to body frame axes. True values of these sensors can be easily computed from true state available for multicopter model. True values are then deliberately corrupted to have similar characteristics like the real sensors of this type.

The same sensor model but with different parameters is used for each of these sensors.

$$x_{OUT} = x_{TRUE} + b_S + b_D + n, \tag{11}$$

where x_{TRUE} is true value, x_{OUT} is modeled senor output, b_S is constant bias, b_D is dynamic bias modeled as first order Gauss-Markov process and n is a white noise. The qualitative characteristics of the sensor model can be seen in Fig. 4, where zero true signal is corrupted by all terms of the sensor model.

4.   Implementation of Model-Based Attitude Estimation

The model-based attitude estimation is not new. Is discussed in some papers including [12], [13]. The new contribution is the identification and investigation of

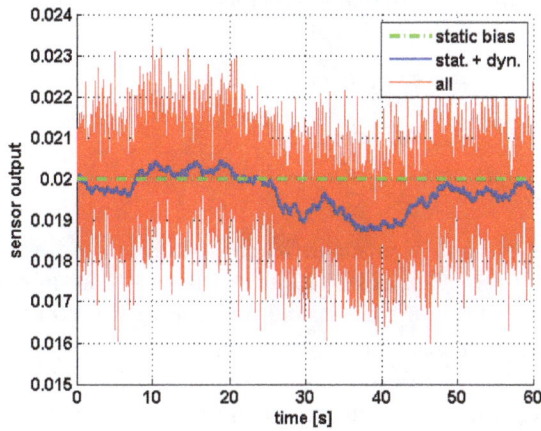

Fig. 4: Output of the sensor model.

main causes which could make the estimation less accurate in real world. The model based attitude estimation for multicopters assumes that this information is available:

- Data from gyroscope, accelerometer and magnetometer.

- Control values for motors.

The traditional algorithms which process the sensors from the first bullet use so-called vector matching method. This method assumes that the accelerometer measures gravitational field only (which holds some information about attitude). But this is generally not true during the whole flight and it is especially not true during aggressive maneuvers and changes of attitude.

On the other hand the model based attitude estimation use mathematical model with modeled aerodynamic drag to predict the true accelerometer measurement. The accuracy of this approach is then theoretically independent of the maneuvers flown by the aerial vehicle.

The model based attitude estimator is based on the mathematical model of multicopter mentioned in the previous chapter. In Fig. 5 there is a scheme of the full multicopter model.

The method used for estimation is extended Kalman filter. It provides the best scalable framework for implementation of such a model-based estimation problem. Extended Kalman filter is an iterative algorithm. Each of its iteration consists of two steps. In prediction step, the next value of the state is predicted based on the inputs and the previous state. In update step the corrections are computed using the measurements and applied to the predicted state.

According to Fig. 5 the full model could be used as a core of a Kalman filter estimator. It means use motor signals and wind speed for prediction of the state

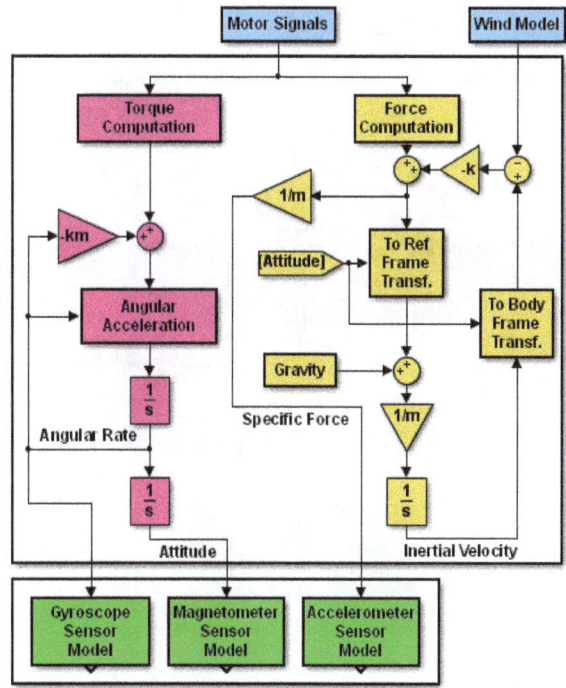

Fig. 5: Scheme of the full multicopter model.

and then use all three sensors (gyroscope, accelerometer and magnetometer) for correction of the predicted state.

However the full model is slightly simplified for model based attitude estimation for various reasons. At first it is assumed that the multicopter is not equipped with any kind of sensor capable of measuring wind speed. For this reason the wind speed input has to be neglected. Next, the motor signals are not used for prediction of angular rate (measured directly by gyroscopes), despite it is possible. This prediction is not used in the presented estimation algorithm, as the accuracy of the prediction is very low compared to the accuracy even of low cost gyroscope. This optimization leads to lower number of states of the filter and in lower computation complexity. Additionally the knowledge of inertia matrix of the multicopter is not required.

Therefore the Kalman filter based on the simplified model uses the gyroscope and motor signal for prediction step and the accelerometer and magnetometer for update step. The scheme of the simplified model is in Fig. 6. The estimated wind is assumed to be zero in the first version. Chapter 6 discuss the effect of estimation of the wind speed.

The general equations of discrete extended Kalman filter are:

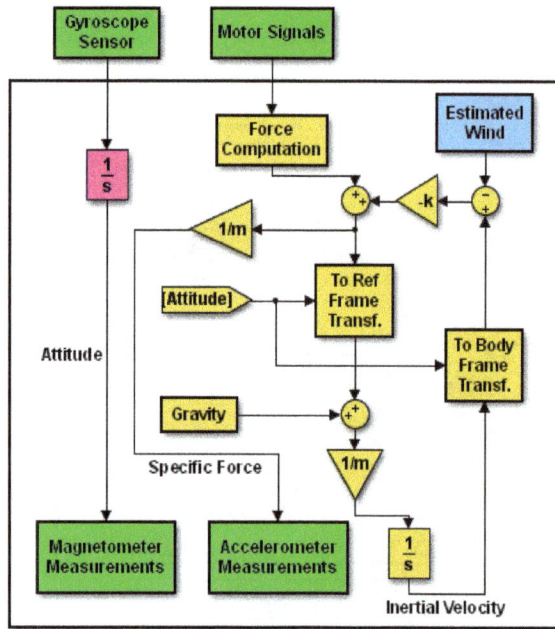

Fig. 6: Scheme of the simplified model used for attitude estimation.

1) Prediction Step

$$\vec{x}(k+1|k) = f(\vec{x}(k|k), \vec{u}(k)),$$
$$\mathbf{P}(k+1|k) = \mathbf{F} \cdot \mathbf{P} \cdot \mathbf{F}^{\mathrm{T}} + \mathbf{Q}. \qquad (12)$$

2) Update Step

$$\mathbf{K} = \frac{\mathbf{P}(k+1|k) \cdot \mathbf{H}^{\mathrm{T}}}{\mathbf{H} \cdot \mathbf{P}(k+1|k) \cdot \mathbf{H} + \mathbf{R}},$$
$$\delta\vec{z}(k+1) = \vec{z}(k+1) - h(\vec{x}(k+1|k)),$$
$$\vec{x}(k+1|k+1) = \vec{x}(k+1|k) + \mathbf{K}(\delta\vec{z}(k+1)),$$
$$\mathbf{P}(k+1|k+1) = (\mathbf{I} - \mathbf{K} \cdot \mathbf{H}) \cdot \mathbf{P}(k+1|k),$$
$$\mathbf{F} = \frac{\partial f}{\partial \vec{x}} \Big|_{\vec{x}(k|k), \vec{u}(k)}, \qquad (13)$$
$$\mathbf{H} = \frac{\partial h}{\partial \vec{x}} \Big|_{\vec{x}(k+1|k)},$$

where $\vec{x}$, $\vec{u}$, $\vec{z}$ are the state, input and measurement vectors respectively, $\mathbf{P}$, $\mathbf{Q}$, $\mathbf{R}$ are the covariance matrices of state, inputs and measurements respectively, $\mathbf{F}$ and $\mathbf{H}$ are Jacobian matrices of functions f and h and finally k is the sample time index.

The estimation algorithm utilizing extended Kalman filter is uniquely defined by the functions f and h and by covariance matrices $\mathbf{Q}$ and $\mathbf{R}$. The state, input and measurements vectors for here presented model based

attitude estimation are defined as follows.

$$\vec{x} = \begin{bmatrix} \vec{q} & \vec{v}_I & \vec{b} \end{bmatrix},$$
$$\vec{u} = \begin{bmatrix} \vec{\omega}_g & \vec{S} \end{bmatrix}, \qquad (14)$$
$$\vec{z} = \begin{bmatrix} \vec{a}_B & \vec{m}_B \end{bmatrix},$$

where $\vec{q}$ is the four element attitude quaternion, $\vec{v}_I$ is the velocity vector expressed in inertial frame, $\vec{b}$ is the gyroscope bias vector, $\vec{\omega}_g$ is the angular rate vector sensed by gyroscope, $\vec{S}$ is the vector of motor control inputs, $\vec{a}_B$ is the specific force vector sensed by accelerometer and $\vec{m}_B$ is the magnetic field sensed by magnetometer. In this case, of model based attitude estimation algorithm, the function f consists of the following subparts:

- Discrete quaternion integration based on Eq. (9) where the angular rate vector has the following format to incorporate estimated gyroscope bias.

$$\vec{\omega} = \vec{\omega}_g - \vec{b}. \qquad (15)$$

- Inertial acceleration computation and discrete integration to obtain inertial velocity which are based on Eq. (1), Eq. (3), Eq. (4) and Eq. (6).

- Bias propagation defined by Gauss-Markov process [11].

$$\vec{b}_{k+1} = \beta_D \vec{b}_k. \qquad (16)$$

The function h which predicts the measurement values have the following parts.

- Specific-force prediction based on Eq. (1), Eq. (3), Eq. (4) and Eq. (6) where gravity is excluded.

- The prediction of magnetic field in body frame.

$$\vec{m}_b = \mathbf{R}_{\mathrm{IB}} \vec{m}_I = \mathbf{R}_{\mathrm{BI}}^T \vec{m}_I, \qquad (17)$$

where $\mathbf{R}_{\mathrm{IB}}$ is transformation matrix from inertial to body frame and $\vec{m}_I$ is supposed to be known Earth magnetic field vector expressed in inertial frame. The measurement model for magnetometer assumes that magnetometer is measuring the Earth magnetic field only which is generally not true and care must be taken when using especially in indoor environments.

The covariance matrices $\mathbf{Q}$ and $\mathbf{R}$ are defined as follows [13].

$$\mathbf{Q} = \begin{bmatrix} \mathbf{Q}_q & 0 & 0 \\ 0 & \mathbf{Q}_V & 0 \\ 0 & 0 & \mathbf{Q}_B \end{bmatrix},$$

$$\mathbf{Q}_q = \mathbf{B}\mathbf{Q}_g\mathbf{B}^T,$$

$$\mathbf{B} = \frac{\partial f}{\partial \vec{\omega}}|_{\vec{x},\vec{\omega}},$$

$$\mathbf{Q}_g = diag(\vec{\sigma}_g^2), \qquad (18)$$

$$\mathbf{Q}_V = diag(\vec{\sigma}_V^2),$$

$$\mathbf{Q}_B = diag(\vec{\sigma}_B^2),$$

$$\mathbf{R} = diag(\vec{\sigma}_A^2, \vec{\sigma}_M^2),$$

where $\vec{\sigma}_g$ is standard deviation vector of gyroscope white noise, $\vec{\sigma}_V$ is standard deviation vector representing the accuracy in computation of the velocity increment, $\vec{\sigma}_B$ is standard deviation vector of white noise driving the Gauss-Markov process for gyroscope biases, $\vec{\sigma}_A$ is standard deviation of accelerometer which overbounds all error sources of this sensor and $\vec{\sigma}_M$ is standard deviation vector of magnetometer which overbounds all error sources of this sensor.

5.   Simulations

The computer simulations of model-based attitude estimation algorithm, described in the previous chapter, were targeted to find the limits of use of such algorithm in real conditions. With respect to this aim, the effects of wind and the inaccurate model parameters on accuracy of attitude estimates were investigated.

At first the parameters of the full multicopter model was defined. The summary of the model parameters are in the Tab. 1. This model was then used for generation of two, 60 s long, testing trajectories. First is low speed normal trajectory with tilts less than 45 degrees. The second one is high speed aggressive trajectory with tilts up to 85 degrees. The estimation errors for attitude recomputed to Euler angles for both trajectories are in Fig. 7 and Fig. 8. The error is computed as a difference between estimated value and the true value. Red lines in the figures are $1 - \sigma$ accuracy of the estimated state directly computed from the Kalman filter covariance matrix. The model parameters used in estimation algorithm match the parameters used in the generation of testing data. Wind is not present in these cases.

5.1.   Wind Effect

The reason that one can assume that wind should have negative impact on accuracy of attitude estimation

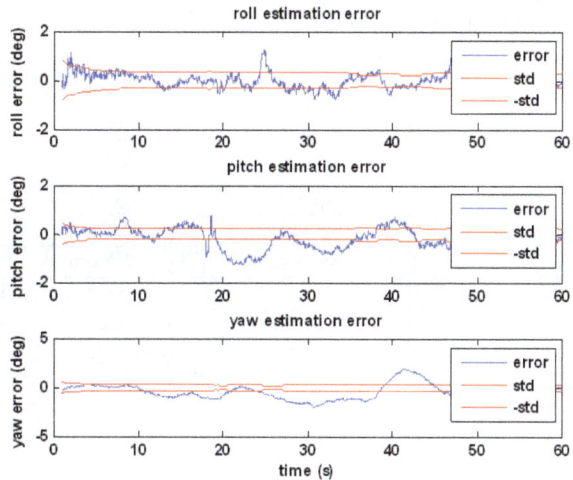

Fig. 7: Estimation error for normal trajectory.

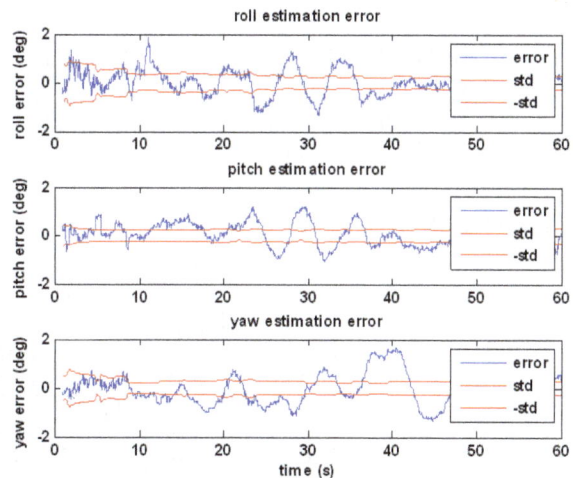

Fig. 8: Estimation error for aggressive trajectory.

comes from the fact, that in real world the aerodynamic drag is proportional to relative velocity with respect to wind while the estimation algorithm assumes it is proportional to inertial speed.

It is appropriate to assume that constant wind will have no effect on the accuracy of estimation because with respect to attitude the constant wind will cause no change of measurements in any used sensor. However the changes of wind can cause significant change in accelerometer measurements which are not predicted by the model. This can obviously lead to decreased accuracy. The following simulations are performed to verify above hypothesis. On both testing trajectories, four different settings of wind were simulated. The measure of negative impact is the RMSE error of estimated attitude expressed in Euler angles. The different settings of wind parameters are in Tab. 2. The resulting RMSE values are in Tab. 3.

Tab. 1: Full multicopter model parameters.

Model parameters			
Parameter	**Value**	**Parameter**	**Value**
Mass	$m = 1$ kg	Force drag z	$k_z = 0.8$ Ns·m^{-1}
Inertia tensor	$I/10^3$ kg·m^2	Motor constant	$c_T = 6.9 \cdot 10^{-5}$ N
Torque drag	$k_M = 0.2$ Nm	Torque constant	$c_R = 0.05$ Nm
Force drag x	$k_x = 2$ Ns·m^{-1}	Arm length	$L = 0.2$ m
Force drag y	$k_y = 2$ Ns·m^{-1}		
Sensor Parameters			
Parameter	**Gyroscope**	**Accelerometer**	**Magnetometer**
Static bias std	0.05 rad·s^{-1}	0 m·s^{-2}	0 mT
GM time constant	300 s	300 s	300 s
GM std	0.013 rad·s^{-1}	0.041 m·s^{-2}	0.041 mT
White noise std	0.014 rad·s^{-1}	0.14 m·s^{-2}	0.022 mT

Tab. 2: Wind parameters for different scenarios.

	Static wind part velocity [m·s^{-1}]			Dynamic wind part	
	x	**y**	**z**	**Time constant [s]**	**Standard deviation of GM process [m·s^{-1}]**
Wind 1	0.5	-0.5	0.5	7.0	0.2
Wind 2	5.0	-5.0	5.0	7.0	0.2
Wind 3	0.5	-0.5	0.5	7.0	2.0
Wind 4	5.0	-5.0	5.0	7.0	2.0

Tab. 3: Resulting RMSE values of wind impact simulations.

	Normal trajectory RMSE (deg)			Aggressive trajectory RMSE (deg)		
	Roll	**Pitch**	**Yaw**	**Roll**	**Pitch**	**Yaw**
No wind	0.33	0.46	0.91	0.50	0.46	0.64
Wind 1	0.47	0.56	0.90	0.73	0.52	0.64
Wind 2	0.74	0.59	0.89	1.06	0.54	0.71
Wind 3	4.09	2.81	1.68	4.31	1.94	2.05
Wind 4	3.89	2.82	1.62	4.17	1.92	2.00

Tab. 4: List of parameters and their corruptions.

	Mass	**Force drag x**	**Force drag y**	**Force drag z**	**Motor constant**	**Arm length**
	Parameter corruption factor (%)					
Scenario 1	-	70	-	-	-	-
Scenario 2	-	130	-	-	-	-
Scenario 3	-	-	-	70	-	-
Scenario 4	-	-	-	130	-	-
Scenario 5	-	-	-	-	70	-
Scenario 6	-	-	-	-	130	-

Tab. 5: Resulting RMSE values of inaccurate model parameters impact simulations.

	Normal trajectory RMSE (deg)			Aggressive trajectory RMSE (deg)		
	Roll	**Pitch**	**Yaw**	**Roll**	**Pitch**	**Yaw**
No wind	0.33	0.46	0.91	0.50	0.46	0.64
Scen. 1	0.88	0.75	1.02	2.11	1.25	1.46
Scen. 2	0.58	0.56	0.93	1.18	0.82	0.85
Scen. 3	1.13	0.65	0.93	3.42	2.33	1.27
Scen. 4	0.68	0.46	0.92	1.90	1.47	1.09
Scen. 5	3.27	1.55	1.23	6.70	5.51	2.76
Scen. 6	3.99	1.62	1.28	5.27	4.80	3.02

Tab. 6: Resulting RMSE values for algorithm with or without the wind vector state.

Aggressive	Algorithm					
	With wind in state RMSE [deg]			Without wind in state RMSE [deg]		
	Roll	Pitch	Yaw	Roll	Pitch	Yaw
Wind 1	0.80	0.63	0.67	0.73	0.52	0.64
Wind 2	0.91	0.62	0.69	1.06	0.54	0.71
Wind 3	2.11	0.74	1.38	4.31	1.94	2.05
Wind 4	0.94	0.62	0.76	4.17	1.92	2.00

The values in Tab. 3 indicate how the wind affects accuracy of the attitude estimates. It can be concluded that static part of the wind has almost no effect on estimation accuracy while dynamic part (the wind speed changes) causes significant decrease of the accuracy. Another conclusion is that the type of trajectory has a little influence on accuracy when wind disturbance is introduced.

5.2. Model Parameters

The inaccurate model parameters can also have negative impact on accuracy of attitude estimates. In order to determine the quantitative information about this effect, another set of simulations was performed.

The model parameters required in estimation algorithm are listed in Tab. 4, where also the corruption scenarios of selected parameters are defined. For each scenario the corresponding parameter was corrupted in estimation algorithm from its true value by the factor shown in Tab. 4 and then the RMSE value for both trajectories was computed which are listed in Tab. 5. Mass and arm length are excluded from simulations, since these quantities can be with ease very accurately measured. In other words it is not expected that these parameters would be inaccurate. The *force drag y* parameter is excluded because it is assumed that the behaviour in x and y axis will be very similar.

From Tab. 5 many interesting conclusions can be made. At first the accuracy of estimation strongly depends on the type of trajectory when some of the parameter is known inaccurately. Further the accuracy of the *force drag z* parameter is more critical than the ones for x and y axes. Finally the most critical parameter is constant relating the motor control signal to motor thrust.

The results of this chapter show the accuracy degradation effects which can be expected when using the model-based estimation algorithm in real application. The wind and model parameters are only a part of all effect which can degrade the accuracy. Even the model equations could be just approximations of real world behavior (which is obviously true) but here mentioned results can help suggesting on which part the deep research should be performed to gain better accuracy.

6. Wind Effect Mitigation

As shown in section 5.1 the wind has negative impact on attitude estimation accuracy using dynamic model approach. One possible way how to inhibit this effect is to include the wind speed vector into the estimated states. In the previous case the model assumed that velocity with respect to wind equals to inertial velocity. In this case they differ by amount of the estimated wind speed. The estimated wind speed is expressed in inertial frame. By rewriting Eq. (4) one gets.

$$\vec{F}_{BA} = \begin{bmatrix} \vec{v}_B^i - \mathbf{R}_{\mathrm{IB}} \vec{v}_I^w \end{bmatrix}^T \begin{bmatrix} -k_x \\ -k_y \\ -k_z \end{bmatrix}, \qquad (19)$$

where $\vec{v}_{iB}$ is velocity vector with respect to inertial frame expressed in body frame, $\mathbf{R}_{\mathrm{IB}}$ is transformation matrix from inertial to body frame and $\vec{v}_I^w$ is estimated wind speed vector expressed in inertial frame. Additionally the individual components of wind velocity vector are modeled as Gauss Markov process. In the following simulations the parameters of Gauss-Markov process for estimated state is the same as the one used for generating the wind speed vector. Results of simulations are in Tab. 6. The results show that including the wind speed vector into estimated state space significantly reduce the RMSE values for last two rows where dynamic wind is present.

7. Conclusion

This paper reveals some properties of so-called model based attitude estimation for multicopters. The standard model of multicopter with induced drag forces was presented as a baseline for designing the estimation algorithm. In simulations the performance of the filter was verified on data generated by the full model with advanced sensor models. Further the effect of wind and inaccurate model parameters were investigated. The results can be effectively used for focusing future research on the key parts which have the biggest influence on estimation accuracy, namely the wind state estimation and online parameter estimation.

Acknowledgment

The completion of this paper was made possible by the grant No. FEKT-S-14-2429 -„The research of new control methods, measurement procedures and intelligent instruments in automation" financially supported by the internal science fund of Brno University of Technology.

References

[1] POUNDS, P., R. MAHONY and P. CORKE. Modelling and Control of a Quad-Rotor Robot. In: *2006 Australasian Conference on Robotics and Automation, ACRA 2006*. Auckland: Australian Robotics & Automation Association, 2006, pp. 501–510. ISBN 978-095875838-3.

[2] BOUABDALLAH, S., A. NOTH and R. SIEGWART. PID vs LQ control techniques applied to an indoor micro quadrotor. In: *2004 IEEE/RSJ International Conference on Intelligent Robots and Systems (IROS)*. Sendai: IEEE, 2004, pp. 2451–2456. ISBN 0-7803-8463-6. DOI: 10.1109/IROS.2004.1389776.

[3] BOUABDALLAH, S. and R. SIEGWART. Backstepping and Sliding-mode Techniques Applied to an Indoor Micro Quadrotor. In: *Proceedings of the 2005 IEEE International Conference on Robotics and Automation*. Barcelona: IEEE, 2005, pp. 2247–2252. ISBN 0-7803-8914-X. DOI: 10.1109/ROBOT.2005.1570447.

[4] BOUABDALLAH, S. and R. SIEGWART. Full control of a quadrotor. In: *2007 IEEE/RSJ International Conference on Intelligent Robots and Systems*. San Diego: IEEE, 2007, pp. 153–158. ISBN 978-1-4244-0912-9. DOI: 10.1109/IROS.2007.4399042.

[5] DUNFIED, J., M. TARBOUCHI and G. LABONTE. Neural network based control of a four rotor helicopter. In: *2004 IEEE International Conference on Industrial Technology, 2004. IEEE ICIT '04*. Hammamet: IEEE, 2004, pp. 1543–1548. ISBN 0-7803-8662-0. DOI: 10.1109/ICIT.2004.1490796.

[6] DUCARD, G. and R. D'ANDREA. Autonomous quadrotor flight using a vision system and accommodating frames misalignment. In: *2009 IEEE International Symposium on Industrial Embedded Systems*. Lausanne: IEEE, 2009, pp. 261–264. ISBN 978-1-4244-4109-9. DOI: 10.1109/SIES.2009.5196224.

[7] MARTIN, P. and E. SALAUN. The true role of accelerometer feedback in quadrotor control. In: *2010 IEEE International Conference on Robotics and Automation*. Anchorage: IEEE, 2010, pp. 1623–1629. ISBN 978-1-4244-5038-1. DOI: 10.1109/ROBOT.2010.5509980.

[8] SOLC, F. Modelling and Control of a Quadrocopter. *Advances in Military Technology*. 2010, vol. 5, iss. 2, pp. 29–38. ISSN 1802-2308.

[9] GREWAL, M. S. and A. P. ANDREWS. *Kalman filtering: theory and practice using MATLAB*. 2nd ed. New York: Wiley, 2001. ISBN 04-713-9254-5.

[10] MAHONY, R., V. KUMAR and P. CORKE. Multirotor Aerial Vehicles: Modeling, Estimation, and Control of Quadrotor. *IEEE Robotics*. 2012, vol. 19, iss. 3, pp. 20–32. ISSN 1070-9932. DOI: 10.1109/MRA.2012.2206474.

[11] ROGERS, R. M. *Applied mathematics in integrated navigation systems*. 2nd ed. Reston: American Institute of Aeronautics and Astronautics, 2003. ISBN 15-634-7656-8.

[12] LEISHMAN, R., J. MACDONALD, S. QUEBE, J. FERRIN, R. BEARD and T. MCLAIN. Utilizing an improved rotorcraft dynamic model in state estimation. In: *2011 IEEE/RSJ International Conference on Intelligent Robots and Systems*. San Francisco: IEEE, 2011, pp. 5173–5178. ISBN 978-1-61284-454-1. DOI: 10.1109/IROS.2011.6094922.

[13] MACDONALD, J., R. LEISHMAN, R. BEARD and T. MCLAIN. Analysis of an Improved IMU-Based Observer for Multirotor Helicopters. *Journal of Intelligent and Robotic Systems*. 2014, vol. 74, iss. 3–4, pp. 1049–1061. ISSN 1573-0409. DOI: 10.1007/s10846-013-9835-5.

About Authors

Radek BARANEK was born in 1987. He received his master degree from Palacky University in Olomouc in 2011. His research interests include state estimation of flying robots, especially multicopters.

Frantisek SOLC was born in 1940. He received his Ph.D. from Technical University of Brno in 1977. His current research activity is focused on modeling and control of dynamic systems, including flying robots.

21

Contribution to the Management of Traffic in Networks

Filip CHAMRAZ, Ivan BARONAK

Institute of Telecommunications, Faculty of Electrical Engineering and Information Technology,
Slovak University of Technology, Ilkovicova 3, 812 19 Bratislava, Slovak Republic

filip.chamraz@gmail.com, ivan.baronak@stuba.sk

Abstract. *The paper deals with Admission control methods (AC) in IMS networks (IP multimedia subsystem) as one of the elements that help ensure QoS (Quality of service). In the paper we are trying to choose the best AC method for selected IMS network to allow access to the greatest number of users. Of the large number of methods that were tested and considered good we chose two. The paper compares diffusion method and one of the measurement based method, specifically „Simple Sum". Both methods estimate effective bandwidth to allow access for the greatest number of users/devices and allow them access to prepaid services or multimedia content.*

Keywords

AC methods, IMS, IP.

1. IMS

IP multimedia subsystem (IMS) was created in 1999 as standard 3GPP. This standard was the first try of the creation of convergent network and creation of a single platform to provide multimedia services. It was originally designed to ensure IP connectivity in UMTS. Brought change from circuit-switched technology, which was used in older generations of the systems, to packet-switched technology. IMS guarantees QoS and brings a number of benefits, technical and economical, for the service provider and customer too. The biggest advantage is cooperation with a previous generation of networks by built-in gates and strong standardization. It is used for all types of services, radio, fixed and cable.

IMS testing operation started in 2006 in Japan and Korea and 2007 in the United States. Today IMS is already fully developed in Slovakia (for example telecommunications operators O2, Telekom, Orange). IMS is used to provide a wide range of services, for example. VoIP, IPTV, video communication, transfer of data services and others [1], [2].

2. AC Methods

Admission control methods are used by creating a new connection to decide that a new connection will be accepted or rejected. AC methods are based on probability theory and mathematical statistics. Have the task keep the balance between the use of network resources and previously agreed on connection parameters. It is the first act to be carried out in the allocation of network resources for a particular connection. AC methods are the first protection against redundancy in network. New connection is allowed only if there is guaranteed QoS, otherwise the connection is refused. QoS must also be observed for the existing connections in the network. If it is not met the new connection will be allowed.

AC method solves the problem when the N connections in multiplex with a total capacity C, the probability that the sum of the immediate bit rate $r_i(t)$ of all connections in multiplex exceeds the total capacity C, is less than a given value ε. This probability can be expressed as [3], [4]:

$$P\left[\sum_{i=1}^{n} r_i(t) \geq C\right] < \varepsilon. \tag{1}$$

AC methods should satisfy three main conditions:

- Effectively allocate bandwidth to utilize maximally of telecommunications network.

- Manage a telecommunications network to meet all requirements of QoS.

- Does not allocate the entire bandwidth so that no overload on the network node is [3].

Network node

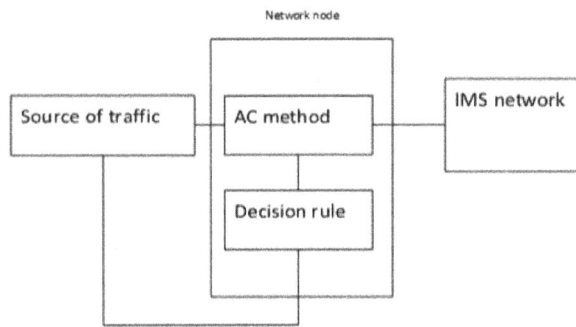

Fig. 1: AC methods.

2.1. Distribution of AC Methods

AC methods can be classified of several parameters. It depends on the view, or parameters which they work and under which requirements are evaluated. The first way to divide these methods is to divide them on the basis of traffic parameters, obtained from pre-defined values in the descriptor service or used online measurement of network. We can divide them through the use of buffer or parameter PLR (Packet Loss Ratio) or effective bandwidth and more. We know several tens of AC methods that are intended to be used in particular networks or in some nodes of telecommunication networks. With many of them we worked well in our Institute of Telecommunications Faculty of Electrical Engineering and Information Technology at Slovak University of Technology in Bratislava. The most frequently used methods are Gaussian approximation method, method of effective bandwidth, diffusion method, convolution method and others. This paper compares two methods, because it could not be possible to compare all known methods and even those most frequently implemented. For simulation the two methods were chosen, diffusion and Simple sum" algorithm. These methods were chosen because after trying many others, these came out better than the others. And of course these two methods are used in similar traffic models what is the reason why we were interesting about those methods right from very beginning.

The future of AC methods is expected in the use of the methods that used online measurement of network, fuzzy logic and neural networks [3], [4], [6], [7].

3. Diffusion Method

In terms of distribution diffusion method uses data from the traffic descriptor. Diffusion processes in this case are more precise approximation of continuous systems with discrete line slot. Method assumes total capacity of the output link C and buffer size B. The process is determined by the maximum speed of the source

R, average speed of the source r and the average cluster size b. We will use diffusion method based on two relations that describe the required bandwidth for the models with finite (FB – finite buffer) and infinite (IB – infinite buffer) capacity of line slot [3].

Packet loss probability of a final buffer size can be defined as:

$$P_{FB} = \frac{1}{\sqrt{2\pi}} e^{\frac{2B}{\alpha}(\lambda - C) - \frac{(\lambda - C)^2}{2\sigma^2}}, \qquad (2)$$

and the probability of packet loss with infinite buffer size as:

$$P_{IB} = \frac{\sigma}{\lambda\sqrt{2\pi}} e^{\frac{2B}{\alpha}(\lambda - C) - \frac{(\lambda - C)^2}{2\sigma^2}}, \qquad (3)$$

aggregated mean value of the bit rate:

$$\lambda = \sum_{i=1}^{N} r_i, \qquad (4)$$

σ^2 the aggregate variance bit rate:

$$\sigma^2 = \sum_{i=1}^{N} \sigma_i^2, \qquad (5)$$

$$\sigma_i^2 = r_i(r_i - R_i). \qquad (6)$$

Immediate variance of arrival packet α calculated as:

$$\alpha = \sum_{i=1}^{N} r_i CV_i^2, \qquad (7)$$

where

$$CV_i^2 = \frac{1 - (1 - \beta_i T_i)^2}{(\beta_i T_i + \gamma_i T_i)}, \qquad (8)$$

$$T_i = \frac{1}{R_i}, \qquad (9)$$

$b_i = \dfrac{1}{\beta_i}$ is the mean value of the active period of the source, $\dfrac{1}{\gamma_i}$ is the mean value of the inactive period.

Relations for calculating statistical bandwidth for FB and IB model, where the packet loss rates is less than ε, can be expressed as:

$$C_{FB} = \lambda - \delta + \sqrt{\delta^2 - 2\sigma^2\omega_1}, \qquad (10)$$

$$C_{IB} = \lambda - \delta + \sqrt{\delta^2 - 2\sigma^2\omega_2}, \qquad (11)$$

$$\omega_1 = \ln(\varepsilon\sqrt{2\pi}), \qquad (12)$$

$$\omega_2 = \ln(\varepsilon\lambda\sqrt{2\pi}) - \ln(\varepsilon), \qquad (13)$$

$$\delta = \frac{2B}{\alpha}\sigma^2. \tag{14}$$

The resulting statistical effective bandwidth is then determined as:

$$C_{df} = \max\{C_{FB}, C_{IB}\}. \tag{15}$$

Decision algorithm for accepting or rejecting new connection can then be summarized as follows:

- At each point of time are monitored parameters λ, σ^2, α.

- After the arrival of a new connection values of parameters λ, σ^2, α are recalculated, so as to include a new connection.

- Calculate the value C_{df}.

- If $C_{df} \leq C$ new connection is accepted.

- Else, a new connection is rejected and the previous values of the parameters λ, σ^2, α are restored.

Diffusion method is conservative with respect to packet loss, but more economic in bandwidth allocation. At the same time provides greater opportunities for homogeneous and heterogeneous telecommunication traffic. This method is easy to implement, technically and economically too.

4. Measurement Based AC Methods

The methods are based on on-line measurements of traffic passing through the switch and the new connection requires only a minimum of information. But the additional information improves efficiency of AC method. Initial estimate of bandwidth is made of the available parameters and further adjusted according to the measurement results. On-line measurement must be fast enough. It applies that the shorter measuring period then more connections can be served. AC method based on the measurement can't be used directly by the current packet loss rates. Therefore use a simpler and more efficient way and measurement of bandwidth.

If N connections passing through the switch use the bandwidth C, we try to estimate the minimum bandwidth $C(N)$. $C(N)$ is bandwidth that these connections need to be able to guarantee predetermined parameters of packet loss rate [3], [5].

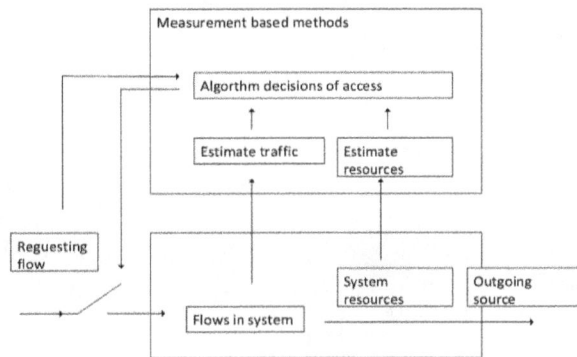

Fig. 2: Measurement based AC methods.

4.1. Algorithm „Simple sum"

It is one of measurement based AC algorithm and it is used the aforesaid principle.

$$C_r + r_{n+1} < C, \tag{16}$$

C the capacity of the line, C_r is the sum of n bit rates and connection r_{n+1}, r_{n+1} is the bit rate connection, requesting for permission.

Most often it is implemented in switches and routers where we don't expect too much load [6].

5. Simulation

Simulations and all the necessary calculations for the individual compared for all methods were developed in Matlab (R210). All results of the individual simulations are shown through specific graphs for their better readability and follow much easier interpretation. Because of all the necessary calculations and their results should clearly not choose a more appropriate method and the results should lose its importance.

5.1. Traffic Model and Parameters

For the simulation (and all the necessary calculations had to be performed at each of the compared methods) were defined traffic parameters. It was necessary calculate with these parameters. And in evaluating the results, taking to consider some limits, so that we can clearly determine the appropriateness of the method.

Defined was:
C capacity of the line,
B buffer size,
ε maximum value of a loss of packets.

Parameters were defined as follows:
$C = 100$ Mbit·s^{-1},
$B = 45$ packets,
ε from 10^{-6} to 10^{-5} packets·s^{-1}.

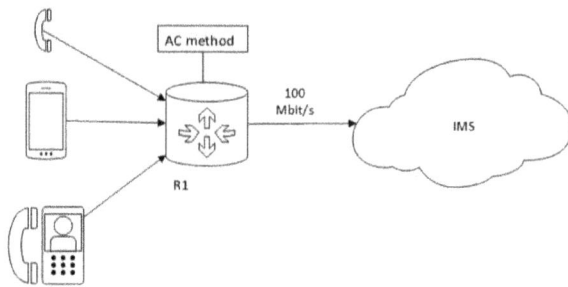

Fig. 3: Model of traffic for simulations.

As a source of traffic was used randomly generated traffic matrix on the size of $n \times T$, were $n = 300$ represented the number of used resources and $T = 300$ represented the number of time cycles, when the traffic was simulated. Traffic matrix represented requirements of network (or users) for connection (user's access to his subscription services). Individual network requirements ranged from 1 Mbit·s^{-1} to 6 Mbit·s^{-1} represented specific multimedia devices (smartphone, tablet, telephone - VoIP), which requested access, to download data (whether from different cloud or web sites), watching video (Youtube) and VoIP communication.

5.2. The Simulation Results for Diffusion Method

The first selected and simulated was diffusion method. In Fig. 4 we can see a graphical simulation result of simulation for the resultant statistical effective bandwidth C_{df}. For access asks 200 users. As we can see, access is granted "only" to 170 users/acceding devices. On further attempts to create a new connection by any other device would already crossed maximum capacity of line $C = 100$ Mbit·s^{-1}. And another user will already have access denied.

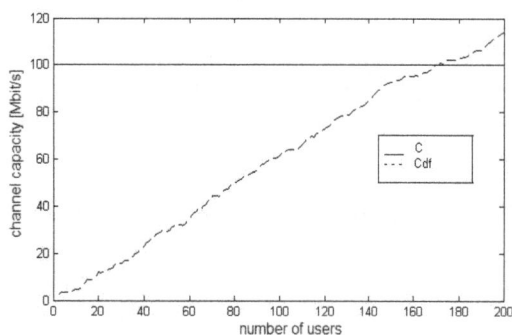

Fig. 4: The resultant statistical effective bandwidth C_{df}.

For a comparison of the resulting statistical effective bandwidth in Fig. 5 we can see waveforms for statisti-

cal bandwidth with finite C_{FB} and infinite C_{IB} buffer size that were simulated (calculated) according to given relations for diffusion method. It was necessary to calculate them, because the resulting statistical effective bandwidth C_{df} is calculated using these relations. Directly as the maximum of the values C_{IB} and C_{FB}, as was mentioned in the description of the method above. We can see, that C_{FB} obtained with the given parameters the maximum capacity of the line rather than C_{IB}. C_{IB} maximum capacity of the line in case of 200 users doesn't reached. It would reach in the case of 255 users. There would also exceed the parameter of loss of packets ε (would be lost to too many packets, about 10^{-3} [packets·s^{-1}]). This is why we take the resulting statistical effective bandwidth C_{FB}, the maximum value of this two, because it keeps all the parameters within the given prescribed values for by us simulated network

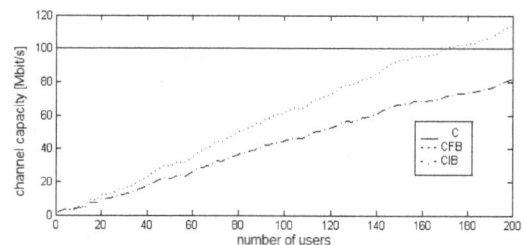

Fig. 5: Statical bandwidth with finite C_{FB} and infinite C_{IB} buffer size.

In Fig. 6 we can see how changes the probability of packet loss with a finite buffer size P_{FB}. Up to 144 incoming connections from users is probability of the loss of packets about 10^{-11} [packets·s^{-1}]. For 170 users, reaching the maximum capacity of the line is the probability of loss of packets about 10^{-9} [packets·s^{-1}], which does not exceed the maximum permissible loss of packets ε (from 10^{-6} to 10^{-5} [packets·s^{-1}]) and do not come close to these values. Even when exceeding the maximum line capacity $C = 100$ Mbit·s^{-1} do not comes to exceeding the maximum acceptable value of loss of packets ε, it is about 10^{-8} [packets·s^{-1}].

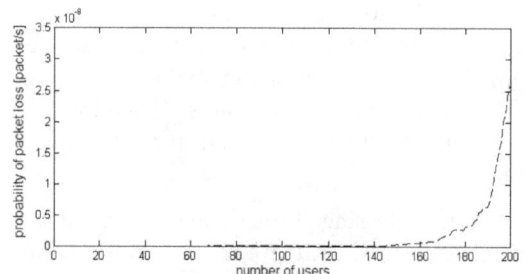

Fig. 6: P_{FB} - The probability of packet loss with a finite buffer size.

In Fig. 7 we can see how changes the probability of packet loss with infinite buffer size P_{IB}. Up to 140 incoming connections from users is probability of the loss of packets about 10^{-8} [packets·s^{-1}]. For 170 users, reaching the maximum capacity of the line is the probability of loss of packets about 10^{-6} [packets·s^{-1}], which does not exceed the maximum permissible loss of packets ε (from 10^{-6} to 10^5 [packets·s^{-1}]). Even when exceeding the maximum line capacity $C = 100$ Mbit·s^{-1} do not comes to exceeding the maximum acceptable value of loss of packets ε, it is about 10^{-5} [packets·s^{-1}]. It's been a while for the imaginary border of loss, but still within the range of the parameter.

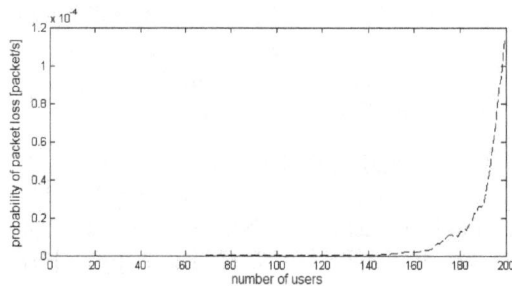

Fig. 7: P_{IB} - The probability of packet loss with an infinite buffer size.

5.3. The Simulation Results for „Simple sum" Algorithm

In Fig. 8 we can see the simulation result for algorithm „Simple sum", which is measurement based. We can see that the maximum line capacity is exceeded for 49 acceding users. This small number of users that access is granted may be caused by the simplicity of the algorithm that uses this method. Therefore the effect of the advantages of online measurements with this method not show fully.

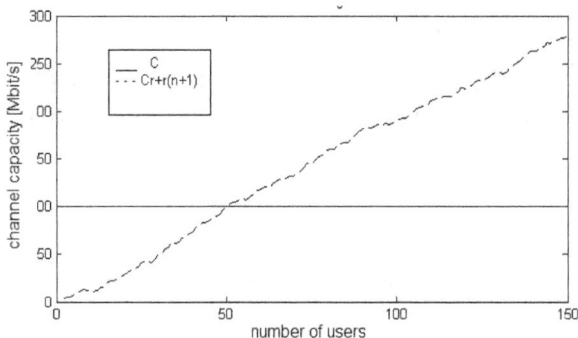

Fig. 8: The simulation results for „Simple sum" algorithm.

5.4. Comparison of Diffusion Method and „Simple sum" Algorithm

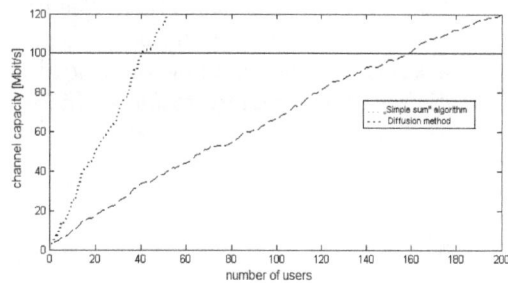

Fig. 9: The simulation results for „Simple sum" algorithm and diffusion method.

In Fig. 9 we can see simulation results for „Simple sum" algorithm and diffusion method. From this direct comparison of diffusion method and „Simple sum" algorithm it is clearly that diffusion method is much better for our traffic model. As was mentioned „Simple sum" algorithm allows access to 49 users, but diffusion method allows access to 170 users.

6. Conclusion

From the simulation results and from the graphs presented it is clear to see that it is preferable in our network node use diffusion method and not the algorithm „Simple sum".

Algorithm „Simple sum" is easier to implement from the economic point cheaper and requires less computing capacity of the system but it is simplicity is in this case more loss. Therefore as was already mentioned is mainly used for switches and routers where we don't expect too much load. In contrast diffusion AC method is more difficult by computing capacity, handling all of the necessary calculations to the decision to allow a new connection takes time computing resources (about μs or ns). Diffusion method is more difficult even from the economic point but still a reasonable and profitable as possible. Comparing the financial resources that we use to implement this method in our network and benefit from a number of service users/customers in a very short time we will return our investments. In the very core of the algorithm is indeed more complex but allow network access to multiple users while respecting maximum packet loss rates ε. Diffusion method allows access 3 - times more users than „Simple sum" algorithm.

Small and cheap solution to improve the method or system may implement certain warnings or queues. From the view of the setting of the user device and the settings of the network itself in reaching maximum capacity, the device may be reminded to wait until the available capacity. Capacity may be after released allocated and notified equipment about availability of or as it is now device can be rejected. In the last case, the device will have to try to new connection. The first cases should preferably wait and order in the queue and would rather be serviced. This would of course lead to greater user satisfaction but mainly to more efficient distributing of the source.

Acknowledgment

Paper was created with the support of MSVVaS SR within the OP Research and Development projects: "Support the completion of the Centre of Excellence for Smart Technologies, Systems and Services II" - ITMS 26240120029 and "University Science Park STU Bratislava" - ITMS 26240220084, co-financed from the resources of the European Regional development.

References

[1] YEGANEH, H., A. H. DARVISHAN and M. SHAKIBA. NGN functional architecture for resource allocation and admission control. In: *9th International Conference on Telecommunication in Modern Satellite, Cable, and Broadcasting Services*. Nis: IEEE, 2009, pp. 533–539. ISBN 978-1-4244-4382-6. DOI: 10.1109/TELSKS.2009.5339452.

[2] BARONAK, I., R. TRSKA and P. KVACKAJ. CAC – Connection admission Control in ATM Networks. *Journal of Electrical Egineering*. 2005, vol. 56, iss. 5–6, pp. 162–164. ISSN 1335-3632.

[3] YERIMA, S. Y. Implementation and Evaluation of Measurement-Based Admission Control Schemes Within a Converged Networks QoS Management Framework. *International journal of Computer Networks*. 2011, vol. 3, iss. 4, pp. 137–152. ISSN 975-2293. DOI: 10.5121/ijcnc.2011.3410.

[4] PISTEK, M. Porovnanie metod riadenia pristupu zalozenych na merani. *Posterus* [online]. 2012, vol. 5, no. 9. ISSN 1338-0087.

[5] CHROMY, E., M. JADRON, M. KAVACKY and S. KLUCIK. Admission Control in IMS Networks. *Advances in Electrical and Electronic Engineering*. 2013, vol. 11, no. 5, pp. 373–379. ISSN 1336-1376. DOI: 10.15598/aeee.v11i5.875.

[6] BARONAK, I. and M. VOZNAK. CaC in ATM - the Diffuse Method. *Acta Polytechnica: Journal of Advanced Engineering*. 2006, vol. 46, no. 3, pp. 57–63. ISSN 1805-2363.

[7] LIAO, J., J. WANG, T. LI, J. WANG and X. ZHU. A token-bucket based notification traffic control mechanism for IMS presence service. *Computer Communications*. 2011, vol. 34, iss. 10, pp. 1243–1257. ISSN 0140-3664. DOI: 10.1016/j.comcom.2010.12.017.

[8] CAO, Z., C. CHI, R. HAO and Y. XIAO. User Behavior Modeling and Traffic Analysis of IMS Presence Servers. In: *IEEE GLOBECOM 2008 - 2008 IEEE Global Telecommunications Conference*. New Orleans: IEEE, 2008, pp. 1–5. ISBN 978-1-4244-2324-8. DOI: 10.1109/GLOCOM.2008.ECP.474.

[9] VOZNAK, M. and F. REZAC. Web-based IP Telephony Penetration System Evaluating Level of Protection from Attacks. *WSEAS Transactions on Communications*. 2011, vol. 10, iss. 2, pp. 66–76. ISSN 1109-2742.

[10] CHROMY, E., J. DIEZKA, M. KOVACIK and M. KAVACKY. Traffic Analysis in Contact Centers. In: *11th International Conference Knowledge in Telecommunication Technologies and Optics, KTTO 2011* Ostrava: VSB–Technical University of Ostrava, 2011, pp.22–24. ISBN 978-80-248-2399-7.

[11] BARONAK, I. and P. KVACKAJ. Statical CAC Methods in ATM. *Radioengineering*. 2006, vol. 15, no. 2, pp. 53—56. ISSN 1210–2512.

[12] KARAMDEEP, S. and K. GURMEET. Connection Admission Control Methods Based on Fuzzy Logic. In: *6th International Multi Conference on Intelligent Systems and Nanotechnology*. Klawad: ResearchGate, 2012, pp. 68–70.

About Authors

Filip CHAMRAZ was born in Malacky, Slovakia, on March 1992. Nowadays he is a student at the Institute of Telecommunications, Faculty of Electrical Engineering and Information Technology at Slovak University of Technology in Bratislava. He completed bachelor degree in 2014. He focuses on problems of traffic management, IMS and admission control methods.

Ivan BARONAK was born in Zilina, Slovakia, on July 1955. He received Master of Science degree (electrical engineering) from Slovak Technical

University Bratislava in 1980. Since 1981 he has been a lecturer at Department of Telecommunications Slovak University of Technology in Bratislava. Nowadays he works as a professor at Institute of Telecommunications of Faculty of Electrical Engineering and Information Technology at Slovak University of Technology in Bratislava. Scientifically, professionally and pedagogically he focuses on problems of digital switching systems, Telecommunication Networks, Telecommunication management (TMN), IMS, VoIP, QoS, problem of optimal modeling of private telecommunication networks and services.

22

MULTIPATH TCP IN LTE NETWORKS

Ondrej VONDROUS, Peter MACEJKO, Zbynek KOCUR

Department of Telecommunication Engineering, Faculty of Electrical Engineering,
Czech Technical University in Prague, Technicka 2, 166 27 Prague, Czech Republic

ondrej.vondrous@fel.cvut.cz, peter.macejko@fel.cvut.cz, zbynek.kocur@fel.cvut.cz

Abstract. *The results of experiments with Multipath TCP (Multipath Transmission Control Protocol) in LTE (3GPP Long Term Evolution) networks are presented. Our results show increased resiliency and availability of network Multipath TCP connection. Using multiple sub-flows over diverse network paths inside Multipath TCP session results in greatly increased bandwidth and availability. When Multipath TCP is used the behavior of this protocol is consistent in situation of adding new sub-flow or removing flow on a faulty line. Different strategies for path selection are possible with great impact on Multipath TCP session performance. Network structure consisting of few connections with similar parameters is optimal for full mesh Multipath TCP path topology. In this case the behavior of Multipath TCP predictable.*

Keywords

LTE, multipath, TCP.

1. Introduction

Despite the fact that the world of communications is rapidly changing now, the TCP is still main protocol for Internet communication today. It is well designed and very reliable, and major applications all over the world depend on them. There is a great increase in a number of devices connected to Internet nowadays and what is more important these devices typically have more than one active interface connected to Internet network at a time. First one is WiFi (Wireless Fidelity) interface and second one is 3G/LTE modem. With decreasing price for mobile broadband data and increasing FUP (Fair User Policy) limits it obvious that there are new possibilities to improve connection stability and network throughput by utilizing multiple interfaces at a time. It is also necessary to increase throughput and preserve fairness among data flows in large Data centers, which is possible through replacing TCP by Multipath TCP as described in [1].

At this point, it is necessary to mention big limitation of TCP. This protocol is not able to take advantage of multiple active interfaces, which means that TCP can utilize only one interface at a time. There are some other protocols that were meant to replace TCP, but it has not happene yet, and TCP is still main protocol for Internet communication. One of that protocols is SCTP (Stream Control Transmission Protocol), [2], [3]. It is one of the newer protocols that were developed with multihoming in mind. SCTP has many interesting features like fail-over and possibility to use multiple paths at the same time. However, SCTP is still not widely used. This situation is caused mainly by the fact that Internet applications have to be redesigned to implement SCTP and NAT (Network Address Translation) is also problematic with SCTP.

In the new era of modern communications, we find Multipath TCP very promising. This new protocol is capable of using multiple interfaces and splitting one TCP session to multiple ones utilizing different network paths.

In this paper, we would like to demonstrate results of performance measurement of Multipath TCP over real mobile broadband networks. We focused our measurement especially on LTE networks.

The paper is organized as follows. At first we introduce Multipath TCP protocol and its specific implementation. Then we describe the methodology of our measurement and result evaluation. Then we show results of measurements on real LTE network.

2. Multipath TCP

Multipath TCP [4] is an extension of TCP, which allows single connection to be split over multiple network paths. At first a single network connection is established to the destination host. When the initial connection is established and source host is aware of destination host IP (Internet Protocol) addresses then source host can finally open additional sub-flows to the destination host. Every sub-flow is treated as a standard TCP connection when transported through the network. It is possible to use different IP addresses or different port numbers with single IP address on source host or destination host side of communication. Multipath TCP has its advantages and disadvantages like any other network protocol. Following list summarize typical advantages and disadvantages of Multipath TCP implementation.

Advantages of Multipath TCP:

- Reliability and connection persistency: it is possible to add or remove paths without connection loss.

- Increased throughput: every single connection through diverse path increase amount of available throughput.

- Transparent for applications: applications communicate the same way as on TCP.

Disadvantages of Multipath TCP:

- Fixed topology: Only client server communication.

- TCP options problem: TCP/IP headers are changed on many devices in network (NAT (Network Address Translation), IPS (Intrusion Prevention System), Firewalls etc.).TCP/IP packet fields which could be modified through network devices are shown in the Fig. 1.

- Protocol modification: Multipath TCP requires modified TCP/IP stack.

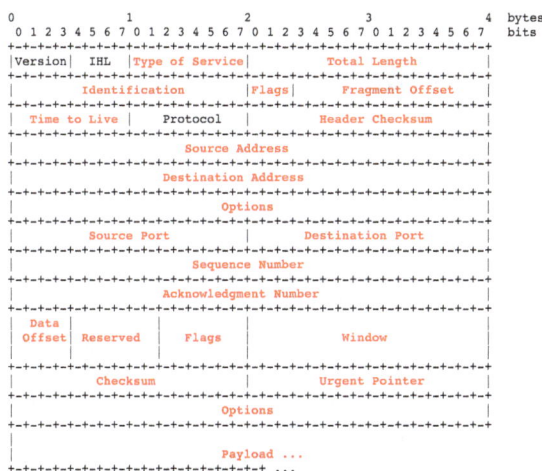

Fig. 1: TCP/IP packet fields (in red) which could be modified through network devices.

Multipath TCP session initialization: At first Multipath TCP connection is initialized in the same way as a standard TCP connection by the three-way handshake. SYN, SYN/ACK, ACK packets are exchanged on a single path. Difference between Multipath TCP and TCP is that Multipath TCP sends packet with options. In this case each packet contains the MP_CAPABLE TCP option. This option ensures receiver that the sender of this packet is capable of communicating via Multipath TCP and wishes to establish such connection. More detailed description of the session initialization, address exchange, connection security can be found in corresponding RFC 6824 [4] or we suggest this article [5] as a starting point for Multipath TCP study.

When initial connection is established and both hosts are aware of opposing host IP addresses, it is possible to open new sub-flows utilizing unused addresses. New sub-flows are typically initialized on host, which initiated original communication, but it is not limited to only opening connections by initiator. Another host can initiate new sub-flows too. These additional connections are established using MP_JOIN option in TCP header and special bit indicates if newly created sub flow is backup flow or if it becomes part of the connection immediately. This is further described in section 3.2 of RFC 6824 [4]. Sub flows can be established between any combinations of hosts IP addresses. If any sub-flow fails, there is no connection drop or packet loss in case that at least one sub-flow is still alive and working.

One of the points of interest in our research is congestion avoidance and control in Multipath TCP. Because of utilizing multiple flows algorithm for congestion avoidance and control, it is different from the algorithm used in standard TCP. Congestion control for standard TCP protocol is described by Alg. 1. Additionally, at the start of a connection and after retransmission timeout, an exponential increase is used in all further presented algorithms.

The variables used in the pseudocode are defined as follows:

- CW: congestion window size.
- CW_P: congestion window size for path P.
- PS: packet size.
- n: number of paths.

Different implementations of Multipath TCP use different algorithms. Let's take a closer look on two of them. The first one is Equally-Weighted TCP (EWTCP). In this scenario, each flow runs its own AIMD (Additive Increase/Multiplicative Decrease) algorithm. This algorithm is modified because increase

Algorithm 1 Regular TCP - AIMD.

1: **for** every incoming ACK **do**
2:     increase the congestion windows CW by $\frac{PS}{CW}$;
3: **end for**
4: **for** every each loss **do**
5:     decrease CW by $\frac{CW}{2}$;
6: **end for**

constant is dependent on a number of active sub-flows. More active sub flows lead to smaller constant. Equally-Weighted TCP is described by Alg. 2.

Algorithm 2 Equally-Weighted TCP - EWTCP.

1: $\alpha = \frac{PS}{\sqrt{n}}$;
2: **for** every incoming ACK from path P **do**
3:     increase CW_P by $\frac{\alpha}{CW_P}$;
4: **end for**
5: **for** every each loss on path P **do**
6:     decrease CW_P by $\frac{CW_P}{2}$;
7: **end for**

The second one is The Coupled Congestion Control. In this scenario, the congestion windows are not maintained for each sub-flow separately but there is one common congestion window for all sub-flow belonging to one connection. Coupled Congestion Control is described by Alg. 3.

Algorithm 3 Coupled Congestion Control.

1: CW is Global Congestion Window size maintained for all sub flows.
2: **for** every incoming ACK from path P **do**
3:     increase increase CW by $\frac{PS}{CW}$;
4: **end for**
5: **for** every each loss on path P **do**
6:     decrease CW by $\frac{CW}{2}$;
7: **end for**

More about design, implementation and evaluation of congestion control can be found in [6]. Another point of interest in Multipath TCP can be an algorithm, which is used to choose paths for Multipath TCP session. Proper path selection for Multipath TCP sub-flows can dramatically influence connection performance metrics. To ensure specific latency, jitter or bandwidth it is crucial to select proper paths, which can fulfill desired requirements on network connection.

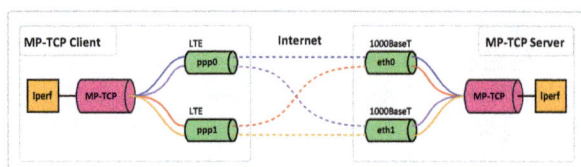

Fig. 2: Multipath TCP measurement topology.

In our measurement, we used full mesh topology because of its simplicity and because of the underlying network topology we used for measurement Fig. 2. We used server with two Gigabit Ethernet interfaces connected to Internet over network with more than sufficient capacity. On the client side, we used two LTE modems. Multipath TCP was setup for utilizing all combinations of IP addresses available. IP addresses were configured as one IP address per interface. As a traffic source and traffic destination, we used Iperf tool [7].

There are also other approaches to path selection like Adaptive Path Selection for Multipath Transport Protocols in the Internet [8] or Multipath TCP Path Selection using PCP [9]. Advanced algorithm for path selection takes its place in more complex topologies with different capacity and quality paths.

2.1.   Methodology

Compared to the traditional TCP there are no tools to analyze Multipath TCP data flows. Therefore, the simple methodology was chosen. The methodology analyzes each data stream and then determine the overall transfer rate. This approach is based on the methodology published in [10]. It uses generated TCP data flow by program named Iperf. Flow is generated between a client and a server. Connection of the server to the Internet must be at least the sum of the individual transmission speeds of all tested technologies. On the client, it is actual data flow captured over all used interfaces (LTE modems). After that program tcptrace [11] was used to obtain detailed information about transmission speed, delay, sequence numbers and various statistics from the actual TCP sessions. The results obtained from the tcptrace analysis and program Iperf were then processed by the analysis tool UDA (Universal Data Analyzer).

1)   Universal Data Analyzer Framework

UDA is a modular software tool for processing and analyzing data obtained from the measurement and simulation. Supported data sources include various tools (tcpdump, tcptrace, python, bash) and simulation tool OMNeT++ [12]. UDA framework is being developed under a grant TA02011015 and will soon be released under one of the open license. For the analysis of Multipath TCP protocol, the module in UDA framework was created. This module uses information about each contribution flow and creates an image of total throughput over Multipath TCP connection. Also, it performs basic statistical analysis. The actual output of this module is then presented in this work as graphs and additional statistical information.

Tab. 1: Table of measurement parameters.

Parameter	Value
Measurement location	Technická 2, 166 27 Prague 6, 8th floor
Measurement tool	Iperf 2.0.5
Measurement type	TCP Upload, Window size 64k
Multipath TCP version	GNU Linux 3.11.10, Multipath TCP ver. 0.88
Mobile Network Technology	LTE FDD (1800MHz band)
LTE terminal type	Cinterion PLS8-E
Internet Service Provider	Telefonica O2
$RSRP_{ppp0}$	−59 dBm
$RSRQ_{ppp0}$	−70 dB
$RSRP_{ppp1}$	−50 dBm
$RSRQ_{ppp1}$	−75 dB

2.2. Results

The aim of the measurement is to verify the behavior of the current implementation of the protocol Multipath TCP in an environment of LTE mobile network. The first objective was to test the ability of the combining individual channels in varying environment of mobile networks. The second objective was to test the ability to regenerate after failure of one of the used channels. All presented measurements were performed in the up-link direction. For a data flow generation, we used the TCP protocol (Iperf). The experiment was set in a way that one of the modems had worse signal then other. This is reflected in a difference in the delay profiles of the individual transmission channels. Parameters of the radio interface and the place of measurement are summarized in Tab. 1 (where RSRP is Reference Signal Received Power and RSRQ is Reference Signal Received Quality). The measurement process was divided into three parts:

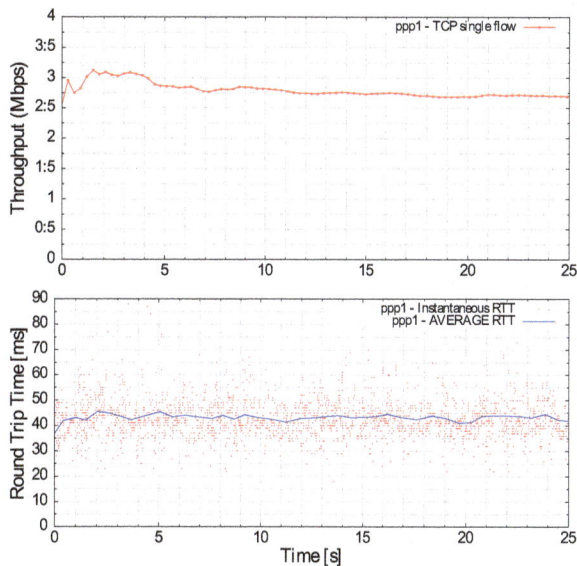

Fig. 3: Throughput and round trip time measurement of single TCP flow.

1) Common TCP Flow Measurement

In the first part we verified the maximum throughput of the data channel without the participation of Multipath TCP protocol. The results are shown in Tab. 2 and in the Fig. 3.

Tab. 2: Summary results of the measurement of the single common TCP flow.

Parameter	Value
$\overline{Throughput}_{ppp1}$	2.74 Mb·s^{-1}, $\sigma = 0.09$ Mb·s^{-1}
$\overline{RTT}_{ppp1}$	43.25 ms, $\sigma = 1.94$ ms
Measurement duration	25 s

The results correspond to measurements carried out in a given place using other applications on mobile phones and computers. Measurements were performed at the interface *ppp1* which had a better signal. The measured data confirms the behavior of LTE technology, which due to significantly lower delays (compared to technologies GPRS/EDGE or older generation of UMTS) can create much better conditions for data

transfer using TCP protocol. Also the transmission speed and delay have very low dispersion.

2) Multipath TCP Throughput Measurement

Verification of the ability to combine different transmission channels was performed in the second measurement. This measurement used the same configuration of the measuring data stream as in the first measurement. The results obtained are shown in Tab. 3 and in the Fig. 4.

In the results can be seen that interface *ppp0* had worse signal compared to *ppp1* which resulted in the higher values of RTT while maintaining similar levels of total throughput. The Fig. 5 also shows the behavior of individual transmission rates for each channel. The total transfer rate corresponds to the sum of the contributing data streams. Even with worsened conditions on the physical layer of channel *ppp0*, Multipath TCP was able to bring together the different data

Tab. 3: Summary results of the measurement of Multipath TCP flows.

Parameter	Value
$\overline{Throughput}_{ppp0}^{1st}$	1.85 Mb·s^{-1}, $\sigma = 0.07$ Mb·s^{-1}
$\overline{Throughput}_{ppp0}^{2nd}$	1.78 Mb·s^{-1}, $\sigma = 0.11$ Mb·s^{-1}
$\overline{Throughput}_{ppp1}^{1st}$	2.04 Mb·s^{-1}, $\sigma = 0.25$ Mb·s^{-1}
$\overline{Throughput}_{ppp1}^{2nd}$	1.42 Mb·s^{-1}, $\sigma = 0.13$ Mb·s^{-1}
$\overline{RTT}_{ppp0}^{1st}$	105.84 ms, $\sigma = 26.29$ ms
$\overline{RTT}_{ppp0}^{2nd}$	106.82 ms, $\sigma = 25.62$ ms
$\overline{RTT}_{ppp1}^{1st}$	37.87 ms, $\sigma = 3.71$ ms
$\overline{RTT}_{ppp1}^{2nd}$	38.71 ms, $\sigma = 3.66$ ms
Measurement duration	25 s

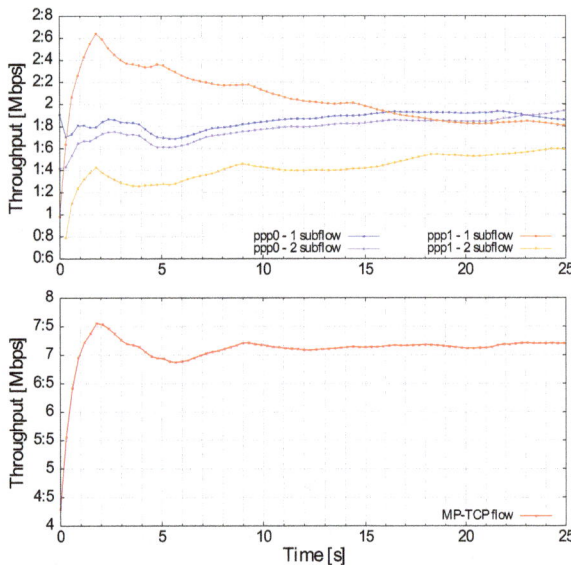

Fig. 4: Throughput measurement of Multipath TCP flows.

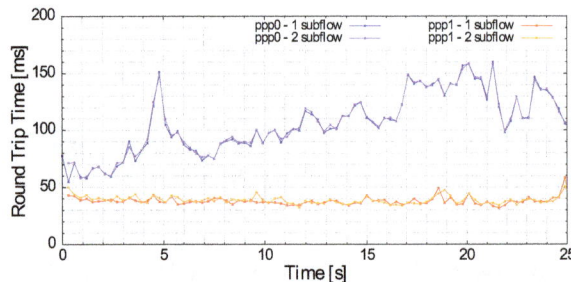

Fig. 5: Round trip time measurement of Multipath TCP flows.

streams. There is also an increase in radio resource utilization via MP-TCP. When utilizing MP-TCP the final throughput at single modem interface is higher than the throughput of single TCP Flow as shown on Fig. 2 and Fig. 4.

3) Multipath TCP Disconnecting Measurement

Ability to recover from the failure of one of the channels was verified in the third measurement. The configuration of measuring data stream was same as in the previous two cases. During the measurements the antenna from channel number 1 (*ppp1* interface) were randomly disconnected (2×). This disconnection resulted in signal loss, and therefore the connection was interrupted. The results of the measurements are shown in Tab. 4 and Fig. 6 and Fig. 7. The graph with total transfer rate Fig. 6 has also transfer rate obtained from the Iperf tool.

Tab. 4: Summary results of the measurement of the Multipath TCP flows with disconnection of the *ppp1* interface.

Parameter	Value
$\overline{Throughput}_{ppp0}^{1st}$	1.50 Mb·s^{-1}, $\sigma = 0.49$ Mb·s^{-1}
$\overline{Throughput}_{ppp0}^{2nd}$	1.83 Mb·s^{-1}, $\sigma = 0.40$ Mb·s^{-1}
$\overline{Throughput}_{ppp1}^{1st}$	1.96 Mb·s^{-1}, $\sigma = 0.19$ Mb·s^{-1}
$\overline{Throughput}_{ppp1}^{2nd}$	1.08 Mb·s^{-1}, $\sigma = 0.10$ Mb·s^{-1}
$\overline{RTT}_{ppp0}^{1st}$	87.68 ms, $\sigma = 32.38$ ms
$\overline{RTT}_{ppp0}^{2nd}$	88.68 ms, $\sigma = 32.85$ ms
$\overline{RTT}_{ppp1}^{1st}$	43.63 ms, $\sigma = 5.01$ ms
$\overline{RTT}_{ppp1}^{2nd}$	39.35 ms, $\sigma = 3.41$ ms
Measurement duration	120 s
Disconnection time 1	in 16 s
Disconnection time 2	in 67 s

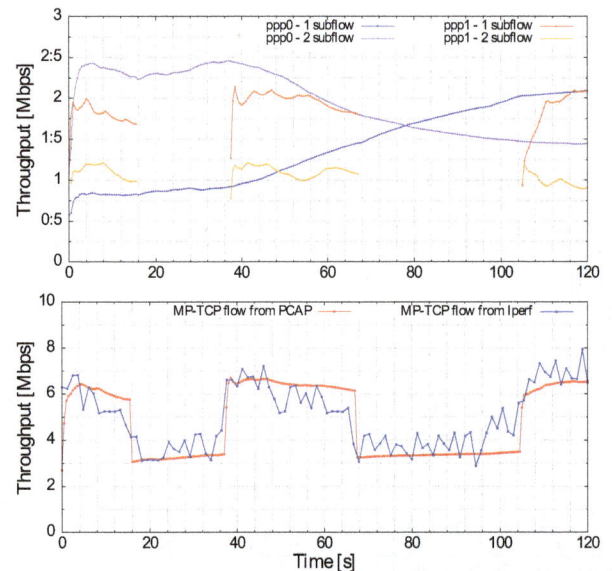

Fig. 6: Throughput measurement of Multipath TCP flows with disconnection of the ppp1 interface.

Progression of throughput in Fig. 6 confirms proper operation of the activation and deactivation of channels. When one of the channels is disabled, the established TCP connection is not terminated, but throughput fall to sum of available connections. The activation of a new or previously disabled transmission chan-

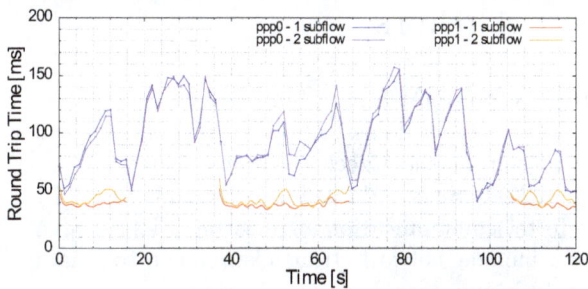

Fig. 7: Round trip time measurement of Multipath TCP flows with disconnection of the ppp1 interface.

nel has no impact on the stability of existing connection. Delays in each channel (see Fig. 7) corresponds to the behavior that has been found in previous measurements. The total throughput of Multipath TCP obtained by our methodology is identical to the throughput obtained from the output of Iperf.

3. Conclusion

The presented measurements demonstrate capability of Multipath TCP to utilize multiple network paths effectively and provide increased availability and bandwidth. The experiments prove that this protocol is mature now and is ready for production use.

Our experiments show that this protocol has minimum overhead, and final bandwidth is the sum of sub-flows bandwidth. If there is diversity among network sub-flow path, then bandwidth is maximized. Multipath TCP also reacts very well on link fault, where neither packet loss is detected nor is latency increased. If faulty line is back in operating state Multipath TCP adds new sub-flow almost instantly, thus increasing bandwidth to its original state. Because of increased resiliency and availability by Multipath TCP connection it would be convenient to deploy this protocol into industrial networks where these features are demanded.

As future work, we will focus on Multipath TCP protocol analysis improvement mainly on RTT and jitter. We will also invest in testing Multipath TCP with other tools, namely tools for measuring systems of realtime. We would like to use methodologies for measurement and analysis presented for VoIP networks [13] and industrial networks [14].

Acknowledgment

This work was supported by Ministry of the Interior of the Czech Republic Grant no. VG20122014095, "Development of adaptable and data processing systems for high-speed, secure and reliable communication in extreme conditions", and the research was carried out in cooperation with CERTICON. Also this work was supported by Students grants at Czech technical university in Prague SGS13/200/OHK3/3T/13 and SGS12/186/OHK3/3T/13.

References

[1] RAICIU, C., S. BARRE, Ch. PLUNTKE, A. GREENHALGH, D. WISCHIK and M. HANDLEY. Improving datacenter performance and robustness with multipath TCP. In: *Conference on SIGCOMM*. New York: ACM Press, 2011, pp. 266–277. ISBN 978-1-4503-0797-0. DOI: 10.1145/2018436.2018467.

[2] RFC 2960. *Stream Control Transmission Protocol* [online]. IETF, 2000. Available at: `https://www.ietf.org/rfc/rfc2960.txt`.

[3] RFC 3286. *An Introduction to the Stream Control Transmission Protocol (SCTP)* [online]. IETF, 2002. Available at: `https://www.ietf.org/rfc/rfc3286.txt`.

[4] RFC 6824. *TCP Extensions for Multipath Operation with Multiple Addresses* [online]. IETF, 2013. Available at: `https://tools.ietf.org/html/rfc6824`.

[5] BARRE, S., Ch. PAASCH and O. BONAVENTURE. MultiPath TCP: From Theory to Practice. In: *10th International IFIP TC 6 Networking Conference*. Valencia: Springer, 2011, pp. 444–457. ISBN 978-3-642-20756-3. DOI: 10.1007/978-3-642-20757-0_35.

[6] WISCHIK, D., C. RAICIU, A. GREENHALGH and M. HANDLEY. Design, implementation and evaluation of congestion control for multipath tcp. In: *8th USENIX Conference on Networked Systems Design and Implementation, ser. NSDI'11*. Berkeley: USENIX Association, 2011, pp. 99–112. ISBN 978-931971-84-3.

[7] Iperf: tools to measure network performance. [online]. 2011. Available at: `https://iperf.fr`.

[8] CHEN, X. Y. Y. and X. WU. *Maps: Adaptive path selection for multipath transport protocols in the internet* [online]. 2011. Available at: `https://www.cs.duke.edu/~yuchen/papers/maps.pdf`.

[9] MPTCP. *Multipath TCP (MPTCP) Path Selection using PCP draft-wing-mptcp-pcp-00* [online]. 2013. Available at: `http://tools.ietf.org/html/draft-wing-mptcp-pcp-00`.

[10] KOCUR, Z., P. CHLUMSKY, P. MACEJKO, M. KOZAK, L. VOJTECH and M. NERUDA. Measurement of Mobile Communication Devices on the Testing Railway Ring. In: *15th International Conference on Research in Telecommunication Technologies*. Bratislava: Slovak University of Technology in Bratislava, 2013, pp. 34–37. ISBN 978-80-227-4026-5.

[11] Tcptrace: tcp/ip analysis tool. [online]. 2014. Available at: http://www.tcpdump.org/.

[12] Omnet++: framework home page. [online]. 2014. Available at: http://www.omnetpp.org/.

[13] VOZNAK, M. and J. ROZHON. Approach to stress tests in SIP environment based on marginal analysis. *Telecommunication Systems*. 2013, vol. 52, iss. 3, pp. 1583–1593. ISSN 1018-4864. DOI: 10.1007/s11235-011-9525-1.

[14] CEPA, L., Z. KOCUR and Z. MULLER. Migration of the IT Technologies to the Smart Grids. Elektronika ir Elektrotechnika. 2012, vol. 123, iss. 7, pp. 123–128. ISSN 1392-1215. DOI: 10.5755/j01.eee.123.7.2390.

About Authors

Ondrej VONDROUS was born in Czech Republic in 1981. He received his M.Sc. degree in electrical engineering from the Czech Technical University in Prague in 2011. Since 2011 he has been studying Ph.D. degree in telecommunication engineering. His research interests include network transmission control, data flow analysis and data flow optimization.

Peter MACEJKO was born in Czech Republic in 1980. He received his M.Sc. degree in electrical engineering from the Czech Technical University in Prague in 2006. He is teaching networking technologies and distributed systems. His research is focused on scheduling in distributed systems and data flow and protocol analysis. He is currently actively involved in projects focused on high speed data transmission from fast moving objects.

Zbynek KOCUR was born in 1982. He received his M.Sc. degree in electrical engineering from the Czech Technical University in Prague in 2008 and Ph.D. degree in electrical engineering in 2014. He is teaching communication in data networks and networking technologies. His research is focused on wireless transmission and data flow analysis, simulation and optimization.

23

Impact of Nodal Centrality Measures to Robustness in Software-Defined Networking

Tomas HEGR, Leos BOHAC

Department of Telecommunications, Faculty of Electrical Engineering,
Czech Technical University in Prague, Technicka 2, 166 27 Prague, Czech Republic

tomas.hegr@fel.cvut.cz, bohac@fel.cvut.cz

Abstract. *The paper deals with the network robustness from the perspective of nodal centrality measures and its applicability in Software-Defined Networking (SDN). Traditional graph characteristics have been evolving during the last century, and numerous of less-conventional metrics was introduced trying to bring a new view to some particular graph attributes. New control technologies can finally utilize these metrics but simultaneously show new challenges. SDN brings the fine-grained and nearly online view of the underlying network state which allows to implement an advanced routing and forwarding. In such situation, sophisticated algorithms can be applied utilizing pre-computed network measures. Since in recent version of SDN protocol OpenFlow (OF) has been revived an idea of the fast link failover, the authors in this paper introduce a novel metric, Quality of Alternative Paths centrality (QAP). The QAP value quantifies node surroundings and can be with an advantage utilized in algorithms to indicate more robust paths. The centrality is evaluated using the node-failure simulation at different network topologies in combination with the Quality of Backup centrality measure.*

Keywords

Centrality, network robustness, OpenFlow, software-defined networking, topology.

1. Introduction

The concept of Software-Defined Networking (SDN) revealed countless new possibilities in the area of packet forwarding in data networks, packet handling and network security. Even though the concept is oriented to the network control centralization, it does not mean that the network reliability, i.e. provided service availability, is automatically decreased. The view of the events in the network and the online accessibility of the traffic statistics gives a possibility to install forwarding rules (also referred as matching rules) with respect to variety network measures. One of the network areas which can be improved at the flow level is a host-to-host communication reliability, i.e. the installed network path resilience, which is important for mission-critical applications.

In SDN, a host-to-host communication path is, usually, established via installation of engineered rules into network nodes along the path. Considering the OF protocol, these rules are usually installed in a reactive way, when the controller needs every time to evaluate the given traffic pattern. To keep the packet latency low, it is necessary to provide rules as fast as possible. An SDN application has to compute a network path keeping a low latency which can be challenging especially in complex networks. First and very understandable way is to optimize the SDN applications built up to the controller to reach the low latency of the network path computation. Another way is to provide pre-computed measures to these algorithms, and so reduce the problem complexity or increase quality of results.

In this paper, we focus on the path robustness in terms of the aforementioned communication reliability and node failure resilience. It is generally true that path protection is better than path restoration from the latency view. In the OF area of SDN, the protection can be realized via the set of equal matching rules with different execution priority and dissimilar output action. The same can be simply established thanks to the *Group Table* architecture and the *fast failover* feature included in recent versions of the OF specification [7]. In such case, the first forwarding action is executed from a set of actions at which the output port is alive for the particular flow in OF. Since the management of rules itself is demanding, there must be assured proper expressiveness, efficiency and correctness, and it should be utilized an appropriate framework as for example is FatTire [14] by Reitblatt et al.

Although the technical means are already available to implement the protected path, the issue with right path selection in complex networks still remains. This is particularly important in high-demanding low-latency networks where the connection protection is vital. We suggest utilizing a graph centrality measures during the path generation process to solve the issue. The graph centrality concept is well known in graph theory and from the area of social network analysis. In the next sections we present and evaluate new centrality named Quality of Alternative Paths (QAP). The centrality reflects an average fitness of node's alternative paths and thus it can be used not only during the path generation process but also for network analysis. The evaluation of the centrality measure impact to network robustness is done through the node failure simulation together with the another centrality measure called Quality of Backup published by Shavitt and Singer in [18].

2.   Related Works

The mechanism of failure recovery was in general extensively investigated from the beginning of the telecommunication age. In the area of SDN and particularly OF, there have been published several papers on this topic in recent years. Beside the FatTire [14] framework, there was presented a runtime system for emulating controller flow installations allowing programmers to implement failure-agnostic code in [8]. The way of handling high controller load during a failure in the network and the appropriate fast restoration algorithm were proposed in [15] by Sharma et al. The authors showed in [17] that using the *Group Table* and two well-known mechanisms of failure recovery, i.e. restoration and protection, the sub-50 ms recovery latency can be achieved at carrier-grade networks. Since the results are encouraging the authors showed in [16] that the OF in-band control, thus utilizing same channel for both control and payload traffic, remains problematic from the latency viewpoint. The optimized controller placement and control-traffic routing can greatly improve the SDN architecture resilience as is presented in [3].

The node degree is one of the basic centrality metrics in graph theory. While it simply shows the node's elementary attribute, it could not encompass all information about the node's surroundings. Therefore, many new metrics were introduced targeting to particular node's attribute such as closeness, betweenness and others [5]. The centrality measures describing network robustness were introduced in [18] where authors describes two new centrality measures, Quality of Backups and Alternative Path Centrality (APC). While QoB is a normalized centrality, APC quantify

the topological contribution of a node to the network functionality in graph-individual way. Beyond these robustness centrality measures, even other measures were also as, for example, advanced compound length-constrained Connectivity and Rerouting Centrality (l-CRC) designated for wireless sensor networks [19].

To evaluate the impact of centrality measures to the path robustness, we decided to apply the node failure simulation. This method is well-known, and it was frequently used for describing relations between nodes and graph components in complex networks in famous work [1]. Focusing particularly on network robustness several papers were published by Manzano et al. In [10], the authors present a robustness analysis of real-world network topologies under different failure scenarios. These scenarios are divided into Static and Dynamic node impairments. Likewise, there were published works analyzing contemporary data-center topologies in [9] and in [6]. As the Static random failure test shows to be one of the most useful simulation test it was selected as a counterexample to our targeted static node failure simulation described in section 4.

Topologies for the QAB metric evaluation were chosen from several different networking areas. The random graphs have been obtained using three models Erdos-Renyi [4], Watts and Strogatz as the small-world representative [20] and Barabasi-Albert as the scale-free one [2]. Traditional random graphs have been supplemented with real-world carrier-grade networks accessible at the SDNlib project [13].

3.   Centrality Measures

In this section, we describe the graph theory background and two centrality measures. The first is QoB and the second is the newly proposed QAP.

3.1.   Graph Theory Background

Every communication network can be represented by a graph $G = (V, E)$ where V is a set of nodes and E is a set of edges. Every pair of two connected nodes are said to be adjacent. A walk of length k between any two nodes $(u, v) \in V$ is a sequence of edges $e_1 e_2 ... e_k$ such that e_i and e_{i+1} are adjacent. If we denote the walk from u to v in k steps as $u \xrightarrow{k} v$ than the distance $\delta(u, v) = \min\{k | u \xrightarrow{k} v\}$ is the length of the shortest path in an unweighted graph. By convention, $\delta(u, v) = \infty$ if there does not exist a walk from u to v and $\delta(u, u) = 0$. If we look for a shortest path from u to w bypassing node v, i.e. the path is not going through the node v, we denote such distance $\delta_v(u, w)$. Both centrality measures involve this vital function.

Moreover, both measures QAP and QoB use specific node sets to express the node centrality. The set of nodes adjacent to node v are called neighbors of node v and similarly $N_G(v) = \{u \in G | (v,u) \text{ or } (u,v) \in E\}$ is named the neighborhood of v. Then the set $C_v = \{u \in G | (v,u) \in E\}$ is the set node v's direct children, thus nodes accessible through v, and the set $P_v = \{u \in G | (u,v) \in E\}$ is the set node v's direct parents. It is clear that in undirected graphs sets P_v and C_v are equal.

3.2. Quality of Backup

The Quality of Backup centrality introduced in [18] quantifies backup efficiency of a given node by examination the cost of re-routing paths from the set of parents to the set of children. The expression Eq. (1) shows the QoB function capturing the backup efficiency:

$$\rho(v) = \frac{\displaystyle\sum_{u \in P_v} \sum_{w \in C_v} \frac{1}{\max\{\delta_v(u,w) - 1, 1\}}}{|P_v| \cdot |C_v|}. \quad (1)$$

QoB describes node's dispensability in the given graph in the normalized way, thus $\rho : V \to [0,1]$, and so is possible to compare different graphs between each other. If a node with perfect backup $\rho = 1$ fails, the network functionality is not reduced. In this case, neither the network connectivity nor the path lengths in the network are affected. On the contrary if $\rho = 0$ the node has no backup.

3.3. Quality of Alternative Paths

While QoB gives a comprehensive centrality measure in terms of node backup quality using the path lengths, it does not consider number of such alternative paths, thus the number of those paths bypassing the evaluated node. The presented Quality of Alternative Paths centrality is a complementary measure to QoB dealing with this gap. This auxiliary centrality measure provides additional information when analyzing network robustness or generating paths in a network.

The QAP has been inspired by QoB as can be seen in expression Eq. (2). In this expression, it is used a newly defined function α_v which assigns a weight to the connections in $P_v \times C_v$, i.e. the connections between all parent and child nodes. This weight takes into account number of alternative paths and consequently it can be used to prioritize nodes with higher load-balancing potential avoiding bottlenecks in case of the evaluated node failure. The final QAP value is given as the average weight of all alternative paths:

$$\eta(v) = \frac{\displaystyle\sum_{u \in P_v} \sum_{w \in C_v} \alpha_v(u,w)}{|P_v| \cdot |C_v|}. \quad (2)$$

$$\alpha_v(u,w) = \begin{cases} 1 & \text{if } \delta_v(u,w) = 0 \\ \displaystyle\sum_{p \in D} \frac{1}{\delta_v(u,w)} & \text{else} \\ 0 & \text{if } \delta_v(u,w) = \infty \end{cases} \quad (3)$$

The newly proposed weight function α expressed in Eq. (3) has the following rationale. Every node-disjoint path between the given parent's node and child's node bypassing the examined node contribute to the total weight by the reverse value of the path's length. This is valid if the nodes are connected, else the weight is 0. In case $u = w$, the weight is 1 to keep the proper η average value. In the α expression, only contribution of node-disjoint paths is considered. Such approach gives the number of effective alternative paths, and it prefers better node connectivity. The set of node-disjoint paths between given nodes from P_v and C_v in Eq. (3) is marked as D. An example network is shown in Fig. 1.

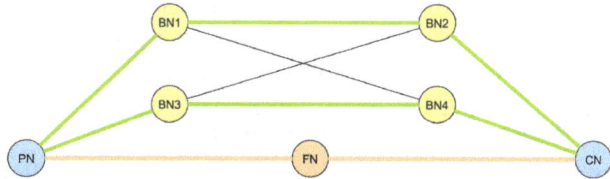

Fig. 1: An example scenario with two node-disjoint paths bypassing node FN between nodes PN and CN going through BN nodes.

The transmission latency is, usually, introduced by interconnecting nodes along the transmission path nowadays. QAP was designed to prefer the shorter length of the path to the number of paths in order to prioritize lower latency. This characteristic is obvious from Fig. 2. In case the parent's node and the child's node are directly connected, the path weight is 1, consequently QAP for all nodes in Full graph reaches 1.

Even though in today's conventional networks links have assigned cost or weight, in QAP we consider only two aforementioned graph properties, number of paths and the path's length, and the application only on unweighted graphs. This can be modified by extending α function by adding an appropriate cost variable according to the chosen network environment.

The basic version of the algorithm for QAP computation is written below in Alg. 1 in form of pseudocode. All failover shortest paths are stored in list marked $\bar{\delta}_v$. These paths are found by BFS algorithm. According to Menger's theorem [11] the minimum vertex cut between two given nodes is equal to the maximum

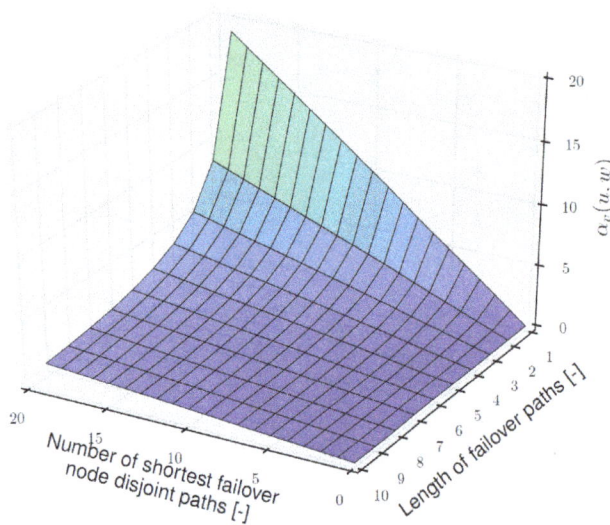

Fig. 2: Model of α_v function showing the relation between lengths and number of shortest failover node-disjoint paths. It shows steep increase of the weight for shorter paths, i.e. preferring shorter paths to number of paths.

number of pairwise node-independent paths between them. This can be done in $O(|V|^3)$. In our approach, the node-disjoint failover paths are found from a set of all shortest failover paths between given nodes by the process of iterative removing most node-intersect paths until no node intersections are present.

Algorithm 1 QAP centrality pseudocode.
---

function QAP(v,G)
  $\eta \leftarrow 0$
  for u **in** P_v **do**
    $\bar{\delta}_v(u) \leftarrow BFS_v(u, G, allpaths)$
    for w **in** C_v **do**
      $\bar{\delta}_v(u,w) = $ Find D for w in $\bar{\delta}_v(u)$
      if $\delta_v(u,w) = 0$ **then**
        $\eta \leftarrow \eta + 1$
      else if $\delta_v(u,w) \neq \infty$ **then**
        $\eta \leftarrow \eta + |\bar{\delta}_v(u,w)| \frac{1}{\bar{\delta}_v(u,w)}$
  $\eta \leftarrow \frac{\eta}{|P_v||C_v|}$

---

4.   Centrality Evaluation

This section describes the impact of complementary centrality measures QoB and QAP to the network under the static node-failure simulation. Since the usability proof of centrality measures by following all paths in the network during the failure simulation would be exceedingly complex, we decided to apply static failure simulation known from topology robustness tests as was mentioned in section 2. In the evaluation, we use All Terminal Reliability (ATR) [12] as a pointer to the beginning and the end of evaluation interval

which described later. We compare network diameter D, average shortest path length δ and total count of all shortest path in the network at both simulation scenarios in the evaluation interval.

Selected topologies and a set of computed classical robustness characteristics is listed in Tab. 1. All topologies were constructed as undirected and un-weighted graphs.

The simulations were performed in two different scenarios. The first is Static Random node-failure simulation (RS) which was repeatedly evaluated to reach desired confidence level of 95 %. The second is Targeted Static node failure simulation (TS). The TS simulation is based on the value of QoB and QAP measures. The higher priority is assigned to QoB in TS since it is the preferred path robustness to quality of alternative paths. QAP is used as a complementary criterion. TS simulations were carried out in descending order of QoB and on the second place in ascending order of QAP. This means that nodes are removed from the less-significant one with the worst quality of alternative paths to the best one. We assume that better results of TS in observed graph parameters give a presumption for usability of QoB and QAP in the selection of more resilient paths with higher quality of alternative paths.

The evaluation interval is in this paper defined as the difference in portion of removed nodes at the moment the ATR falls down under 1 for the first time during the node removing process in both scenarios. This means the moment when the evaluated graph is unconnected for the first time during the simulation run. Since RSs were carried repeatedly, we focused on the average portion of removed nodes. All numbers of removed nodes are expressed as the portion of removed nodes to the total number of nodes in the graph.

The comparison of both simulation scenarios was done at the begging of the evaluation interval (RS removed nodes) and in the mid-interval. Even though, it was possible to analyze also the end of the evaluation interval, the results were not relevant in most cases because of the small residual graph size.

An example histogram showing distribution of QAP for the nobel-eu is depicted in Fig. 3. In this particular case, all QAP values are in the range from 0.45 to 0.75. This is because most nodes have similar nodal degree. The colored plot of the topology is shown in Fig. 4, where the colors match to the computed QAP values. The QAP value has no upper bound and can be significantly higher at different topologies.

The results are presented in the form of percentage comparison between TS and RS scenarios at the begging and in the mid-interval. The first part is the diameter comparison depicted in Fig. 5 which is an important graph parameter giving information about

Tab. 1: Main characteristics of constructed graphs. First three are random graphs with 400 nodes and similar average nodal degree close to 12. Their names are as follow: ern400d12 for Erdos-Renyi, swn400d12 for Watts and Strogatz and ban400d12 for Barabasi-Albert. The rest of graphs is based on real-world telecommunication topologies from SNDlib project [13].

Characteristic	ban400d12	ern400d12	swn400d12	geant	germany50	india35	nobel-eu	norway	pioro40
$\|V\|$	400	400	400	22	50	35	28	27	40
$\|E\|$	2379	2381	2400	36	88	80	41	51	89
Diameter	3	4	4	5	9	7	8	7	7
Average short-est path length	2.16	2.68	2.68	2.53	4.05	2.94	3.56	3.13	3.31
StdDev	0.44	0.59	0.58	0.99	1.75	1.25	1.65	1.47	1.48
Average nodal degree	11.90	11.91	12.00	3.27	3.52	4.57	2.93	3.78	4.45
StdDev	22.31	3.53	3.17	1.70	1.05	1.69	0.86	1.19	0.50
Maximum de-gree	183	27	24	8	5	9	5	6	5
Minimum de-gree	6	3	4	2	2	2	2	2	4
Edge density	0.03	0.03	0.03	0.16	0.07	0.13	0.11	0.15	0.11
Edge per vertex	5.95	5.95	6.00	1.64	1.76	2.29	1.46	1.89	2.23
Heterogeneity	0.09	0.01	0.01	0.11	0.04	0.06	0.05	0.06	0.02

Tab. 2: Observed moments when the ATR falls below 1 for the first time in the simulation run and the mid-interval between those moments at TS and RS.

Topology	TS removed nodes [%]	RS removed nodes [%]	Mid interval [%]
geant	72.7	28.9 ± 1.9	50.8
germany50	52.1	19.1 ± 1.0	35.6
india35	62.9	32.5 ± 1.6	47.7
nobel-eu	22.2	18.5 ± 1.0	20.3
norway	44.4	29.0 ± 1.1	36.7
pioro40	47.5	33.1 ± 1.3	40.3
ern400d12	74.8	50.2 ± 2.6	62.5
ban400d12	97.8	46.4 ± 3.5	72.1
swn400d12	69.3	53.0 ± 3.3	61.1

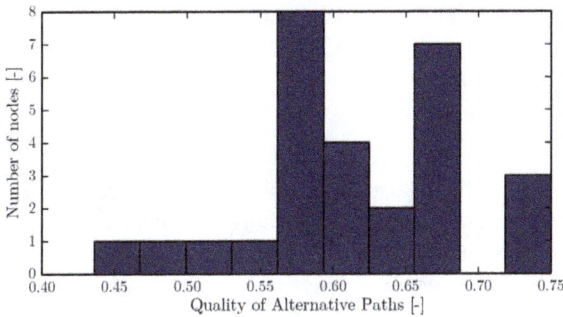

Fig. 3: An example QAP distribution for the nobel-eu topology. Since the topology is not excessive and nodal degree of most nodes is similar the histogram shows low values of QAP up to 0.75.

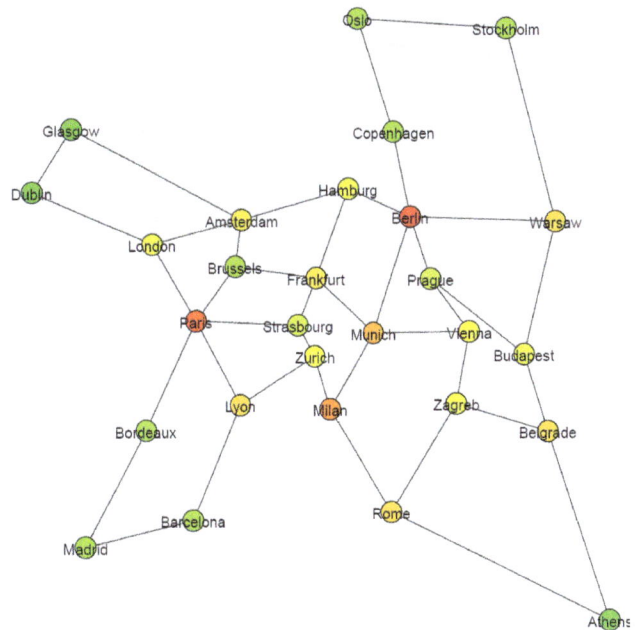

Fig. 4: The plotted nobel-eu topology shows QAP values depicted in Fig. 3. The color range from green to red represents QAP values in range from 0.45 to 0.75.

maximum path lengths in the graph. Results show that the TS scenario gives in almost all cases smaller graph diameter especially at the begging of the interval, where the improvement reaches almost 52 % in case of ban400d12. The state in the mid-interval is unbalanced at real-world showing that engineered topologies are in case removing higher number of unimportant nodes degrade considerably faster.

The second observed characteristic is the average shortest path length which gives another information about path lengths to diameter. The results depicted in Fig. 6 almost correlate in many cases with results for di-

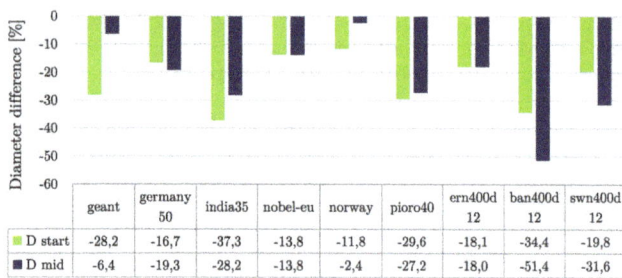

Fig. 5: The bar chart shows a percentage difference in graph diameter between RS and TS for evaluated topologies. The TS shows that diameter at all cases is smaller up to 52 % as in case of ban400d12 in the mid-interval.

ameter. Values reach 23 % shorter paths in case of the india35 topology at the begging of the interval that is lower than the improvement at graph diameter. Higher improvement than india35 reaches again random graph ban400d12 at 27 %. One can observe more significant reduce of improvement in case of randomly generated graphs. This is highly probably due to the different graph structure than at the engineered real-world telecommunication topologies. The improvement drop is most significant at norway network in mid-interval as in the case of diameter.

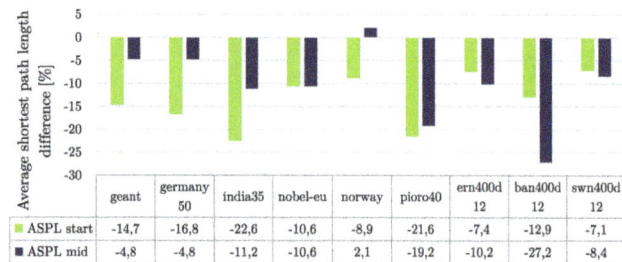

Fig. 6: The value of average shortest path length shows roughly the same results as in case of diameter. The percentage difference reaches up to 23 % for india35 topology at the beginning of the evaluation interval and 27 % for ban400d12 in the mid-interval.

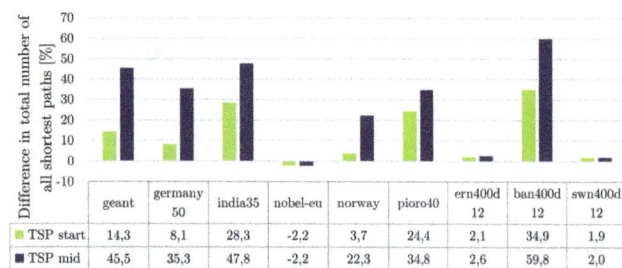

Fig. 7: Total number of all shortest paths shows the positive impact of evaluated centrality measures especially in mid-interval. The TS gives in several cases from 30 % to 40 % more paths.

The last characteristic reflecting the impact of QAP is the total number of all shortest paths between all pairs of nodes. Results depicted in Fig. 7 show that the general improvement is predominantly in mid-interval.

More paths exist between nodes more possibly the path protection can be implemented. While the improvement for average path lengths for norway network was negative in mid-interval, in case of the total number of all shortest paths is gives more than 22 % better results in the same interval. The opposite trend shows nobel-eu and also random graphs. The total number of paths for TS reaches up to 60 % in case of ban400d12 random graph and keeps above 20 % in half of all real-world topologies.

5.   Conclusion

The SDN era brings new possibilities to the network control also covering the path protection by failover mechanism. The application built at top of the SDN controller can now take into account a plethora of statistics and network state information during the generation of forwarding rules for a particular path while preserving it more robust. Moreover, these applications are enabled to use the surplus computing capacity available in data centers and run optimization algorithms to obtain advanced and complex measures there. Such measure can help to construct paths with higher resilience to the node or link failure in complex networks.

In this paper, we suggest using centrality measures for the path protection purpose: Quality of Backup complemented by the newly proposed Quality of Alternative Paths. These centrality measures describe paths bypassing a particular failed network node. The impact of these measures was demonstrated by the method of Static failure simulation.

It was performed random static node-failure simulation and targeted static node-failure simulation taking into account both centrality measures. The simulations were carried out on three different random graphs and six real-world telecommunication topologies. The results show that on undirected and unweighted graphs targeted simulation can give better results up to tens of percent for graph diameter, average path length and also for total number of all shortest paths in case of the massive node failure.

The simulations can be in the future work extended to follow and quantify the impact of centrality measures to every single path between all node pairs. Moreover, the proposed metric can be improved to consider weights of links.

Acknowledgment

This work was supported by Students grants at Czech Technical University in Prague

SGS13/200/OHK3/3T/13 and by grant of Ministry of the interior of the Czech Republic number VG20132015104.

References

[1] ALBERT, R., H. JEONG and A.-L. BARABASI. Error and attack tolerance of complex networks. *Nature*. 2000, vol. 406, iss. 6794, pp. 378–382. ISSN 0028-0836. DOI: 10.1038/35019019.

[2] BARABASI, A.-L. and R. ALBERT. Emergence of scaling in random networks. *Science*. 1999, vol. 286, no. 5439, pp. 509–512. ISSN 0036-8075. DOI: 10.1126/science.286.5439.509.

[3] BEHESHTI, N. and Y. ZHANG. Fast failover for control traffic in Software-defined Networks. In: *2012 IEEE Global Communications Conference (GLOBECOM)*. Anaheim: IEEE, 2012, pp. 2665–2670. ISBN 978-1-4673-0920-2. DOI: 10.1109/GLOCOM.2012.6503519.

[4] BOLLOBAAS, B. *Modern graph theory*. New York: Springer, 1998. ISBN 0-387-98488-7.

[5] BRANDES, U. On variants of shortest-path betweenness centrality and their generic computation. *Social Networks*. 2008, vol. 30, iss. 2, pp. 136–145. ISSN 0378-8733. DOI: 10.1016/j.socnet.2007.11.001.

[6] COUTO, R. S., M. E. M. CAMPISTA and L. H. M. K. COSTA. A reliability analysis of datacenter topologies. In: *2012 IEEE Global Communications Conference (GLOBECOM)*. Anaheim: IEEE, 2012, pp. 1890–1895. ISBN 978-1-4673-0920-2. DOI: 10.1109/GLOCOM.2012.6503391.

[7] OpenFlow 1.2. In: *Open Networking Foundation* [online]. 2011. Available at: https://www.opennetworking.org.

[8] KUZNIAR, M., P. PERESINI, N. VASIC, M. CANINI and D. KOSTIC. Automatic failure recovery for software-defined networks. In: *Proceedings of the second ACM SIGCOMM workshop on Hot topics in software defined networking*. New York: ACM Press, 2013, pp. 159–160. ISBN 978-145032056-6. DOI: 10.1145/2491185.2491218.

[9] MANZANO, M., K. BILAL, E. CALLE and S. U. KHAN. On the Connectivity of Data Center Networks. *IEEE Communications Letters*. 2013, vol. 17, iss. 11, pp. 2172–2175. ISSN 1089-7798. DOI: 10.1109/LCOMM.2013.091913.131176.

[10] MANZANO, M., J. L. MARZO, E. CALLE and A. MANOLOVAY. Robustness analysis of real network topologies under multiple failure scenarios. In: *2012 17th European Conference on Networks and Optical Communications*. Vilanova i la Geltru: IEEE, 2012, pp. 1–6. ISBN 978-1-4673-0949-3. DOI: 10.1109/NOC.2012.6249941

[11] MENGER, Karl. Zur allgemeinen Kurventheorie. *Fundamenta Mathematicae*. 1927, vol. 10, iss. 1, pp. 96–115. ISSN 0016-2736.

[12] NEUMAYER, S. and E. MODIANO. Network Reliability With Geographically Correlated Failures. In: *2010 Proceedings IEEE INFOCOM*. San Diego: IEEE, 2010, pp. 1–9. ISBN 978-1-4244-5836-3. DOI: 10.1109/INFCOM.2010.5461984.

[13] ORLOWSKI, S., R. WESSALY, M. PIORO and A. TOMASZEWSKI. SNDlib 1.0-Survivable Network Design Library. *Networks*. 2009, vol. 55, iss. 3, pp. 276–286. ISSN 0028-3045. DOI: 10.1002/net.20371.

[14] REITBLATT, M., M. CANINI, A. GUHA and N. FOSTER. FatTire: Declarative fault tolerance for software-defined network. In: *Proceedings of the second ACM SIGCOMM workshop on Hot topics in software defined networking*. New York: ACM Press, 2013, pp. 109–114. ISBN 978-145032056-6. DOI: 10.1145/2491185.2491187.

[15] SHARMA, S., D. STAESSENS, D. COLLE, M. PICKAVET and P. DEMEESTER. Enabling fast failure recovery in OpenFlow networks. In: *2011 8th International Workshop on the Design of Reliable Communication Networks (DRCN)*. Krakow: IEEE, 2011, pp. 164–171. ISBN 978-1-61284-124-3. DOI: 10.1109/DRCN.2011.6076899.

[16] SHARMA, S., D. SAESSENS, D. COLLE, M. PICKAVET and P. DEMEESTER. Fast failure recovery for in-band OpenFlow networks. In: *2013 9th International Conference on the Design of Reliable Communication Networks (DRCN)*. Budapest: IEEE, 2013, pp. 52–59. ISBN 978-1-4799-0049-7.

[17] SHARMA, S., D. STAESSENS, D. COLLE, M. PICKAVET and P. DEMEESTER. Open-Flow: Meeting carrier-grade recovery requirements. *Computer Communications*. 2013, vol. 36, iss. 6, pp. 656–665. ISSN 0140-3664. DOI: 10.1016/j.comcom.2012.09.011.

[18] SHAVITT, Y. and Y. SINGER. Beyond Centrality - Classifying Topological Significance Using Backup Efficiency and Alternative Paths. In: *6th International IFIP-TC6 Networking*

Conference. Atlanta: Springer Berlin Heidelberg, 2007, pp. 774–785. ISBN 978-3-540-72605-0. DOI: 10.1007/978-3-540-72606-7_66.

[19] SITANAYAH, L., K. N. BROWN and C. J. SREENAN. Fault-Tolerant Relay Deployment Based on Length-Constrained Connectivity and Rerouting Centrality in Wireless Sensor Networks. In: *9th European Conference, EWSN 2012*. Trento: Springer Berlin Heidelberg, 2012, pp. 115–130. ISBN 978-3-642-28168-6. DOI: 10.1007/978-3-642-28169-3_8.

[20] WATTS, D. J. and S. H. STROGATZ. *Nature*. vol. 393, iss. 6684, pp. 440–442. ISSN 0028-0836. DOI: 10.1038/30918.

About Authors

Tomas HEGR received his M.Sc. in computer science at the Czech Technical University in Prague in 2012. He participate in teaching activities at the department of telecommunication engineering. His research interests involves industrial networks based on Ethernet and Software-Defined Networking in all research areas.

Leos BOHAC received the M.Sc. and Ph.D. degrees in electrical engineering from the Czech Technical University, Prague, in 1992 and 2001, respectively. Since 1992, he has been teaching optical communication systems and data networks with the Czech Technical University, Prague. His research interest is on the application of high-speed optical transmission systems in a data network.

24

MODELLING PACKET DEPARTURE TIMES USING A KNOWN PDF

Stanislav KLUCIK, Martin LACKOVIC, Erik CHROMY, Ivan BARONAK

Institute of Telecommunications, Faculty of Electrical Engineering and Informational Technology, Slovak University of Technology in Bratislava, Illkovicova 3, 812 19 Bratislava, Slovak Republic

klucik@ut.fei.stuba.sk, lackovic@ut.fei.stuba.sk, chromy@ut.fei.stuba.sk, baronak@ut.fei.stuba.sk

Abstract. *This paper deals with IPTV traffic source modelling and describes a packet generator based on a known probability density function which is measured and formed from a histogram. Histogram based probability density functions destroy an amount of information, because classes used to form the histogram often cover significantly more events than one. In this work, we propose an algorithm to generate far more output states of random variable X than the input probability distribution function is made from. In this generator is assumed that all IPTV packets of the same video stream are the same length. Therefore, only packet times are generated. These times are generated using the measured normalized histogram that is converted to a cumulative distribution function which acts as a finite number of states that can be addressed. To address these states we use an ON/OFF model that is driven by an uniform random number generator in (0, 1). When a state is chosen then the resulting value is equal to a histogram class. To raise the number of possible output states of the random variable X, we propose to use an uniform random number generator that generates numbers within the range of the chosen histogram class. This second uniform random number generator assures that the number of output states is far more larger than the number of histogram classes.*

Keywords

H.264, IPTV, MPEG2 TS, probability density function, traffic generators.

1. Introduction

Traffic generators are mainly used in simulators to test predefined scenarios. They are often used in real network environments for stress testing of paths, devices or QoS (Quality of Service) [1], [2] implementations within tested NGN networks.

This paper deals with an IPTV (Internet Protocol Television) traffic generator that can be applied in above-mentioned scenarios. It describes the proceeding how to derive and construct such a generator of packet times. In our scenario, we developed it as a MatLab function that uses a PDF and desired sequence length in seconds as inputs and inter-generation times are the output. Subsequently we preprogrammed this function to Java that uses a write function to generated packets of identical arbitrary content at the network interface of a network interface card. This generator generates only one IPTV stream. This means no streams are multiplexed together. In more detail, we propose this IPTV generator with following attributes: VBR (Variable Bit-rate), H.264, MPEG2 TS (Moving Pictures Experts Group) (Transport Stream). These properties are the result of examined traffic that we have received from our national IPTV provider. Because our provider uses only CBR (Constant Bit-rate) traffic and we are interested in VBR, only pre coded VBR video records were used. This means that the coding scheme was left to H.264 main profile and Full HD (High Definition) resolution and only one channel per multiplex was examined. Then the recorded video samples were stream streamed out using the following encapsulations: MPEG2 TS/RTP (Real-time Transport Protocol)/UDP (User Datagram Protocol)/IP (Internet Protocol)/Ethernet. With this proceeding, we obtain IPTV VBR traffic with the coding used as by the IPTV provider. Results obtained from this generator are compared to real traffic. Both, real and generated traffic are streamed to our simulator which was created within MatLab environment. Measurements were made at the Ethernet layer.

Traffic generators have to be constructed as simple as possible, but they also have to be precise in comparison with real traffic. Therefore, it is always a decision between the complexity of the generator and the resulting computer load. In this paper, we propose a one-level packet generator that imitates a defined source. It uses a known PDF to generate only packet times. This approach minimizes the computer CPU and RAM usages.

Within the scope of the project „Support of Center of Excellence for SMART Technologies, Systems and Services II" funded by structural funds of European union we have built a modern IP Multimedia Subsystem [3] lab at the Institute of Telecommunications. In this lab, we use this generator to simulate IPTV traffic from several sources.

The rest of this paper is organized as follows. The second section describes the current state of the art. The third chapter is devoted to the generator itself. The fourth chapter is the conclusion.

2.   State of the Art

Packet generators can be divided into several modeling approaches. Many packet generators are based on several mathematical models like semi-Markovian chains [4], wavelets [5], multifractal analyses [6] or combined Markovian models [7]. These models can be used to generate packet sizes, times between generating of packets, absolute times or they can even aim only a specific part of the whole generator. When the real traffic has some specific properties, the traffic model has to take these properties into account, but some of them are omitted. In [8] authors published a survey on different VBR traffic models. Following models were examined: Markov Modulated Gamma (MMG) model, the Discrete Autoregressive (DAR) model, the second order Autoregressive AR(2) model, and a wavelet-based model These models are compared to each other, IPTV and video conference traffic. In [9] authors use frame oriented generator that mix Normal and Log-normal distributions for frame sizes generation. In the case of H.264 video Burr or Gamma distributions fit better frame sizes distributions of real traffic. In our case, we use a recorded PDF, therefore, the resulting PDF of the generator matches the input PDF and there are no needs to examine frame structures.

IPTV VBR traffic of one stream is a time variant self-similar process, which can be described at several levels. In this paper we propose a one level IPTV packet generator, which does not take the self-similarity parameter into account at all. We assume the case where the self-similarity goes to the background because of flow switching. When tenths of IPTV flows are aggregated together and streamed out, the effect of self-similarity of one flow reduces. We studied this behavior using our MatLab simulator where 125 real and 125 generated switched streams were compared against each other. This simulator is a programmed function of a one network node that can simulate some queueing systems, works as a time driven simulator, and its ac-

curacy was proven with analytically expressed Markovian chains or using other methods. Several nodes can create a network of nodes. Real traffic has its own self-similarity because it consisted from 125 separate recorded IPTV streams that were switched together to a one IPTV stream. The same was made with the generated traffic, where the generator we used was based on the Burr distribution function. We compared results from real and generated traffic using the same way as in our previous paper [10]. One-way delay, packet loss and jitter were observed. Results are a bit more correlated like the results in sub-chapter 3.3. Because of this partial success we decided to create a one-level packet generator that uses a known PDF (Probability Density Function) that is not analytically expressed. Proposed algorithm is also usable in many other applications where finite number of states at the input has to be extended to nearly unlimited number of output states without using any analytically expressed function.

3.   One Level Packet Generator

This chapter describes procedures to create a packet generator from a measured PDF of times between generation of packets. It is assumed that the all generated packets have the same size. Therefore, only packet times are needed to be generated. The MPEG2 TS layer creates 7×188 bytes long packets that are encapsulated into RTP, UDP, IP and Ethernet layer. Therefore the generated packet is in real an Ethernet frame with the total length of $7 \times 188 + 12 + 8 + 20 + 38 = 1394$ B. This length is usable in our network simulator that works nearly on the L2/L1 RM OSI (Reference Model Open System Interconnection) layer.

3.1.   Input – Known Probability Density Function

PDF that we use as an input to our generator was measured in our experimental laboratory where we streamed the traffic obtained from our IPTV provider. Inter-arrival times were observed. Because measuring time using a personal computer is not very precise, we had to filter the results, and the resulting PDF can be seen in Fig. 1. CDF (Cumulative Distribution Function) of this PDF is showed on Fig. 2. To stream out traffic this time we used the VLC player software. Time is a continuous variable. When generating time stamps, real values have limits defined by the variable type used within programming environment. In MatLab, we used the default double-precision floating-point data type. This data type is more precise than the floating-point values needed to represent time stamps.

From filtered results, we created a histogram using defined classes. Then we normalize the absolute counts of occurrences to the sum of occurrences, so the sum of resulting occurrences is equal to one. In this way, we meet the rule that the integral or sum of a PDF over the y axis has to be equal to one, where y axis represents the probability of occurrence of a given class. In using a recorded PDF, we do not need to look for a usable analytical expressible probability distribution function. However, using a PDF that was made from a histogram with limited number of classes decreases the absolute count of states that can be generated using a known recorded PDF.

Creating frame times on an Ethernet line with a defined clock where time is a continuous variable using a histogram with only about thousand of classes may not be enough. Therefore, we propose a method wherewith we do not need to look for an analytically expressible PDF but we can create nearly continuous states of this random variable described by the recorded PDF.

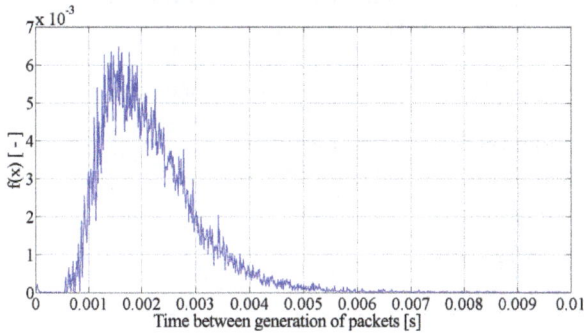

Fig. 1: PDF obtained from measurements on real IPTV traffic.

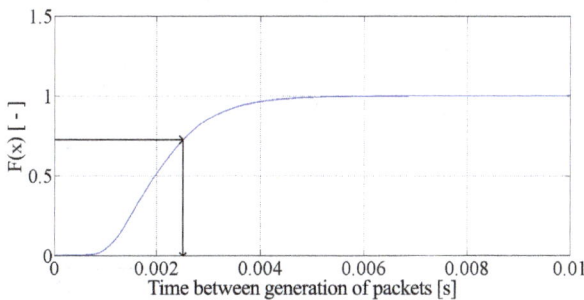

Fig. 2: CDF obtained from measurements on real IPTV traffic.

3.2. Defining the Generator

In this section, the process of finding the appropriate x value from a generated y value is described. When generating numbers according to a known PDF, firstly we create a CDF from this known PDF. Then we use an uniform pseudorandom generator with defined limits between $\langle 0, 1 \rangle$ (MatLab rand() function). When the pseudorandom generator generates a value as a double-precision floating-point data type, it represents a real

value with a limited number of numbers behind the comma sign. So this function generates a real number that is used to find the best value on the $F(x)$ vector representing the CDF y axis. Then a value from the x axis is assigned to the equivalent y value. This procedure is outlined in Fig. 1 and Fig. 2. As we already mentioned above, the PDF or CDF is a normalized histogram that uses a finite number of classes. If the histogram has 1001 boundaries, 1000 classes are created. Using the process of finding the appropriate class for an uniform random generated value representing a value on the $F(x)$ we only obtain 1000 finite times when a packet or frame is generated. These times are states that the resulting random variable can achieve. So we propose an algorithm to create nearly unlimited number of states distributed according to a recorded PDF.

Table 1 shows a PDF example that is created from a histogram. The first column shows centers of histogram classes, $f(x)$ are the function values or probability of class occurrence, and $F(x)$ are CDF values. Centers of classes are used because histograms are often created with uniformly distributed bins. Using this PDF or CDF would result in only 7 states of random variable. We added a new step after the corresponding class was found. When a value with the center of 3.5 is created and the boundaries are from 3 to 4 (only uniform distributed classes are considered) an uniform distributed number is generated with limits $\langle 3, 4 \rangle$. This generated number replaces the number representing the found class.

Tab. 1: Example PDF created from a histogram.

Center value of class (x)	f(x)	F(x)
0.5	0.1	0.1
1.5	0	0.1
2.5	0	0.1
3.5	0.5	0.6
4.5	0.2	0.8
5.5	0.1	0.9
6.5	0.1	1.0

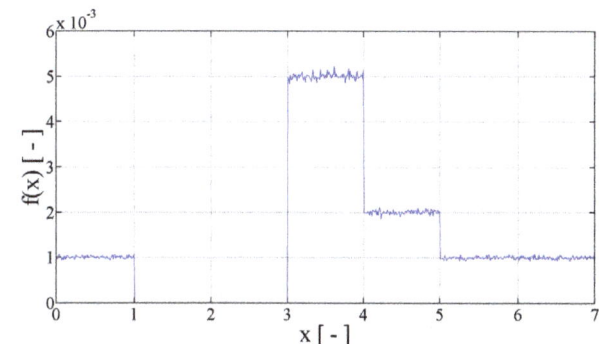

Fig. 3: PDF generated according to Tab. 1.

1) The Algorithm Presets

Input variables to our algorithm are the PDF and time. PDF consists of two columns where the first column contains the classes defined by the center values, and the second column contains probability of occurrences of equivalent classes. Classes are representing times between generations of IPTV packets. Time defines the length of the generated IPTV stream in seconds.

Algorithm is as follows. From the PDF, the mean value of time between the generation of packets is computed and from this value the number of output elements is computed according to Eq. (1):

$$E = \frac{time}{\mu}, \qquad (1)$$

where, E is the number of elements to be generated and E also controls the number of cycles for generating elements, $time$ is the requested time of the generated stream in seconds and μ represents the mean value of inter-generation time of packets. Then the CDF is computed from the input PDF using a cumulative sum of elements from the second column that contains the probability values. Assuming that classes have the same width, class width is computed according to Eq. (2):

$$I = x(2) - x(1), \qquad (2)$$

where I is an interval or class width, $x(1)$ and $x(2)$ are the centers of two consecutive classes or the centers of the first and second class.

2) Algorithm - Searching for Classes

Generating values uses a simple for cycle with a nested search algorithm to search for best CDF values, which are used to find the class to generate time from. Every step a random number from an uniform interval of $\langle 0, 1 \rangle$ is generated. This number represents the y axis of the CDF. Algorithm for searching the best $F(x)$ according to the generated y uses the condition defined in Eq. (3):

$$F(x) < y, \qquad (3)$$

where, $F(x)$ is a concrete probability value computed from the input PDF and y is the generated random value in $\langle 0, 1 \rangle$.

In our generator PDF or CDF values are defined as matrixes. If there is no information about the placement of y values whether they are pointing to the middle or beginning of a corresponding class within previous used histogram, we have to decide what the

meaning of these values is. This information is needed because the generator has to decide what class should be used for elements (packet or time) generation. The condition in Eq. (3) denotes that the concrete $F(x)$ value points to the beginning of the current class.

There is no use for finding the nearest neighbor of a corresponding class. Figure 4 shows what will happen when the nearest neighbor would be searched. This figure corresponds to the PDF defined in Tab. 2. The X on the Fig. 4 between the first and second classes represents a generated value of approximately 0.3 which was generated from the uniform interval of $\langle 0, 1 \rangle$. If this value X is nearer from the point of view of y axis to the value $F(x) = 0.2$ as to the value $F(x) = 0.4$ then the resulting area would cover 1.5 classes as a limit from left (down), when searing the nearest value would be used. On the other hand, the last class would be used only in $1/2$ of generated values that should correspond to that class. Therefore, it would be better to define the PDF and CDF as matrixes that are pointing to the beginning of corresponding classes. Also, the condition in Eq. (1) must be met.

Tab. 2: Example PDF – defining the condition of interest.

Center value of class (x)	Beginning value of class (x)	f(x)	F(x)
0.5	0	0.2	0.2
1.5	1	0.2	0.4
2.5	2	0.2	0.6
3.5	3	0.2	0.8
4.5	4	0.2	1.0

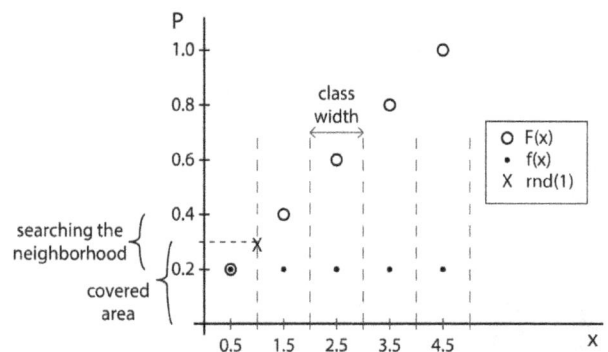

Fig. 4: PDF and CDF from Tab. 2.

Now is the class to generate from known, then the second generation of the current resulting time has to be done. The resulting time is generated between boundaries of two classes and is described in Eq. (4):

$$t = a + (b - a) \cdot rand(1), \qquad (4)$$

where, t is the current generated time, a and b are boundaries of the current class and are defined in Eq. (5) and $rand(1)$ generates a random variable in

$\langle 0, 1 \rangle$:

$$a = x(j) - I/2,$$
$$a = x(j) + I/2, \qquad (5)$$

where $x(j)$ is the center of the current class j and $I/2$ is the one-half of the computed interval defined in Eq. (2).

Figure 3 shows the resulting PDF after the uniform random number generator was applied to the input values from Tab. 1. Even when the input PDF uses 7 classes the resulting number of states is restricted only to data type used for the variable, which contains the generated uniform random value in limits of the concrete chosen class. Certainly the resulting curve of the PDF contains stairs, but when more classes are used, the shape at all should not be a problem when generating IPTV packets. This process of creating packets includes the need of generating 2 values using the uniform pseudo-number generator and a find function to find the best $F(x)$. Finding the best $F(x)$ value depends on the representation of x axis for the $F(x)$.

After generating an appropriate time representing the inter-generation time of packets or frames, a final customization of this generator must be used. MPEG2 TS uses a PCR (Program Clock Reference) that has to be sent every 100 ms. Within this 100 ms interval, the inter-generation time has to be constant. So if a real number is generated, the number of packets that have to be sent out with this time is computed in Eq. (6):

$$NP = \left\lceil \frac{100 \cdot 10^{-3}}{t} \right\rceil, \qquad (6)$$

where NP represents the number of packets that have to be sent out with the same time between the generation of packets and t represents the time between the generation of packets itself. When that time is smaller, more packets have to be sent out within the interval between sending of PCR samples. Because of this repetition in times the overall computer load stays low.

This type of packet generator does not take the self-similarity of an IPTV traffic source into account. Because of using a real PDF, the resulting mean values like mean output speed or variance in comparison with the real traffic are nearly the same. Therefore, we do not show any plots supporting this statements. However, when using this generator in a real network, the resulting packet loss or jitter would be smaller. It is because this generator does not consider sending I frames, which have a bigger size in comparison to B or P frames. Both frames have to be sent out with the same time interval. Therefore, this type of generator is more suitable for testing of a large set of IPTV streams that are multiplexed within the network. In using many streams, the resulting variance of streams smoothes-up and there is no such big difference in packet loss and jitter.

3.3. Results

Because we used a recorded PDF, the resulting mean and variance are nearly the same and slightly alter between separate usages of this generator. There is no point to show them. Therefore, higher level results are a better approach how to compare proposed generator with the recorded traffic.

In this tests VLC player was used to stream traffic, therefore, the PDF originates from this streaming. Using VLC brings a new issue into this chapter. VLC player does not meet the rule that every 100 ms is the bit-rate of the video constant, but rather uses its own rules. These rules are showed in Fig. 5.

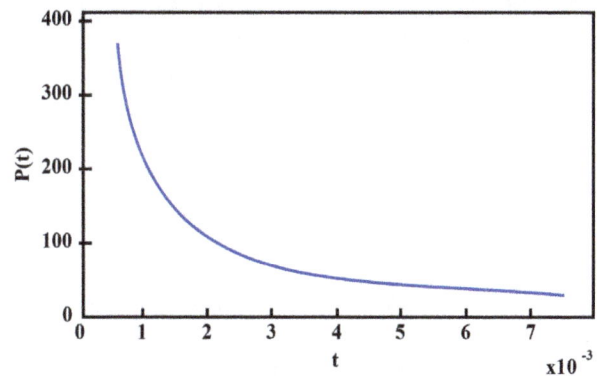

Fig. 5: Number of packets $P(t)$ to be generated with the same inter-generation time as a function of the resulting generated inter-generation time t.

Normally the number of packets to be generated with the same time between the individual generations of packets would be computed according to Eq. (6). Because we used the VLC player, we had to adapt this step and we used an analytically expressed form of the curve represented in Fig. 5. This curve denotes that the 100 ms interval is not constant and depends on the current inter-generation time. Otherwise, this curve would be a straight line. This is only a problem of the current usage of this streamer, therefore, there is no point to concern about this issue more.

In our tests, no packet losses were observed and therefore we present only the resulting packet jitter within our tested network (Fig. 6). The results confirm the assumption that the generator produces a bit small jitter then the original traffic. As already mentioned, this is because the aggregated traffic that consists of 100 IPTV streams has no dependence to the self similarity of a real IPTV stream. However, aggregating traffic reduces this effect quite visible.

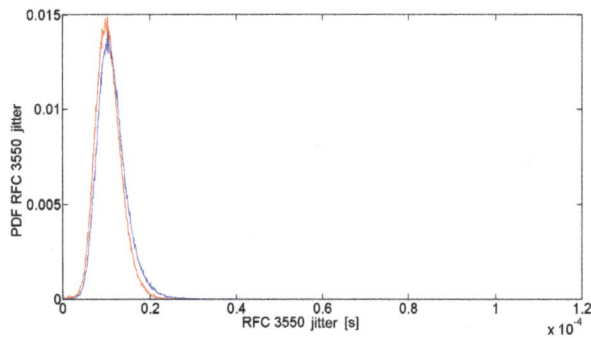

Fig. 6: RFC 3550 jitter in test network of recorded traffic (blue) and generated traffic (red).

This type of generator is very little computing intensive, because only two number generation per packet has to be done and a one searching function is used. However, using MPEG2 TS ensures that this generation is called approximately every $N \times 100^{th}$ packets depending on the input PDF. This generator is also slower in comparison to a generator that uses an analytically expressed function. However, in our case, the recorded PDF is desired as an input.

4. Conclusion

In this paper, we propose a one level packet generator used to generate IPTV packet times between generation of packets or inter-arrival times. This generator uses a known recorded PDF to generate packets from. The input PDF to the proposed algorithm is expressed as a matrix and has its origins in a histogram that uses classes within defined ranges. PDF with uniform classes reduces the number of usable states to generate from. Therefore, we propose an algorithm to generate nearly continuous states that are representing time. Resulting IPTV generator does not meet the rule of IPTV stream self-similarity and therefore it is usable as a generator for aggregating streams to higher speeds so the self-similarity does not apply so much. As the results show, the resulting generator is less compute intensive then our four-level IPTV generator and can delivers nearly similar results, but only when multiplexing is used.

Proposed algorithm is also applicable in other generators that have to use a recorded PDF with defined classes because the PDF originates from a histogram.

In further work, we intend to create an adjustable IPTV H.264 packet generator with the ability to choose the desired output bitrate of the IPTV stream. This generator will use a known recorded PDF and apply in this article introduced model to raise the number of output states of the generator.

Acknowledgment

This article was created with the support of the Ministry of Education, Science, Research and Sport of the Slovak Republic within the Research and Development Operational Programme for the project "University Science Park of STU Bratislava", ITMS 26240220084, co-funded by the European Regional Development Fund. This article is a part of research activities conducted at Slovak University of Technology Bratislava, Faculty of Electrical Engineering and Information Technology, Institute of Telecommunications, within the scope of the projects "Grant programme to support young researchers of STU - QoS in IMS networks".

References

[1] REZAC, F., M. VOZNAK and F. HROMEK. Delay Variation Model with Two Service Queues. *Advances in Electrical and Electronic Engineering*. 2010, vol. 8, no. 1, pp. 24–28. ISSN 1804-3119. DOI: 10.15598/aeee.v8i1.15.

[2] ROZHON, J., F. REZAC, J. SLACHTA and M. VOZNAK. Monitoring of Speech Quality in Full-Mesh Networks. In: *21st International Conference, CN 2014*. Brunow: Springer, 2014, pp 157–166. ISBN 978-3-319-07940-0. DOI: 10.1007/978-3-319-07941-7_16.

[3] JANEVSKI, T. *NGN architectures, protocols, and services*. United Kingdom: Wiley-Blackwell, 2014. ISBN 978-1-118-60720-6.

[4] KEMPKEN, S. and W. LUTHER. Modeling of H.264 high definition video traffic using discrete-time semi-markov processes. In: *Proceedings of the 20th International Teletraffic Congress, ITC20 2007*. Ottawa: Springer, 2007, pp. 42–-53. ISBN 978-3-540-72989-1. DOI: 10.1007/978-3-540-72990-7_8.

[5] DAI, M., Y. ZHANG and D. LOGUINOV. A Unified Traffic Model for MPEG-4 and H.264 Video Traces. *IEEE Transactions on Multimedia*. 2009, vol. 11, iss. 5, pp. 1010–1023. ISSN 1520-9210. DOI: 10.1109/TMM.2009.2021802.

[6] RELJIN, I., A. SAMCOVIC and B. RELJIN. H.264/AVC video compressed traces: multifractal and fractal analysis. *EURASIP Journal on Advances in Signal Processing*. 2006, vol. 2006, pp. 1–13. ISSN 1687-6180. DOI: 10.1155/ASP/2006/75217.

[7] WAN, F., L. CAI and T. A. GULLIVER. A Simple, Two-Level Markovian Traffic Model for

IPTV Video Sources. In: *IEEE GLOBECOM 2008 - 2008 IEEE Global Telecommunications Conference*. New Orleans: IEEE, 2008, pp. 1–5. ISBN 978-1-4244-2324-8. DOI: 10.1109/GLOCOM.2008.ECP.288.

[8] TANWIR, S. and H. PERROS. A Survey of VBR Video Traffic Models. *IEEE Communications Surveys & Tutorials*. 2013, vol. 15, iss. 4. pp. 1778–1802. ISSN 1553-877X. DOI: 10.1109/SURV.2013.010413.00071.

[9] REAZ, A., D. MURAYAMA, K.-I. SUZUKI, N. YOSHIMOTO, G. KRAMER and B. MUKHERJEE. Synthetic traffic generation for streaming video to model IPTV. In: *2011 Fifth IEEE International Conference on Advanced Telecommunication Systems and Networks (ANTS)*. Bangalore: IEEE, 2011, pp. 1–6. ISBN 978-1-4673-0093-3. DOI: 10.1109/ANTS.2011.6163638.

[10] KLUCIK, S. and M. LACKOVIC. Modelling of H.264 MPEG2 TS Traffic Source. *Advances in Electrical and Electronic Engineering*. 2013, vol. 11, no. 5, pp. 404-409. ISSN 1804-3119. DOI: 10.15598/aeee.v11i5.870.

About Authors

Stanislav KLUCIK was born in Bratislava, Slovakia, in 1984. He received the Master degree in telecommunications in 2009 from Faculty of Electrical Engineering and Information Technology of Slovak University of Technology in Bratislava (FEI STU). In 2012 he submitted a Ph.D. work from the field of Source and QoS traffic parameters modeling in NGN networks and his scientific research is focused on packet generators and modeling of packet networks.

Nowadays he works as research at the Institute of Telecommunications of Faculty of Electrical Engineering and Information Technology of Slovak University of Technology in Bratislava.

Martin LACKOVIC was born in Trnava, Slovakia, in 1989. He received the Master degree in telecommunications in 2013 from Faculty of Electrical Engineering and Information Technology of Slovak University of Technology (FEI STU) Bratislava. His scientific research is focused on modeling of packet generators, traffic parameters and networks.

Erik CHROMY was born in Velky Krtis, Slovakia, in 1981. He received the Master degree in telecommunications in 2005 from Faculty of Electrical Engineering and Information Technology of Slovak University of Technology (FEI STU) Bratislava. In 2007 he submitted Ph.D. work. Nowadays he works as assistant professor at the Institute of Telecommunications of Faculty of Electrical Engineering and Information Technology of Slovak University of Technology in Bratislava.

Ivan BARONAK was born in Zilina, Slovakia, on July 1955. He received the electronic engineering degree from Slovak Technical University Bratislava in 1980. Since 1981 he has been a lecturer at Department of Telecommunications Slovak University of Technology in Bratislava. Nowadays he works as a professor at Department of Telecommunications of Faculty of Electrical Engineering and Information Technology of Slovak University of Technology in Bratislava. Scientifically, professionally and pedagogically he focuses on problems of digital switching systems, ATM, Telecommunication management (TMN), NGN, IMS, VoIP, QoS, problem of optimal modeling of private telecommunication networks and services.

25

Network Degradation Effects on Different Codec Types and Characteristics of Video Streaming

Jaroslav FRNDA[1], *Lukas SEVCIK*[1], *Miroslav UHRINA*[2], *Miroslav VOZNAK*[1]

[1]Department of Telecommunications, Faculty of Electrical Engineering and Computer Science, VSB–Technical University of Ostrava, 17. listopadu 15, 708 00 Ostrava-Poruba, Czech Republic
[2]Department of Telecommunications and Multimedia, Faculty of Electrical Engineering, University of Zilina, Univerzitna 8215/1, 010 07, Zilina, Slovakia

jaroslav.frnda@vsb.cz, lukas.sevcik.st1@vsb.cz, miroslav.uhrina@fel.uniza.sk, miroslav.voznak@vsb.cz

Abstract. *Nowadays, there is a quickly growing demand for the transmission of voice, video and data over an IP based network. Multimedia, whether we are talking about broadcast, audio and video transmission and others, from a global perspective is growing exponentially with time. With incoming requests from users, new technologies for data transfer are continually developing. Data must be delivered reliably and with the fewest losses at such high speed. Video quality as part of multimedia technology has a very important role nowadays. It is influenced by several factors, where each of them can have many forms and processing. Network performance is the major degradation effect that influences the quality of resulting image. Poor network performance (lack of link capacity, high network load...) causes data packet losses or different delivery time for each packet. This work focuses exactly on these network phenomena. It examines the impact of different delays and packet losses on the quality parameters of triple play services, to evaluate the results using objective methods. The aim of this work is to bring a detailed view on the performance of video streaming over IP-based networks.*

Keywords

Delay variation, packet loss, PSNR, SSIM, video quality assessment.

1. Introduction

The growth of the Internet network is using more and more resources for performance analysis. This is simulated using different models. This work compares the performance of the network from an experimental viewpoint. The objective of this work is to analyse various impacts on transmitted video, such as packet loss, jitter, reordering. The work focuses on the presence of video and data packets in the network. It compares the impact of data loss and out of order data. The results show whether it is better to get the packets in a different order or completely lost. It compares static and dynamic video; varying quality of the transmitted video. We compare the impact of the size of the transmitted data.

We deal with objective methods for the evaluation of the quality of videos in the works. There are many attributes of the video image. These can be compared; therefore, to measure the exploits of several of the most well-known methods for the evaluation of image quality. Each method has different procedures and different metric evaluation system. We apply packet loss over predetermined steps on stream. The individual objective methods will evaluate the captured streams. The aim of the paper is to evaluate the impact of loss during transmission using different compression technologies from several different perspectives.

2. State of the Art

The recently growing interest in real-time service (such as audio and video) transfer through packet networks based on IP protocol has led to analyses of these services and their behavior in such networks becoming more intensive. Logically, the greatest emphasis is being put on the transfer of voice, since this service is the most sensitive to the overall network status. But on the other hand, video has become the majority part of all data traffic sent via IP networks. In general, a video service is one-way service (except e.g. video calls) so network delay is not such an important factor as in voice service. Dominant network factors that influence the final video quality are especially packet loss, delay variation and the capacity of the transmission links [1]. Analysis of video quality concentrates on the resistance of video codecs to packet loss in the network, which causes artefacts in the video [2], [3]. On the other hand, a few factors still lack, such as a complex view of video parameters on the final video quality. In our previous works [4], [7], we focused on the quality of the triple play service prediction model implementation, where one part was dedicated to the quality of the video service.

The main motivation behind this work is to extend the mentioned computational model and bring a complex view of all video parameters like codec type, character and resolution, and their influence on negative network factors resistance. In addition to packet loss, we focused on another network disruption phenomenon called delay variation (also known as jitter). This phenomenon is very often overlooked due to de-jitter buffer implementation on the receiving side, but for better process of network situation modeling and prediction, it is good to know how it influences the final video quality.

3. Methodology

3.1. Video Processing

1) Size of Digital Image Data

The volume of digital video data is usually described in the terminology of bandwidth or transfer rate. Bandwidth of a classical digital video transmission without compression is up to hundreds of Mbps. The amount of data of the picture signal is higher with an increase in resolution. The volume of data is a major problem in the transmission, processing, storage and display of video information. Digital video compared with static images is very sensitive to memory needed saving [5].

Standard television broadcasting has a frame rate of at least 25 fps (frames per second), [6]. It is sufficient for the delay in perception of the human eye. Every second of the movie at resolution 1080p (Full HD) of uncompressed video can take up to tens of megabytes of memory. Video typically contains a large amount of redundant data. Those can be removed using the appropriate compression algorithms [5].

2) Video Transmission

To transfer video files, fundamentally unreliable protocols are used. The principle consists in sending and receiving data without feedback between the sender and the recipient.

Factors affecting the video transmission are:

- Latency: This is the time that elapses between sending a message from the source and adoption of the destination node.

- Packet order: Variability in the packet delivery time to the destination node causes incorrect order.

- Packet loss: This is the average number of packets that arrive at the destination node due to the state of the network. It is most often expressed as a percentage.

- Bandwidth: This expresses the capacity of the transmission channel.

- Delay: This is caused by overcrowding the packet queue on the outgoing interface [5].

3) Methods for Evaluating the Quality

In the work, we used the objective methods - PSNR and SSIM. Objective evaluation metric involves the use of the metric's computational methods, which form a "score" of the quality of the investigated video. These methods measure the physical characteristics of the video signal, such as the amplitude, timing and signal-to-noise ratio.

PSNR (Peak signal-to-noise ratio) is the ratio between the maximum signal energy and noise energy. It is expressed on a logarithmic scale because different signals have different dynamic ranges. PSNR in decibels is defined by the formula [7]:

$$\mathrm{PSNR} = 10 \log_{10} \frac{\mathrm{MAX}_I^2}{\mathrm{MSE}} [\mathrm{dB}], \qquad (1)$$

where MAX is the maximum value that the pixel can take (e.g. 255 for 8-bit image) and MSE is the difference between two gay-level images or video sequences. Technically, MSE reflects the diversity of the image, while PSNR expresses its identity. The strongest PSNR method is an easy and fast calculation, which is the reason why it is still used very often in scientific papers although the correlation with the human perception is worse than SSIM [6], [8].

The SSIM (Structural Similarity Index) method includes three components - the similarity of the intensity, the corresponding contrast and the corresponding structure. The combination of these three factors forms one value. This demonstrates the quality of the test video. This method differs by evaluating structural distortion and not error rate. The main reason for this difference is characteristic of the human visual system. This perceived distortion changes in the structure of the frame much better than the error rate. Since the SSIM method achieves a good correlation to the subjective impression, rating is defined in the interval [0–1], where 0 represents the worst value and 1 the best one (identity), [7].

3.2. Video Quality Evaluation

The aim of the measurement was to simulate the effect of packet loss and jitter for the video formats MPEG-2 and MPEG-4, to determine the impact on the resulting image using objective methods for measuring the quality of the video and comparing the results. We made measurements for one static and two dynamic videos of 25 seconds. All the movies were measured at a resolution of 720×576 (PAL), 1280×720 (HD) and 1920×1080 (Full HD). Static video was represented by TV news (slow motion), the first dynamic video by a space shuttle launch and the third video with the highest bitrate (60 Mbps) by an open source animated movie called Big Buck Bunny. The whole process of measuring is shown in the Fig. 1 and Fig. 2. To evaluate the quality we used the methods SSIM and PSNR. SSIM correlates better with the perception of the human eye [6]. We evaluated these methods using MSU VQM Tools. As a first step, we created a stream in the VLC Player. As for the video content, streaming process RTP/UDP/IP method with MPEG2(TS) and H.264(MP4) was used.

This broadcast the video stream on the local computer interface. We captured and saved the stream to disk using another VLC Player. We saved this transfer video and tagged it as the original video. Our testing scenarios reflect the situation that can actually happen in the network. Especially the mobile networks capable of using IP architecture like UMTS and LTE reach high values of packet loss and delay variation [10]. For the purpose of settings of our scenarios, we used Linux tool called Netem. Netem provides Network Emulation functionality for testing protocols by emulating the properties of wide area networks. The current version emulates variable delay, loss, duplication and re-ordering [9].

1) Packet Loss

We set the packet loss to 1 on the interface using Netem and then repeated the whole measurement. Then we repeated this step for packet loss in increments of 1 %, 2 %, ..., 10 %.

2) Jitter

We set that 25 % of packets will be delayed (results of our previous work [4] showed that approximately 25 % of all traffic had different one-way delay). We repeated the measurements for 10, 20, 30, 50, 75 and 100 ms

delay variation. By streaming the videos, we set the value of the de-jitter buffer to 0 in VLC, so that the delays were real.

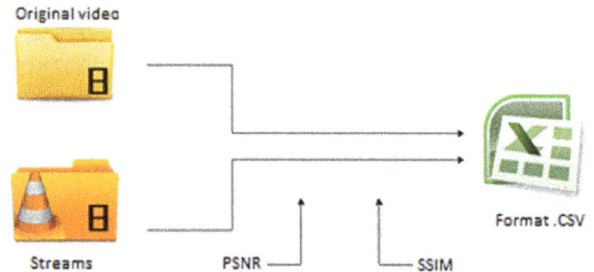

Fig. 1: Measurement procedure.

Tab. 1: Parameters for measurements.

Used codecs	MPEG-2, MPEG-4 - H.264
Video resolution [pixels]	720×576, 1280×720, 1920×1080
Evaluation methods	PSNR, SSIM

Setting packet loss on local interface:

- `#tcqdisc add dev lo root netem loss 1 %`.

Change packet loss on local interface:

- `#tcqdisc change dev lo root netem loss 2 %`.

This causes that 2 % (i.e., 2 out of 100) packets are randomly dropped. Videos for measurement were in these formats, so we did not set any additional transcoding by creating or capturing a stream.

Setting packet delay on local interface:

- `#tcqdisc add dev eth0 root netem delay 10 ms reorder 75 % 50 %`.

In this example, 75 % of the packets (with a correlation of 50 %) will get sent immediately, the others will be delayed by 10 ms. In our case, correlation 50 % means that the delayed part of all data traffic is oscillated around a value of 25 %. This setting simulates the network performance behaviour more exactly.

Change packet delay on local interface:

- `#tcqdisc change dev eth0 root netem delay 20ms reorder 75 % 50 %`.

3)   Evaluation of the Results

By using the program MSU VQMT, we compared the original sample and the tested sample, which included damage caused by our settings. The program exports results into a CSV format, where we can find the value for every compared frame and the total average value for the whole video.

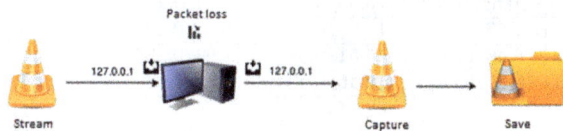

Fig. 2: Comparing stream with original video.

4.   Results

The results of the measurements verified our prediction that not only the type of video codec has a degradation impact on video quality. On the other hand, video resolution was not proven as a significant parameter of video robustness.

The most important factor was shown to be the video code type. Video codec H.264 (MPEG-4 Part 10) is more prone to packet loss rate in the network infrastructure than the older MPEG-2. According to the results of the static video measurements, there is no big difference between the resolutions used. We detected a slight decrease in higher resolution. During the static scene, where changes were very slow, mainly the P and B frames contained approximately the same information regardless of the resolution used [8].

The first tested dynamic video achieved worse results than the static video. Again, the differences between the used resolutions were small. All the GOP frames contain more information, so packet loss significantly affects the picture distortion. That is the reason why dynamic video is generally more sensitive to data loss [5], [8].

The third video has a dynamic character, too, but a very high bitrate compared to the previous two videos. Performance of this video was very poor. High bitrate means a lot of information contained in the I, B and P frames and its loss causes significant degradation of the video quality.

For a better illustration, figures of the number 3 and 4 demonstrate the video quality results for full HD resolution. This paper follows on from our previous work [4] and extends the video prediction model that was used there. All these mentioned result were processed into the following regressive equations.

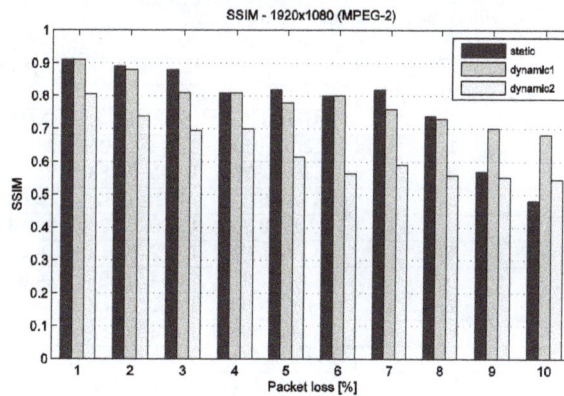

Fig. 3: SSIM results for MPEG-2.

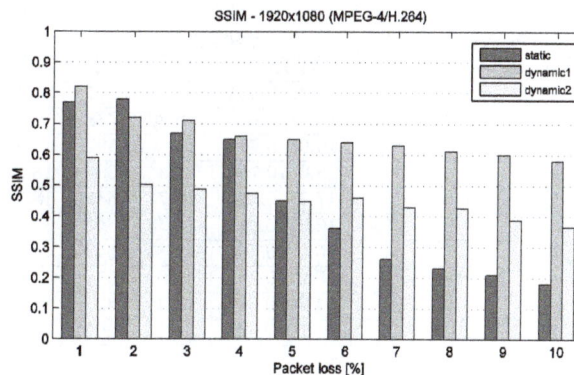

Fig. 4: SSIM results for MPEG-4(H.264).

4.1.   Slow-Motion Video

1)   MPEG-2

$$SSIM = \alpha(a + b \cdot (X^2)) + \beta(a + b \cdot \sqrt{X}) + \\ + \gamma(a + b \cdot (X^2)). \tag{2}$$

2)   MPEG-4

$$SSIM = \alpha\left(\sqrt{a + \frac{b}{X}}\right) + \beta\left(\frac{1}{a + b \cdot X}\right) + \\ + \gamma(exp(a + b \cdot X)). \tag{3}$$

All the necessary coefficients are presented in Tab. 5.

Because measurements were performed on two dynamic videos, the following regressive equations represent a prediction model for both of them.

4.2.   Dynamic Video with Ordinary and High Bitrate

1)   MPEG-2

$$SSIM = \alpha\left(\frac{1}{a + b \cdot X}\right) + \beta(a + b \cdot ln(X)). \tag{4}$$

Tab. 2: Static video measurements results.

Packet loss [%]	PSNR ([dB])/SSIM					
	720×576		1280×720		1920×1080	
	MPEG-2	MPEG-4	MPEG-2	MPEG-4	MPEG-2	MPEG-4
1	29.20/0.914	23.84/0.805	26.25/0.942	17.87/0.742	27.45/0.905	17.63/0.774
2	28.73/0.876	15.05/0.688	25.84/0.894	17.36/0.730	23.54/0.885	16.94/0.776
3	23.320.877	13.21/0.52	22.02/0.839	14.43/0.709	20.86/0.876	15.70/0.673
4	22.89/0.8	13.90/0.529	20.03/0.830	15.13/0.713	21.28/0.821	13.69/0.648
5	23.74/0.77	12.72/0.514	14.29/0.743	12.64/0.555	16.69/0.820	14.59/0.450
6	20.36/0.721	11.34/0.499	14.36/0.726	11.36/0.539	14.84/0.812	11.92/0.362
7	17.98/0.669	12.12/0.486	13.08/0.707	10.34/0.534	15.53/0.800	12.44/0.255
8	15.44/0.536	10.33/0.485	12.43/0.660	9.37/0.483	15.53/0.740	10.86/0.225
9	15.68/0.481	9.96/0.463	13.90/0.658	8.46/0.444	15.51/0.569	11.64/0.205
10	13.40/0.452	8.55/0.362	13.08/0.638	9.77/0.421	12.46/0.481	10.33/0.184

Tab. 3: PSNR and SSIM results for first tested dynamic video.

Packet loss [%]	PSNR ([dB])/SSIM					
	720×576		1280×720		1920×1080	
	MPEG-2	MPEG-4	MPEG-2	MPEG-4	MPEG-2	MPEG-4
1	21.22/0.864	17.71/0.777	21.75/0.859	16.29/0.736	19.00/0.915	15.90/0.817
2	19.08/0.830	16.30/0.738	19.27/0.837	15.89/0.750	17.99/0.875	13.59/0.715
3	19.08/0.746	14.89/0.692	17.90/0.789	14.73/0.745	14.96/0.809	13.25/0.715
4	18.65/0.754	14.84/0.686	18.23/0.766	14.83/0.714	15.86/0.808	11.51/0.657
5	17.38/0.716	14.73/0.683	18.37/0.752	14.69/0.699	14.41/0.781	12.24/0.649
6	16.23/0.711	14.73/0.645	16.07/0.741	15.07/0.663	14.95/0.799	12.62/0.637
7	15.02/0.706	13.83/0.644	14.83/0.717	13.21/0.582	15.29/0.764	12.97/0.632
8	15.27/0.709	12.60/0.651	14.73/0.701	12.66/0.545	14.10/0.728	12.15/0.605
9	14.29/0.693	12.46/0.622	14.83/0.694	10.57/0.502	14.13/0.703	11.46/0.604
10	14.70/0.686	11.65/0.605	14.73/0.678	11.72/0.485	13.20/0.683	12.11/0.582

Tab. 4: PSNR and SSIM results for the second dynamic tested video.

Packet loss [%]	PSNR ([dB])/SSIM					
	720×576		1280×720		1920×1080	
	MPEG-2	MPEG-4	MPEG-2	MPEG-4	MPEG-2	MPEG-4
1	27.0/0.923	21.32/0.695	27.63/0.927	16.10/0.530	24.72/0.806	15.56/0.589
2	26.51/0.9	17.72/0.551	22.79/0.765	15.75/0.511	22.41/0.737	15.16/0.501
3	21.59/0.714	15.36/0.446	22.56/0.745	15.30/0.422	22.0/0.694	14.42/0.487
4	21.26/0.718	15.28/0.447	21.43/0.7	14.92/0.421	22.01/0.699	13.8/0.474
5	20.73/0.705	15.33/0.441	21.10/0.664	14.04/0.428	19.84/0.614	13.37/0.446
6	19.97/0.668	15.21/0.405	21.10/0.597	12.05/0.393	18.72/0.589	13.02/0.458
7	20.07/0.618	14.28/0.338	19.15/0.566	11.46/0.372	18.22/0.564	12.17/0.429
8	19.14/0.575	14.41/0.333	17.90/0.531	10.91/0.353	17.59/0.558	12.09/0.425
9	17.97/0.562	11.08/0.303	17.65/0.528	10.91/0.352	18.28/0.552	11.09/0.387
10	17.81/0.493	10.95/0.326	17.29/0.511	10.87/0.326	18.07/0.545	10.70/0.364

Tab. 5: Coefficients for static video.

Coef.	MPEG-2			MPEG-4 (H.264)		
	720×576	1280×720	1920×1080	720×576	1280×720	1920×1080
a	0.89957	1.08748	0.9216	0.146704	1.1027	0.08596
b	−0.004924	−0.143973	−0.00389	0.528499	0.12312	−0.1839
α	1	0	0	1	0	0
β	0	1	0	0	1	0
γ	0	0	1	0	0	1

2) MPEG-4

$$\text{SSIM} = \alpha \left(a + b \cdot ln\left(X \right) \right) + \beta \left(\frac{1}{a + b \cdot X} \right). \qquad (5)$$

Table 6 and Tab. 7 contain the coefficients for these two equations. All regressive models described here

gained an R-square factor (R^2) higher than 90 %, which represents a high level of veracity.

The second group of measurements led to an analysis of the degradation effect of delay variation – Jitter. The results of the performed tests uncover a critical boundary of 20 ms. Above this value, a significant reduction of final video quality is observed. Due to the process of decompressing and processing the video

Tab. 6: Coefficients for MPEG-2 dynamic videos.

Coef.	Lower bitrate dynamic video			High bitrate dynamic video		
	720×576	1280×720	1920×1080	720×576	1280×720	1920×1080
a	0.858125	0.875076	0.927705	0.9538	0.924954	0.819482
b	−0.076882	−0.080159	−0.00389	0.094724	−0.179784	−0.1216
α	0	0	0	1	0	0
β	1	1	1	0	1	1

Tab. 7: Coefficients for MPEG-4 (H.264) dynamic videos.

Coef.	Lower bitrate dynamic video			High bitrate dynamic video		
	720×576	1280×720	1920×1080	720×576	1280×720	1920×1080
a	0.783421	1.12248	0.800991	0.678364	1.8151	00.5818
b	−0.073105	0.086301	−0.092020	−0.169198	0.122547	−0.0833
α	1	0	0	0	0	0
β	0	1	1	1	1	1

stream on the end user side costing some time, both codecs are tolerant for small delay variation.

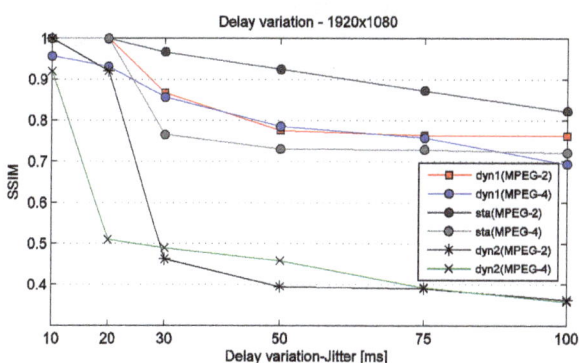

Fig. 5: Results of delay variation measurements, HD resolution.

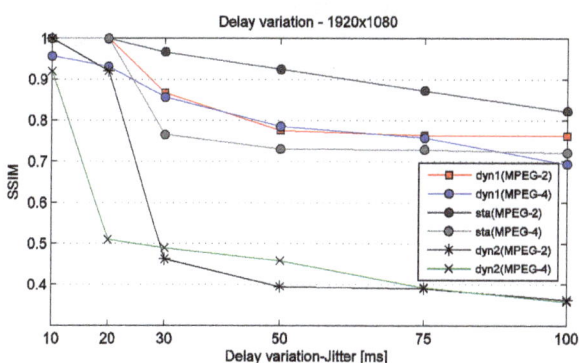

Fig. 6: Results of delay variation measurements, Full HD resolution.

The behaviour of dynamic videos is approximately on the same level, with bigger differences between the MPEG-2 and MPEG-4 codec when the static video was used.

Typically in the real world, de-jitter buffer is used for elimination of this phenomenon, but it is good to know how big a degradation effect on video quality has been caused particularly by the delay variation.

5. Conclusion

The aim of this work was to bring a detailed view of the performance of video streaming over an IP-based network. The measured results showed the relation between the video codec type and bitrate to the final video quality. These results helped us to create and extend our previous mathematical models of video streaming behaviour. The second part of the measurements was dedicated to another adverse network impact on video quality called Jitter. The results proved the importance of De-jitter buffer implementation not only for voice services but also for video streaming services.

Our future works will focus on two directions: Firstly, the new generation of video codecs such as H.265 and VP9. Due to the limitations of our evaluation MSU VQMT program, we are currently awaiting the official support for these new video codecs. The second part will be focused on analysis of the impact of security mechanisms, and on the encryption algorithm implemented to QoS parameters. Security is a highly discussed topic nowadays, and protocols such as IPsec, VPN/SSL and SRTP are becoming more and more frequently used to secure the content of voice or video, so computational mathematical models should handle this new situation.

Acknowledgment

This work was supported by Grant of the SGS No. SP2014/72 and was partially supported by the European Regional Development Fund in the IT4Innovations Centre of Excellence project (CZ.1.05/1.1.00/02.0070) and by the Development of human resources in research and the development of latest soft computing methods and their application in practice project (CZ.1.07/2.3.00/20.0072) funded by the Operational Programme Education for Competitiveness.

References

[1] CHIKKERUR, S., V. SUNDARAM, M. REISSLEIN and L. J. KARAM. Objective Video Quality Assessment Methods: A Classification, Review, and Performance Comparison. *IEEE Transactions on Broadcasting.* 2011, vol. 57, iss. 2, pp. 165–182. ISSN 0018-9316. DOI: 10.1109/TBC.2011.2104671.

[2] YIM, C., A. C. BOVIK, M. REISSLEIN and L. J. KARAM. Evaluation of temporal variation of video quality in packet loss networks: A Classification, Review, and Performance Comparison. *Signal Processing: Image Communication.* 2011, vol. 26, iss. 1, pp. 24–38. ISSN 0923-5965. DOI: 10.1016/j.image.2010.11.002.

[3] FEAMSTER, N. and H. BALAKRISHNAN. Packet Loss Recovery for Streaming Video. In: *12th International Packet Video Workshop.* Pittsburgh: ACM, 2002.

[4] FRNDA, J., M. VOZNAK, J. ROZHON and M. MEHIC. Prediction model of QoS for Triple play services. In: *2013 21st Telecommunications Forum Telfor (TELFOR).* Belgrade: IEEE, 2013, pp. 733–736. ISBN 978-1-4799-1419-7. DOI: 10.1109/TELFOR.2013.6716334.

[5] BOVIK A. *Handbook of image and video processing.* 2nd ed. Burlington: Elsevier Academic Press, 2005. ISBN 01-211-9792-1.

[6] UHRINA, M., J. HLUBIK and M. VACULIK. Correlation Between Objective and Subjective Methods Used forVideo Quality Evaluation. *Advances in Electrical and Electronic Engineering.* 2013, vol. 11, no. 2, pp. 135–146. ISSN 1804-3119. DOI: 10.15598/aeee.v11i2.775.

[7] FRNDA, J., M. VOZNAK and L. SEVCIK. Network Performance QoS Prediction. In: *Proceeding of the First Euro-China Conference on Intelligent Data Analysis and Applications.* Shenzen: IEEE, 2012, pp. 165–174. ISBN 978-3-319-07776-5. DOI: 10.1007/978-3-319-07776-5_18.

[8] UHRINA, M., J. HLUBIK and M. VACULIK. Impact of Compression on the Video Quality. *Advances in Electrical and Electronic Engineering.* 2012, vol. 10, no. 4, pp. 251–258. ISSN 1804-3119. DOI: 10.15598/aeee.v10i4.737.

[9] Netem. *Linux Foundation* [online]. 2009. Available at: http://www.linuxfoundation.org/collaborate/\workgroups/networking/netem.

[10] CHAN, M. C. and R. RAMJEE. TCP/IP performance over 3G wireless links with rate and delay variation. In: *Proceedings of the 8th annual international conference on Mobile computing and networking - MobiCom '02.* New York: ACM Press, 2002, pp. 71–82. ISBN 1-58113-486-X. DOI: 10.1145/570645.570655.

About Authors

Jaroslav FRNDA was born in 1989 in Martin, Slovakia. He received his M.Sc. from the VSB–Technical University of Ostrava, Department of Telecommunications, in 2013, and now he is continuing his Ph.D. study at the same place. His research interests include Quality of Triple play services and IP networks.

Lukas SEVCIK was born in 1989 in Cadca, Slovakia. He received his M.Sc in Informatics from the Faculty of Management Science and Informatics, University of Zilina, in 2013. Now, he is continuing his Ph.D. study at the Department of Telecommunications, VSB–Technical University of Ostrava. His research interests include Quality of Triple play services and IP networks.

Miroslav UHRINA was born in 1984 in Zilina, Slovakia. He received his M.Sc and Ph.D. degrees in Telecommunications from the Department of Telecommunications and Multimedia, University of Zilina, in 2008 and 2012, respectively. Nowadays, he is an assistant lecturer at the mentioned Department of Telecommunications and Multimedia. His research interests include audio and video compression, TV broadcasting (IPTV. DVB-T, DVB-H) and IP networks.

Miroslav VOZNAK is an associate professor at the Department of Telecommunications, VSB–Technical University of Ostrava. He received his Ph.D. in telecommunications, and his dissertation thesis was entitled "Voice traffic optimization with regard to speech quality in networks with VoIP technology" in 2002. Topics of his research interests are Next Generation Networks, IP telephony, speech quality and network security.

Permissions

List of Contributors

Milan TYSLER
Institute of Measurement Science, Slovak Academy of Sciences, Dubravska cesta 9, 841 04 Bratislava, Slovak Republic

Jana LENKOVA
Institute of Measurement Science, Slovak Academy of Sciences, Dubravska cesta 9, 841 04 Bratislava, Slovak Republic

Jana SVEHLIKOVA
Institute of Measurement Science, Slovak Academy of Sciences, Dubravska cesta 9, 841 04 Bratislava, Slovak Republic

Mindaugas KURMIS
Institute of Mathematics and Informatics, Vilnius University, Akademijos St. 4, LT-08663 Vilnius, Lithuania
Department of Informatics Engineering, Faculty of Marine Engineering Klaipeda University, Bijunu St. 17, LT-91225 Klaipeda, Lithuania

Arunas ANDZIULIS
Department of Informatics Engineering, Faculty of Marine Engineering Klaipeda University, Bijunu St. 17, LT-91225 Klaipeda, Lithuania

Dale DZEMYDIENE
Institute of Communication and Informatics, Faculty of Social Policy, Mykolas Romeris University, Ateities St. 20, LT-0830 Vilnius, Lithuania

Sergej JAKOVLEV
Department of Informatics Engineering, Faculty of Marine Engineering Klaipeda University, Bijunu St. 17, LT-91225 Klaipeda, Lithuania

Miroslav VOZNAK
Department of Telecommunications, Faculty of Electrical Engineering and Computer Science, VSB{Technical University of Ostrava, 17. listopadu 15, 708 33 Ostrava, Czech Republic

Darius DRUNGILAS
Department of Informatics Engineering, Faculty of Marine Engineering Klaipeda University, Bijunu St. 17, LT-91225 Klaipeda, Lithuania

Pavla HRUSKOVA
Department of Applied Mathematics, Faculty of Electrical Engineering and Computer Science, VSB–Technical University Ostrava, 17. listopadu 15, 708 33 Ostrava, Czech Republic

Erik CHROMY
Institute of Telecomunications, Faculty of Electrical Engineering and Information Technology, Slovak, University of Technology Bratislava, Ilkovicova 3, Bratislava 812 19, Slovakia

Marcel JADRON
Institute of Telecomunications, Faculty of Electrical Engineering and Information Technology, Slovak, University of Technology Bratislava, Ilkovicova 3, Bratislava 812 19, Slovakia

Matej KAVACKY
Institute of Telecomunications, Faculty of Electrical Engineering and Information Technology, Slovak, University of Technology Bratislava, Ilkovicova 3, Bratislava 812 19, Slovakia

Stanislav KLUCIK
Institute of Telecomunications, Faculty of Electrical Engineering and Information Technology, Slovak, University of Technology Bratislava, Ilkovicova 3, Bratislava 812 19, Slovakia

Arkadiusz BUKOWIEC
Institute of Computer Engineering and Electronics, Faculty of Electrical Engineering, Computer Science and Telecommunications, University of Zielona Gora, Podgorna 50, 65-246 Zielona Gora, Poland

Jacek TKACZ
Institute of Computer Engineering and Electronics, Faculty of Electrical Engineering, Computer Science and Telecommunications, University of Zielona Gora, Podgorna 50, 65-246 Zielona Gora, Poland

Marian ADAMSKI
Institute of Computer Engineering and Electronics, Faculty of Electrical Engineering, Computer Science and Telecommunications, University of Zielona Gora, Podgorna 50, 65-246 Zielona Gora, Poland

Filip DVORAK
Department of radar technology, Faculty of Military Technology, University of Defence, Kounicova 65, Brno, 602 00, Czech Republic

Jan MASCHKE
Department of electrical engineering, Faculty of Military Technology, University of Defence, Kounicova 65, Brno, 602 00, Czech Republic

Cestmir VLCEK
Department of electrical engineering, Faculty of Military Technology, University of Defence, Kounicova 65, Brno, 602 00, Czech Republic

Ameya Anil KESARKAR
Department of Avionics, Indian Institute of Space Science and Technology, Department of Space, Government of India, Valiamala P.O., Thiruvananthapuram – 695 547 Kerala, India

Selvaganesan NARAYANASAMY
Department of Avionics, Indian Institute of Space Science and Technology, Department of Space, Government of India, Valiamala P.O., Thiruvananthapuram – 695 547 Kerala, India

Pavel SKALNY
Department of Applied Mathematics, Faculty of Electrical Engineering and Computer Science, VSB–Technical University of Ostrava, 17. listopadu 15/2172, 708 33 Ostrava Poruba, Czech Republic

Kiattisin KANJANAWANISHKUL
Mechatronics Research Unit, Faculty of Engineering, Mahasarakham University, Kantarawichai District, Khamriang Sub-District, Maha Sarakham, 44150, Thailand

Lionel MAGNIS
The Systems and Control Centre, MINES ParisTech, 60 Bd Saint-Michel, 75272 Paris Cedex, France

Nicolas PETIT
The Systems and Control Centre, MINES ParisTech, 60 Bd Saint-Michel, 75272 Paris Cedex, France

Richard ANDRASIK
Department of Mathematical Analysis and Applications of Mathematics, Faculty of Science, Palacky University Olomouc, st. 17. listopadu 12, 771 46 Olomouc, Czech Republic

Deniss STEPINS
Institute of Radio Electronics, Faculty of Electronics and Telecommunications, Riga Technical University, 16 Azenes Street, LV-1048 Riga, Latvia

Jin HUANG
School of Electrical and Electronic Engineering, Huazhong University of Science and Technology, 1037 Luoyu Road, Wuhan 430074, China

Adam CHROMY
CEITEC - Central European Institute of Technology, Brno University of Technology, Technicka 3058/10, 616 00 Brno, Czech Republic

International Clinical Research Center, St. Anne's University Hospital Brno, Pekarska 53, 656 91 Brno, Czech Republic
Department of Control and Instrumentation, Faculty of Electrical Engineering and Communication, Brno University of Technology, Technicka 3082/12, 616 00 Brno, Czech Republic

Ludek ZALUD
CEITEC - Central European Institute of Technology, Brno University of Technology, Technicka 3058/10, 616 00 Brno, Czech Republic
International Clinical Research Center, St. Anne's University Hospital Brno, Pekarska 53, 656 91 Brno, Czech Republic
Department of Control and Instrumentation, Faculty of Electrical Engineering and Communication, Brno University of Technology, Technicka 3082/12, 616 00 Brno, Czech Republic

Marie HAVLIKOVA
Department of Control and Instrumentation, Faculty of Electrical Engineering and Communication, Brno University of Technology, Technicka 3082/12, 616 00 Brno, Czech Republic

Sona SEDIVA
Department of Control and Instrumentation, Faculty of Electrical Engineering and Communication, Brno University of Technology, Technicka 3082/12, 616 00 Brno, Czech Republic

Zdenek BRADAC
Department of Control and Instrumentation, Faculty of Electrical Engineering and Communication, Brno University of Technology, Technicka 3082/12, 616 00 Brno, Czech Republic

Miroslav JIRGL
Department of Control and Instrumentation, Faculty of Electrical Engineering and Communication, Brno University of Technology, Technicka 3082/12, 616 00 Brno, Czech Republic

Mahmoud Reza SHAKARAMI
Department of Electrical and Electronic Engineering, Engineering Faculty, Lorestan University, Daneshgat Street, 71234-98653 Khoramabad, Lorestan, Iran

Reza SEDAGHATI
Department of Electrical and Electronic Engineering, Engineering Faculty, Lorestan University, Daneshgat Street, 71234-98653 Khoramabad, Lorestan, Iran

Mohammad Bagher HADDADI
Department of Electrical Engineering, School of Engineering, Shiraz University, Bahar Azadi Street, 73159-88553 Kazeroon, Shiraz, Iran

Witthaya MEKHUM
Department of Industrial Management, Faculty of Industrial Technology, Suan Sunandha Rajabhat University, 103 00 Bangkok, Thailand

Winai JAIKLA
Department of Engineering Education, Faculty of Industrial Education, King Mongkut's Institute of Technology Ladkrabang, 105 20 Bangkok, Thailand

Jana NOWAKOVA
Department of Cybernetics and Biomedical Engineering, Faculty of Electrical Engineering and Computer Science, VSB–Technical University of Ostrava, 17. listopadu 15/2172, 708 33 Ostrava-Poruba, Czech Republic

Miroslav POKORNY
Department of Cybernetics and Biomedical Engineering, Faculty of Electrical Engineering and Computer Science, VSB–Technical University of Ostrava, 17. listopadu 15/2172, 708 33 Ostrava-Poruba, Czech Republic

Petr JARES
Department of Telecommunication Engineering, Faculty of Electrical Engineering, Czech Technical University in Prague, Technicka 2, 166 27 Prague 6, Czech Republic

Jiri VODRAZKA
Department of Telecommunication Engineering, Faculty of Electrical Engineering, Czech Technical University in Prague, Technicka 2, 166 27 Prague 6, Czech Republic

Petr CHLUMSKY
Department of Telecommunication Engineering, Faculty of Electrical Engineering, Czech Technical University in Prague, Technicka 2, 166 27 Prague, Czech Republic

Zbynek KOCUR
Department of Telecommunication Engineering, Faculty of Electrical Engineering, Czech Technical University in Prague, Technicka 2, 166 27 Prague, Czech Republic

Jiri VODRAZKA
Department of Telecommunication Engineering, Faculty of Electrical Engineering, Czech Technical University in Prague, Technicka 2, 166 27 Prague, Czech Republic

Tomas KORINEK
Department of Electromagnetic Field, Faculty of Electrical Engineering, Czech Technical University in Prague, Technicka 2, 166 27 Prague, Czech Republic

Radek BARANEK
Department of Control and Instrumentation, Faculty of Electrical Engineering and Communication, Brno University of Technology, Technicka 12, 616 00 Brno, Czech Republic

Frantisek SOLC
Department of Control and Instrumentation, Faculty of Electrical Engineering and Communication, Brno University of Technology, Technicka 12, 616 00 Brno, Czech Republic

Filip CHAMRAZ
Institute of Telecommunications, Faculty of Electrical Engineering and Information Technology, Slovak University of Technology, Ilkovicova 3, 812 19 Bratislava, Slovak Republic

Ivan BARONAK
Institute of Telecommunications, Faculty of Electrical Engineering and Information Technology, Slovak University of Technology, Ilkovicova 3, 812 19 Bratislava, Slovak Republic

Ondrej VONDROUS
Department of Telecommunication Engineering, Faculty of Electrical Engineering, Czech Technical University in Prague, Technicka 2, 166 27 Prague, Czech Republic

Peter MACEJKO
Department of Telecommunication Engineering, Faculty of Electrical Engineering, Czech Technical University in Prague, Technicka 2, 166 27 Prague, Czech Republic

Zbynek KOCUR
Department of Telecommunication Engineering, Faculty of Electrical Engineering, Czech Technical University in Prague, Technicka 2, 166 27 Prague, Czech Republic

Tomas HEGR
Department of Telecommunications, Faculty of Electrical Engineering, Czech Technical University in Prague, Technicka 2, 166 27 Prague, Czech Republic

Leos BOHAC
Department of Telecommunications, Faculty of Electrical Engineering, Czech Technical University in Prague, Technicka 2, 166 27 Prague, Czech Republic

Stanislav KLUCIK
Institute of Telecommunications, Faculty of Electrical Engineering and Informational Technology, Slovak University of Technology in Bratislava, Illkovicova 3, 812 19 Bratislava, Slovak Republic

Martin LACKOVIC
Institute of Telecommunications, Faculty of Electrical Engineering and Informational Technology, Slovak University of Technology in Bratislava, Illkovicova 3, 812 19 Bratislava, Slovak Republic

Erik CHROMY
Institute of Telecommunications, Faculty of Electrical Engineering and Informational Technology, Slovak University of Technology in Bratislava, Illkovicova 3, 812 19 Bratislava, Slovak Republic

Ivan BARONAK
Institute of Telecommunications, Faculty of Electrical Engineering and Informational Technology, Slovak University of Technology in Bratislava, Illkovicova 3, 812 19 Bratislava, Slovak Republic

Jaroslav FRNDA
Department of Telecommunications, Faculty of Electrical Engineering and Computer Science, VSB–Technical University of Ostrava, 17. listopadu 15, 708 00 Ostrava-Poruba, Czech Republic

Lukas SEVCIK
Department of Telecommunications, Faculty of Electrical Engineering and Computer Science, VSB–Technical University of Ostrava, 17. listopadu 15, 708 00 Ostrava-Poruba, Czech Republic

Miroslav UHRINA
Department of Telecommunications and Multimedia, Faculty of Electrical Engineering,
University of Zilina, Univerzitna 8215/1, 010 07, Zilina, Slovakia

Miroslav VOZNAK
Department of Telecommunications, Faculty of Electrical Engineering and Computer Science, VSB–Technical University of Ostrava, 17. listopadu 15, 708 00 Ostrava-Poruba, Czech Republic